# The Physics and Technology of Laser Resonators

# The Physics and Technology of Laser Resonators

Edited by

D R Hall and P E Jackson

*Department of Physics,*
*Heriot-Watt University, Edinburgh*

Adam Hilger, Bristol and New York

*British Library Cataloguing in Publication Data*

The physics and technology of laser resonators
1. Lasers. Resonators
I. Hall, D. R. (Denis R) II. Jackson, P. E. (Paul E)
621.36′6

ISBN 0-85274-117-0

*Library of Congress Cataloging-in-Publication Data are available*



Consultant Editor: **Professor W T Welford**, Imperial College, London

Published under the Adam Hilger imprint by IOP Publishing Ltd
Techno House, Redcliffe Way, Bristol BS1 6NX, England
335 East 45th Street, New York, NY 10017, USA

Printed in Great Britain by JW Arrowsmith Ltd, Bristol

# CONTENTS

## PART TWO *LASER RESONATOR TECHNOLOGY*

## LIST OF CONTRIBUTORS


| | |
|---|---|
| K M ABRAMSKI | Optoelectronics and Laser Engineering, Department of Physics, *Heriot-Watt University*, Edinburgh, Scotland<br>On leave from The Technical University of Wroclaw, Poland |
| M J ADAMS | *British Telecom Research Laboratories*, Martlesham Heath, Ipswich, England |
| H J BAKER | Optoelectronics and Laser Engineering, Department of Physics, *Heriot-Watt University*, Edinburgh, Scotland |
| G BURROWS | *Lumonics Ltd*, Cosford Lane, Swift Valley, Rugby, Warwickshire, England |
| M H DUNN | Department of Physics and Astronomy, *University of St Andrews*, North Haugh, St Andrews, Scotland |
| P E DYER | Department of Applied Physics, *University of Hull*, Hull, England |
| D J DYSON | *Ferranti International*, Dunsinane Avenue, Dundee, Fife, Scotland |
| A I FERGUSON | Department of Physics, *University of Strathclyde*, Glasgow, Scotland |
| W A GOETHALS | *Advanced Production Automation BV*, Veldhoven, The Netherlands |
| D R HALL | Optoelectronics and Laser Engineering, Department of Physics, *Heriot-Watt University*, Edinburgh, Scotland |
| I D HENNING | *British Telecom Research Laboratories*, Martlesham Heath, Ipswich, England |
| C A HILL | *Royal Signals and Radar Establishment*, St Andrews Rd, Malvern, Worcs, England |
| P E JACKSON | Optoelectronics and Laser Engineering, Department of Physics, *Heriot-Watt University*, Edinburgh, Scotland |
| T A KING | Department of Physics, *Manchester University*, Manchester, England |
| V MAGNI | Consiglio Nazionale delle Ricerche, Istituto di Fisica, *Politecnico di Milano*, Milan, Italy |
| I E ROSS | *Ferranti International*, Dunsinane Avenue, Dundee, Scotland |
| M W SASNETT | *Coherent General Inc.*, Palo Alto, California, USA |
| M J SHAW | *Rutherford Appleton Laboratory*, Didcot, Oxon, England |
| P URQUHART | *British Telecom Research Laboratories*, Ipswich, England |
| A C WALKER | Department of Physics, *Heriot-Watt University*, Edinburgh, Scotland |

# PREFACE

The topics covered in this book are central to the design, construction and application of lasers. Most of the unique characteristics of lasers as optical sources are linked to the properties of the laser *resonator*. This applies as much to microscopic semiconductor lasers as to the high power industrial material processing lasers whose volume may be more than $10^{11}$ times larger. The idea for the book has grown from a 1-day *tutorial-style* meeting organised under the auspices of the Quantum Electronics Group of the UK Institute of Physics. Although the contents of the book are based on the presentation given at this meeting considerable efforts have been made, including substantial additional material, to shape the contributions into a coherent whole.

The objectives have been to bring together in a single volume not only an introduction to the underlying physics of laser resonators and laser beams, but also material, not so readily available in text books, on the application of basic principles to the design of resonators for most of the important types of laser in common use today. Although, in some instances particular technologies are described, the overall aim has been to emphasize design principles rather than specific implementations.

Part One begins with a pivotal chapter describing the elementary principles of laser beams and conventional 'free space' Gaussian resonators. It is followed by four chapters which describe the design of certain variants on the basic resonator theme. Thus, Chapter 2–5 cover 'Unstable Resonators, 'Waveguide Resonators', 'Ring Resonators' and 'Multifold Resonators'. In each case, the emphasis is on fundamental principles rather than technology. There has been considerable interest recently in the use of laser mirrors which have spatial variations in reflectivity as a means of controlling the transverse beam intensity profile. This is covered in Chapter 6 describing 'Variable Reflectivity Mirrors'. Recognizing that a number of important resonator types cannot be described by simple Gaussian beam analysis, Chapter 7 has been included to describe computer modelling of diffraction effects in 'Numerical Resonator Calculations'. Many important applications depend on the availability of laser sources of very narrow linewidths leading to the requirement for techniques for 'Frequency Stabilisation of Lasers' as described in Chapter 8.

Many lasers, particularly those operating at high average power produce outputs which are not ideal $TEM_{00}$ beams. Techniques for characterising the 'Propagation of Multimode Beams' and for 'Beam Analysis by Geometrical Optics' are presented in the final two chapters of Part One.

Part Two of the book is devoted to discussion of the principles of laser resonator design as they impact most of the important types of laser. The technology appropriate to these lasers is extremely varied, and although it is covered where appropriate, considerable efforts have been made to describe underlying design goals and principles rather than specific technological solutions. These discussions encompass YAG lasers, dye lasers, optical parameter oscillators, high power and waveguide $CO_2$ lasers, diode lasers, fibre lasers, diode–pumped solid–state lasers and excimer lasers.

The text is primarily intended for use in introductory courses in Lasers and Optoelectronics at the level of final year undergraduate or first year graduate students. It is also hoped that practising scientists and engineers and others looking for an introduction to the field will benefit from the material and design principles described.

We would like to thank a number of people, without whom this book could not have been produced. Most importantly we thank the authors, whose contributions *make* the book, for finding time in busy schedules to adapt and expand their oral presentations to the needs of a text book. They have been extremely patient in the face of our cajoling and editorial efforts. We gratefully offer our thanks to colleagues and graduate students at Heriot–Watt University for responding so patiently and conscientiously to requests to proof–read parts of the manuscript and particularly to Dr Howard Baker for additional editorial assistance; to Mrs Lesley Lumsden for typing part of the manuscript and Mrs Sherron Gilroy who prepared some of the illustrations.

Finally, our gratitude and admiration are due to Alan Colley for his creative wordprocessing and text productions skills, his conscientious attention to detail, amazing capacity for work and unflagging enthusiasm; he is mainly responsible for the high quality of the camera–ready copy.

Edinburgh
August 1989

*Denis R Hall and Paul E Jackson*

# PART ONE FUNDAMENTALS OF RESONATOR PHYSICS

## 1 Laser Resonators and Gaussian Beams

***A C WALKER***

### 1.1 INTRODUCTION

The first laser to be operated – the ruby laser in 1960 – and the vast majority of other lasers developed since, consist of a gain medium placed within an optical resonator (or *cavity*). Although the gain medium, which gives each type of laser its name, determines the output power/energy and ultimate tuning range, it is the *resonator* that controls the spatial quality and spectral linewidth of any particular laser. As these two parameters are the very factors that make lasers unique light sources, with high spatial and temporal coherence, it follows that the design and construction of the resonator deserves close attention.

Over the last 30 years or so, a wide range of resonators have been developed and exploited within commercially available laser systems. These include: free–space *stable* standing–wave and ring cavities, both constructed with discrete mirrors; three dimensional *guided–wave* resonators formed out of the gain medium itself or with external reflectors; and *unstable or non–confining* cavities relying on mirrors of restricted transverse dimension or spatially graded reflectivities. This chapter introduces the basics of laser resonator design, in particular discussing what are called conventional *stable* resonators and the spatial properties of the light inside, and emitted from, such cavities. Subsequent chapters cover the physics of some of the more complex resonators such as rings, unstable and waveguide resonators.

In general, a stable resonator, whether acoustic, electrical or optical, can be defined as an extended environment within which the relevant type of wave or oscillation can circulate in a self–consistent fashion. That is, every time the excitation returns to a particular point in space, its phase and amplitude are the same as the previous occasion. In this way, when driven for times longer than this circulation time, constant amplitude constructive interference occurs with consequent strong coupling between the driving mechanism and the resonator. Thus, in the same way as with musical intruments, where the resonator (*sound–box*) efficiently couples the energy source (eg. vibrating string) to free–space sound waves; a laser optical resonator provides a mechanism by which the gain medium can be coupled to well defined free–space optical modes.

### 1.2 THE FABRY–PEROT RESONATOR

The simplest, and most commonly used, laser resonator is the Fabry–Perot cavity, in which two mirrors oppose each other across the gain medium (See Figure 1.1). Such a resonator provides physical confinement in one dimension only and consequently light simply reflects back and forth along the axis

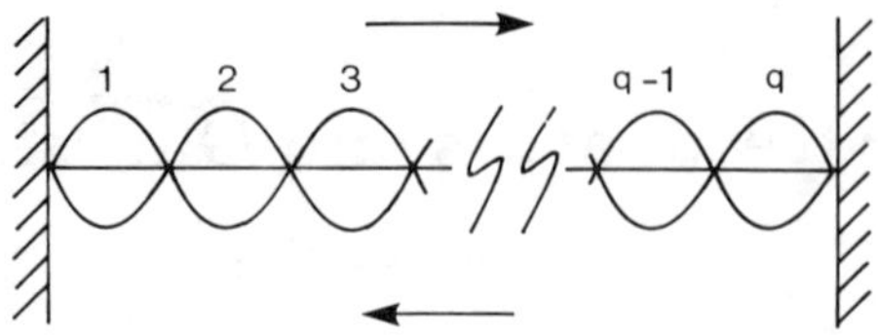


Fig 1.1 A resonant mode of an optical Fabry–Perot cavity. The curve represents the instantaneous standing wave produced by the oscillatory electric field circulating within the resonator. The two mirrors are separated by *q* half-wavelengths.

joining the two mirrors, in a manner determined by their detailed specifications (eg. curvatures, sizes, reflectivities). In such a cavity, the conditions for *stable* resonance can be specified in terms of *three* parameters which must remain unchanged after one complete round-trip:

(i) the phase of the optical wave-front
(ii) the on-axis amplitude of the wave, and
(iii) the amplitude distribution, transverse to the axis

The consequence of each of these conditions can be considered in turn.

**Phase**

In order for the phase to be reproduced on each round-trip, such as to ensure the components of the optical field coupled out of the cavity (through the partially transmitting mirror) all add constructively, the oscillating wavelength of the light must correspond to an integer fraction of the cavity round-trip length.

Thus: $$\lambda = \frac{2nL}{q} \tag{1.1}$$

where $q$ is an integer and $n$ is the average refractive index over the length of the cavity, $L$. As indicated in Figure 1.1, this corresponds to $q$ half-wavelengths fitting between the mirrors and is precisely the same condition as for maximum transmission through a conventional Fabry-Perot interferometer. Each value of $q$ corresponds to a different longitudinal (or "axial") mode of the cavity. In the absence of any additional frequency-selective components inside the cavity, as many modes will oscillate as are compatible with the gain line-width.

The separation between modes is constant in the frequency domain ($\nu = c/\lambda$) and is readily calculated using equation (1.1):

$$\begin{aligned}\Delta\nu &= \nu_{q+1} - \nu_q \\ &= \frac{c}{2nL}(q+1) - \frac{c}{2nL}q\end{aligned}$$

$$\text{i.e.} \qquad \Delta\nu = \frac{c}{2nL} \qquad (1.2)$$

For a conventional Fabry–Perot interferometer the spectral width of each mode is determined by the finesse:

$$F = \frac{\pi(Re^{-\alpha L})^{\frac{1}{2}}}{1 - Re^{-\alpha L}} \qquad (1.3)$$

where $R$ is the geometric mean of the two mirror irradiance reflectivities and $\alpha$ is the internal power loss coefficient. The mode width (FWHM) is then given by:

$$\delta\nu \approx \Delta\nu/F \qquad (1.4)$$

If the loss is replaced by gain and the cavity operated as a cw laser, then at equilibrium, the total round–trip transmission is just unity, i.e. $Re^{-\alpha L} = 1$ (see next section). As a consequence the finesse, given by equation (1.3), theoretically becomes infinite and hence $\delta\upsilon$ should be very small. This is indeed the case; linewidths for a laser operating on a single longitudinal mode can be well below the kHz level (effective finesse $\sim 10^5$). Although the ultimate limit is fixed by quantum fluctuations, the main factor that usually determines the actual mode width is the stability of the overall system. Fluctuations in gain, cavity length or alignment of a cw laser, or the time variation of a pulsed laser, correspond to mode broadening in the frequency domain. Despite these factors, single frequency (single mode) Fabry–Perot resonator lasers achieve extraordinary temporal coherence by comparison with other light sources. This topic is covered in more detail in Chapter 8.

**Amplitude**

For the light flux within the cavity to build up from the initial low level of spontaneous emission, the round trip net gain must be greater than unity. As long as this is the case, the optical field grows exponentially. Clearly, such an amplification cannot proceed indefinitely and in practice, the gain starts to fall or *saturate* as the power–demand placed on the gain medium exceeds the limits imposed on it by the rate of excitation (i.e. the *pump* power). As this happens the resonator round–trip gain falls until it reaches unity. Any further decrease in gain would result in decay of the optical field and recovery of the gain. In a continuously pumped laser, the power circulating within the resonator stabilises at a level at which the saturated gain just balances the internal losses. Thus the resonance condition that the amplitude be reproduced each round–trip is, as with most types of resonator, automatically satisfied.

It is clear from the above discussion that the transmission of the output mirror is critical in determining the power output of the resonator. If this transmission is very high then there is a large cavity loss factor and the build up of internal irradiance, and hence the output power, is limited by the need to maintain a large amplification factor in the gain medium. Conversely, if the output mirror transmission is very low then the available output power is shared between this low loss and other comparable sources of loss in the cavity, eg. absorption, scatter, reflections off internal optical surfaces, etc.

(a)

(b)

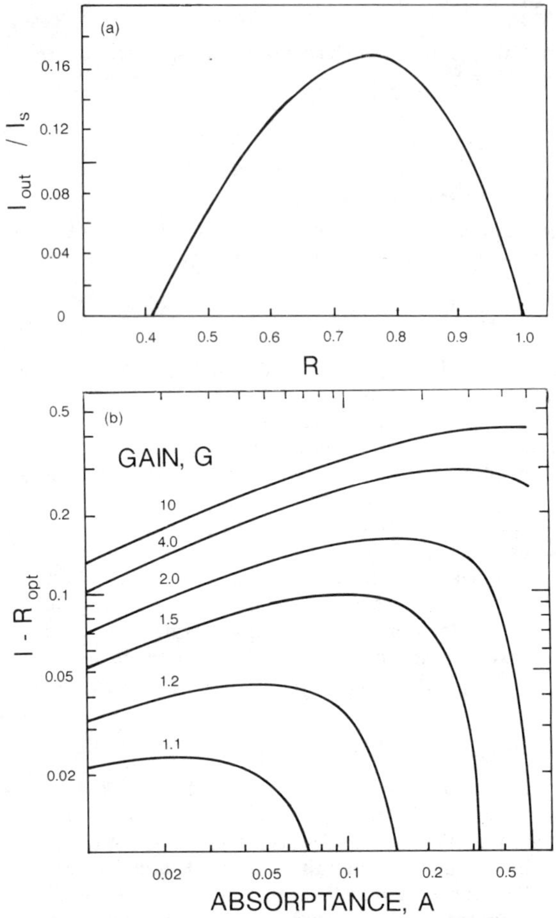


Fig 1.2 (a) Plot showing that to maximise the output irradiance $I_{out}$, from a laser resonator, an optimum output coupler reflectivity $R$, must be selected. Output irradiance has been normalised to the saturation irradiance of the gain medium $I_s$ (b) Curves showing how the optimum reflectivity $R_{opt}$ depends upon the small-signal, double-pass gain factor $G$ and the parasitic round-trip loss factor $A$. Both plots are based on the infinite plane-wave mean-field equation 1.7.

If the background absorptance (loss) is $A$, and the output coupler reflectivity $R$, then at equilibrium:

$$(1 - A)\, R \exp(2gL) = 1 \tag{1.5}$$

Writing the small signal gain factor for a double pass through the gain medium as $G$, and assuming a saturation response appropriate for a homogeneously broadened transition, then this relation becomes:

$$\frac{\ln G}{1 + I/I_s} \approx -\ln[R(1-A)] \tag{1.6}$$

where $I$ is the mean internal cavity irradiance and $I_S$ the saturation irradiance of the gain medium. Using $I_{out} = -\frac{1}{2}I(\ln R)$ then:

$$I_{out} \approx I_s \frac{\ln R}{2} \left[1 + \frac{\ln G}{\ln[R(1-A)]}\right] \tag{1.7}$$

For particular values of the gain and loss factor, $G$ and $A$, equation (1.7) indicates an optimum output coupler reflectivity, $R$, which maximises $I_{out}$. Figure 1.2 shows examples of the dependence of the optimum $R$ upon $G$ and $A$, deduced from equation (1.7).

The mean–field theory approach described above is only strictly valid for low $A$, and $R$ close to unity, when the spatial variation of irradiance along the length of the gain medium is small. Otherwise the spatial dependence of gain saturation, and the positions in the cavity of those components contributing to parasitic losses, should properly be taken into account within a more detailed theoretical description.

**Transverse Amplitude Profile**

In general, a cavity of the type indicated in Figure 1.1, i.e. formed from two plane mirrors, will not act as a stable resonator. Any beam of finite cross–section within such a cavity will diverge due to normal diffraction processes and consequently the transverse amplitude profile of the optical field will not be reproduced on each round trip.

Within this book a variety of techniques for overcoming this divergence are discussed, including: (i) transverse confinement of the optical field (waveguide lasers); (ii) coupling out of the cavity all light beyond some radius (unstable resonators); and (iii) the inclusion of focusing elements inside the cavity of sufficient power to compensate diffraction spreading (stable resonators). The latter approach, usually exploiting focussing cavity mirrors, is one of the most common techniques and is the major topic of the remainder of this chapter.

## 1.3 GEOMETRIC RAY ANALYSIS OF OPTICAL RESONATORS

Although ultimately it is diffraction which causes finite diameter light beams to diverge within a plane–mirror Fabry–Perot cavity, the degree of internal focussing which ensures that this is correctly compensated, can be calculated on the basis of a simple geometrical ray–optics analysis. This is conveniently described using a matrix approach.

### 1.3.1. Ray Transfer Matrices

A generalized ray of light in an optical system, passing through a specific plane, set normal to the optical axis ($z$), can be characterised by its distance from that axis and its angle relative to the plane normal. (See Figure 1.3a). In the paraxial ray approximation, the angle $\rho = dr/dz$ and thus over a distance $L$ along the $z$-axis, and in the absence of any optical elements, the initial values of $r_1$ and $\rho_1$ are transformed to $r_2 = r_1 + L\rho_1$ and $\rho_2 = \rho_1$ (see Figure 1.3b). This relationship between these four parameters can be summarised in a vector/matrix equation of the type:

$$\underline{R}_2 = [M_L]\underline{R}_1 \tag{1.8}$$

$$\underline{R}_1 = \begin{bmatrix} r_1 \\ \rho_1 \end{bmatrix}, \qquad \underline{R}_2 = \begin{bmatrix} r_2 \\ \rho_2 \end{bmatrix} \tag{1.9}$$

$$[M_L] = \begin{bmatrix} 1 & L \\ 0 & 1 \end{bmatrix} \tag{1.10}$$

In this simple example the matrix $M_L$ is the *ray-transfer matrix* corresponding to propagation of the ray over some distance, $L$. Similar ray-transfer matrices can be constructed for any optical transformation including the action of refracting surfaces, thin lenses, spherical mirrors, etc.

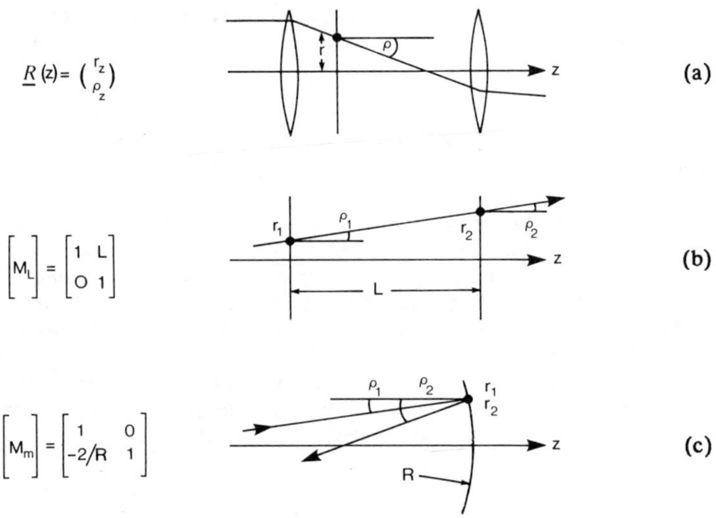


Fig 1.3 (a) Definition of a ray vector, $\underline{R}(z)$, for a ray at position $z$ within an optical system. (b) Ray transfer matrix, $M_L$, for free-space propagation over distance $L$. (c) Ray transfer matrix, $M_m$, for a concave mirror with radius of curvature $R$.

The overall effect on a ray entering an optical system can be computed by multiplying the matrices representing the individual optical components to give a single matrix corresponding to the overall system.

When considering the simplest stable resonator structures, only two types of matrix are required: (i) the matrix $M_L$, above; and (ii) the matrix corresponding to a thin focusing element (lens or mirror). The latter has to leave the ray position unchanged and transform the angle according to the lens equation, thus: $r_2 = r_1$, $\rho_2 = \rho_1 - r_1/f$.

In the case of a concave mirror of curvature radius $R$, the focal length, $f$, can be replaced by $R/2$, giving the matrix for a concave mirror:

$$[\mathrm{M}_m] = \begin{bmatrix} 1 & 0 \\ -2/R & 1 \end{bmatrix} \qquad (1.11)$$

Note that this matrix does not account for the change in direction of the ray on reflection, because in this analysis it is convenient to refer the angles, $\rho$, to the general direction of propagation.

### 1.3.2 Stability Conditions for an Open Resonator

Ray transfer matrices can be used to determine the conditions under which the simplest open resonator, formed by two spherical mirrors, are stable. At this point it should be noted that for a specific ray starting at some arbitrary position and angle, the resonator stability condition *does not* demand that the ray return to this same position and angle on every round trip. It simply requires that the light remain inside some finite boundary, such that all rays contribute to a reproducible amplitude distribution (transverse mode). It is therefore necessary to compute the effect on a ray of an arbitrary large number of cavity round-trips and determine whether the components of the ray vector will remain real and finite.

The general structure assumed for an open resonator is the common arrangement of two spherical mirrors shown in Figure 1.4. (Other elements,

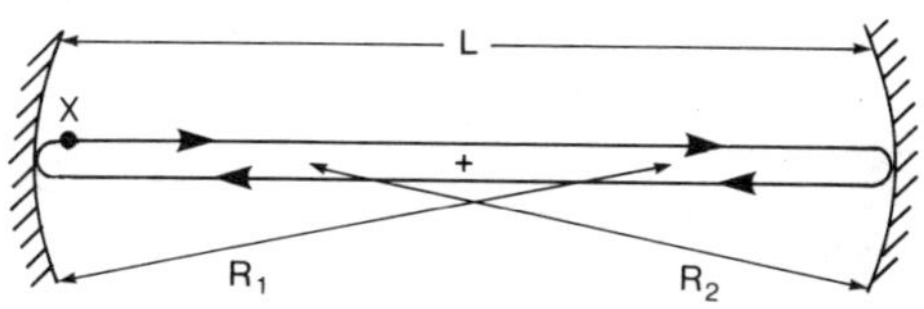


Fig 1.4 An optical cavity of length $L$ formed by two spherical mirrors with radii of curvature $R_1$ and $R_2$.

such as focusing lenses, dielectric gain–media, plane windows, frequency–selective elements, may also be included but the basic principles remain the same). Starting at point X, the matrix for a complete round trip is given by:

$$[\mathrm{M}_R] = [\mathrm{M}_L]\ [\mathrm{M}_{m2}]\ [\mathrm{M}_L]\ [\mathrm{M}_{m1}]$$

$$= \begin{bmatrix} 1-2L/R_2 & 2L-4L/R_2 \\ 4L/R_1R_2-2/R_1-2/R_2 & 1+4L^2/R_1R_2-4L/R_1-2L/R_2 \end{bmatrix} \qquad (1.12)$$

To analyze some number of round-trips, $n$, it is necessary to calculate: $[\mathrm{M}_R]^n$. This can be achieved by applying Sylvester's theorem:

$$\begin{bmatrix} A & B \\ C & D \end{bmatrix}^n = \frac{1}{n\sin\theta}\begin{bmatrix} A\sin(n\theta)-\sin(n-1)\theta & B\sin(n\theta) \\ C\sin(n\theta) & D\sin(n\theta)-\sin(n-1)\theta \end{bmatrix} \qquad (1.13)$$

where, $$\sin\theta = [1-\tfrac{1}{4}(A+D)^2]^{\frac{1}{2}} \qquad (1.14)$$

For $[\mathrm{M}_R]^n$ to be real, so too must be the term $\sin\theta$. Given that $A$, $B$ $C$

and $D$ are all real, the stability criterion is then:

$$\tfrac{1}{4}(A+D)^2 < 1$$

or
$$-1 < \tfrac{1}{2}(A+D) < 1 \qquad (1.15)$$

For $[M_R]$ as calculated in equation (1.12), this corresponds to the condition:

$$-1 < \left[ 1 - \frac{2L}{R_1} - \frac{2L}{R_2} + \frac{2L^2}{R_1 R_2} \right] < 1$$

or
$$0 < (1-L/R_1)(1-L/R_2) < 1 \qquad (1.16)$$

This simple relation (1.16) specifies the conditions under which a cavity is stable. It is sometimes written, using $g_1 = 1-L/R_1$ and $g_2 = 1-L/R_2$, as $0 < g_1 g_2 < 1$.

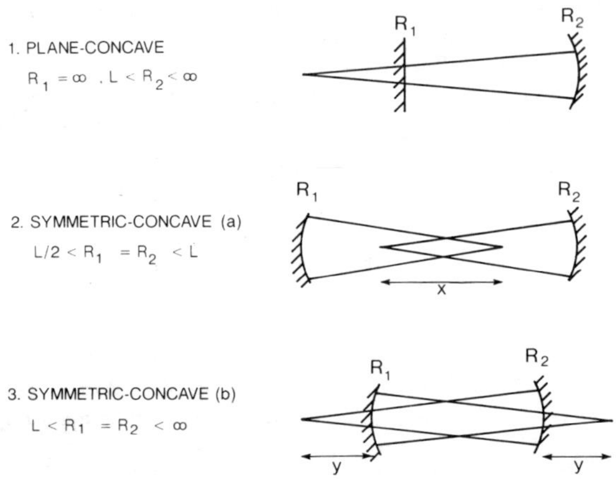


Fig 1.5 Three examples of stable optical resonators formed by two spherical mirrors, radii of curvature $R_1$ and $R_2$, distance $L$ apart.

### 1.3.3 Examples of Stable Open Resonators

Figure 1.5 shows the most common stable resonator geometries. By far the most convenient and frequently used cavity is the plane–concave configuration. One advantage of this arrangement is that the plane reflector can be the partially transmitting mirror and consequently, as both its surfaces are easily made flat, no geometric divergence/convergence is imposed on the output beam – unlike the concave reflector where, when acting as the output coupler, the outer surface has to be polished accurately convex to avoid negative lens effects. A second advantage of the plane–concave cavity is that those frequency selective elements (e.g. reflecting gratings, etalons) that are most readily available, act as planar optical elements and can be conveniently incorporated in this type of cavity as direct replacements for the flat mirror. (In this case the output often *has* to be the curved mirror, with an appropriately correcting outer surface).

The limitations on the values of $R_1$ and $R_2$ that are specified in Figure 1.5 indicate the borderlines between stable and unstable configurations. These borderline cavities are not normally used, since small manufacturing errors or the presence of additional (weak) optical elements inside the cavity (including lens effects in the gain medium itself) could make them unstable. Within the unstable region, the resonator becomes highly lossy as the rays diverge to infinity within a finite number of round trips. Thus for the plane–concave resonator (Figure 1.5), the plane–plane limit, $R_1 = R_2 = \infty$, should be avoided, along with the geometry $R_2 = L$. The latter case corresponds to the concentric cavity limit in the symmetric resonator, $R_1 = R_2 = L/2$ ($x = 0$ in Figure 1.5), as clearly the plane–concave cavities are simply folded equivalents of the symmetric variety.

Cavities near the confocal limit $R_1 = R_2 = L$ ($y = 0$ in Figure 1.5) are only unstable if the symmetry is broken. For example, assuming $\delta$ is small, $R_1 = L + \delta$, $R_2 = L - \delta$ is unstable, while $R_1 = L + \delta$, $R_2 = L + \delta$ is stable. This has no parallel in the plane–concave cavities as these are inevitably always equivalent to symmetric resonators.

The "focal" symmetric cavity in which the focus of each mirror lies on the surface of the opposing mirror, ie $R_1 = R_2 = 2L$, is the only example of this type which is not at a stability boundary. In fact it is a highly stable configuration – well away from the marginal confocal ($R_1 = R_2 = L/2$) and plane–plane ($R_1 = R_2 = \infty$) resonator geometries.

Besides the cavities discussed above their exist many other geometries that satisfy the stability condition of equation (1.16). These include the asymmetric concave mirror combinations and cavities in which one cavity mirror is convex ($R < 0$). Such cavities are used relatively rarely.

## 1.4 DIFFRACTION APPROACH TO STABLE RESONATORS

### 1.4.1 Gaussian Spherical Waves

The stability conditions derived in the previous section by ray analysis remain equally valid when diffraction is included. A detailed diffraction analysis shows that a number of transverse amplitude distributions – transverse *modes* – can resonate stably in any one cavity.

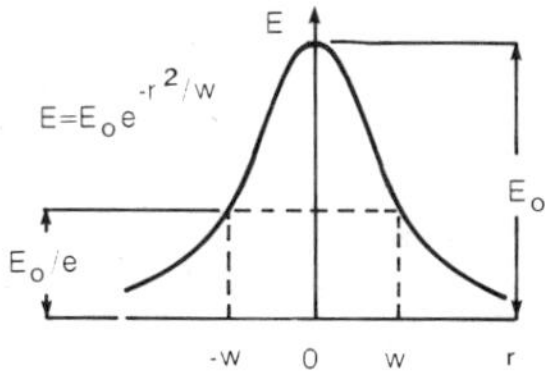


Fig 1.6 The variation of field amplitude, in the transverse direction for a Gaussian beam.

The lowest order mode – the mode that is confined most strongly near the optical axis – has a Gaussian amplitude distribution. That is the electric field amplitude, $E$, varies in the transverse dimension, $r$, as:

$$E = E_0 \exp(-r^2/w^2) \tag{1.17}$$

where $E_0$ is the central amplitude (at $r$=0) and $w$ is a measure of the width of the distribution. Commonly, $w$ is called the spot radius and, as shown in Figure 1.6, when $r=w$, $E = E_0/e$.

Section 1.5 discusses the properties of Gaussian light beams in general. However, in the context of this resonator discussion, *three* properties need be noted immediately:

(i) A plane wave with a Gaussian transverse amplitude distribution ($w = w_0$) will diverge as it propagates, but will always retain the Gaussian distribution.

(ii) The divergence can be characterised by the change in $w$ as the light (wavelength, $\lambda$) propagates along the z-axis, **in either direction (+z or −z)**:

$$w(z) = w_0 \left[1+\left(\frac{\lambda z}{\pi w_0^2}\right)^2\right]^{\frac{1}{2}} \tag{1.18}$$

(iii) The divergence corresponds to the (initially plane) wave-fronts becoming spherical (again with symmetry about $z = 0$, $w = w_0$) with radius of curvature given by:

$$R(z) = z\left[1+\left(\frac{\pi w_0^2}{\lambda z}\right)^2\right] \tag{1.19}$$

### 1.4.2 Lowest-Order Mode Diameters in Stable Resonators

From the above properties of Gaussian spherical waves we can deduce the size of the lowest order transverse mode of any resonator. The relevant boundary condition is that the wave-front curvature must match the curvature of the mirror when reflecting from it. In this way it will simply retrace the same pattern each time it crosses the cavity.

For the symmetric concave cavities, it is clear that the *waist* of the beam − the narrowest, plane-wave part of the beam ($z = 0$, $w = w_0$) − must be located at the centre and the mirrors, therefore, at $z = \pm L/2$ (see Figure 1.7). Using equation (1.19), it can then be concluded that the mirror curvature determines $w_o$ through the relation:

$$R = \tfrac{1}{2}L\left[1+\left(\frac{2\pi w_0^2}{\lambda L}\right)^2\right]$$

and thus, rearranging:

$$w_o = \left(\frac{\lambda L}{2\pi}\right)^{\frac{1}{2}}\left(\frac{2R}{L} - 1\right)^{\frac{1}{4}} \tag{1.20}$$

Fig 1.7 A concave-symmetric cavity with a resonant Gaussian mode indicated by the variation of waist size and wavefront curvature.

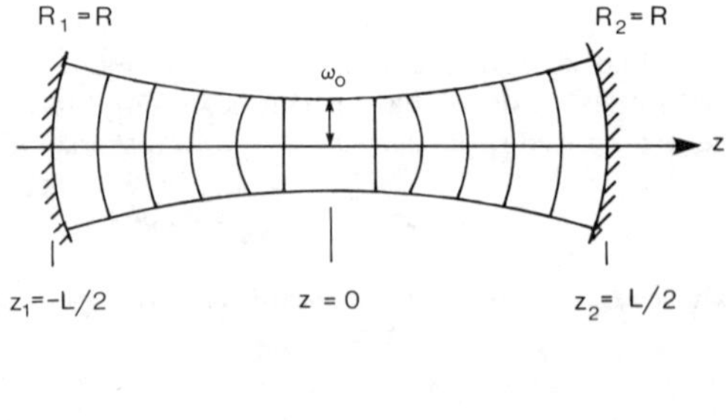

It can be seen that, given a specific resonator length, mirror curvature, and wavelength, the variation of spot radius throughout the cavity is fully specified by equations (1.20) and (1.18).

The plane–concave cavities, as noted in the earlier ray analysis, are effectively folded versions of the symmetric variety. The fact that this is logically consistent with the wave model is apparent by inspection of Figure 1.7, where it is clear that the plane wave fronts at the mode waist would perfectly match a plane reflector placed at this point. In this case, assuming the plane mirror is the output coupler, the waist size corresponds to the output beam dimension. (The output beam then proceeds to diverge in exactly the same way as the reflected component which remains inside the cavity).

In general, $w_0$ is relatively small. For example, for a symmetric stable cavity with $R$ = 2m, L = 1m and $\lambda$ = 1$\mu$m, equation (1.20) indicates $w_0$ = 0.5mm. This is one of the limitations of a stable cavity operating on a single transverse mode – it couples poorly to a *large* cross–section gain medium. Two possible solutions to this, that are discussed in later chapters, are the exploitation of unstable cavities with much larger mode volumes or the use of waveguide resonators in which the gain medium is shrunk to a size equal or less than the free–space mode dimension.

The *general* equation for the waist radius in an asymmetric concave resonator is:

$$w_0 = \left[\frac{\lambda}{\pi}\right]^{\frac{1}{2}} \left[\frac{L(R_1-L)\ (R_2-L)\ (R_1+R_2-L)}{(R_1+R_2-2L)^2}\right]^{\frac{1}{4}} \qquad (1.21)$$

which reduces to equation (1.20) when $R_1$=$R_2$. As before, this equation is derived from the simple assumptions that the resonating mode has a Gaussian transverse amplitude distribution and that the wave–front curvature must match the mirror curvature on reflection. Justification of the choice of a Gaussian mode comes from considering the limitations imposed on the cavity by equation (1.21). For $w_0$ to be real, the numerator within the second brackets must be positive, i.e. :

$$L(R_1-L)(R_2-L)(R_1+R_2-L) > 0$$

Noting that: $(R_1+R_2-L)L/R_1R_2 = 1-(1-L/R_1)(1-L/R_2)$

this can be re–written: $(1-L/R_1)(1-L/R_2)[1-(1-L/R_1)(1-L/R_2)] > 0$

which is equivalent to: $0 < (1-L/R_1)(1-L/R_2) < 1$ (1.22)

Thus it is apparent that the stability condition for a Gaussian mode within a cavity is precisely consistent with that derived from geometric ray analysis – equation (1.16).

When calculating the lowest order transverse mode geometry for a general cavity, only the waist dimension – equation (1.21) – and the waist position are required. The deviation of the latter from the resonator centre, simply follows from the degree of mirror curvature asymmetry. Thus the ratio $-z_1/z_2$

(where the subscripts refer to the mirrors 1 and 2) is given by:

$$\gamma = -\frac{z_1}{z_1} = \frac{R_2 - L}{R_1 - L} \tag{1.23}$$

Finally, it can be deduced that the beam radii on each mirror are:

$$w_1 = \left[\frac{\lambda R_1}{\pi}\right]^{\frac{1}{2}} \left[\frac{\gamma L}{R_1 + R_2 - L}\right]^{\frac{1}{4}} \tag{1.24}$$

$$w_2 = \left[\frac{\lambda R_2}{\pi}\right]^{\frac{1}{2}} \left[\frac{L/\gamma}{R_1 + R_2 - L}\right]^{\frac{1}{4}} \tag{1.25}$$

### 1.4.3 Fresnel Number and Diffraction Losses

Having determined the dimensions of the lowest order transverse modes, there remains the question: how large should the mirrors be to avoid significant losses resulting from light diffracting past their edges? Clearly, such a loss factor can never be completely zero as all Gaussian-mode irradiance distributions theoretically extend to infinity.

A useful dimensionless parameter in discussing this aspect of resonators is the Fresnel number, $N$:

$$N = a^2/L\lambda \tag{1.26}$$

where a refers to the radii of the mirror apertures. $N$ corresponds to the number of Fresnel zones that fit within the mirror diameter, viewed from the centre of the opposing mirror, distance $L$ away. The condition $N = 1$ defines a length $L = a^2/\lambda$ over which a plane wave of radius $a$, will propagate before developing strong diffraction rings. It is reasonable to assume therefore that a cavity of Fresnel number around unity, or greater, will experience minimal diffraction loss.

A more quantitative approach to this problem uses the equations for the cavity spot radii from the previous section. For example, the largest spot radius within a plane-concave resonator (i.e. on the concave mirror) is given by equation (1.24) with $R_2 = \infty$:

$$w = \left[\frac{\lambda L}{\pi}\right]^{\frac{1}{2}} \left[\frac{R/L}{1 - L/R}\right]^{\frac{1}{4}} \tag{1.27}$$

For typical $L/R$ values, the second bracket is of order unity. If $N \sim 1$, then $\lambda L \sim a^2$ and hence $a = w\sqrt{\pi}$.

The power loss for a gaussian beam of spot-radius $w$, reflecting from a mirror of radius $a = w\sqrt{\pi}$ is just 0.2% (see section 1.5.3) and it can therefore be concluded that the condition $N \geqslant 1$ is indeed generally sufficient to ensure negligible diffraction losses.

Only when the plane-plane cavity condition is approached, $L/R \rightarrow 0$, does the final term in equation (1.27) lead to significant round-trip diffraction losses in an $N = 1$ cavity. In this regime, where the Gaussian mode volume ultimately diverges to infinity at $R_1 = R_2 = \infty$, the resonator modes are no longer Gaussian but are instead determined dominantly by the mirror dimensions. In practice, whatever the cavity parameters, if a significant part of the ideal Gaussian mode extends beyond the mirror radii (or any other limiting aperture) the mode that actually oscillates is a truncated Gaussian with somewhat lower loss than might otherwise be expected.

### 1.4.4 Higher Order Transverse Modes And Their Suppression

As already indicated, the fundamental Gaussian mode is only the lowest-order of a set of spherical-wave transverse modes that all match the resonator stability requirements. The higher-order modes also propagate with fixed amplitude distribution imposed on (diverging/converging) spherical wave-fronts, with identical curvature to the Gaussian mode. However, the power distribution across a specific wave-front has a greater transverse extent than the Gaussian. The higher-order modes can be described: (i) in a rectangular coordinate system by products of the Gaussian function with Hermite polynomials (Hermite-Gaussian modes), or (ii) in a cylindrical coordinate system by products of the Gaussian function and Laguerre polynomials (Laguerre-Gaussian modes). The two descriptions are formally equivalent.

The transverse irradiance distributions that these functions describe are multi-lobed; i.e. they correspond to a series of separated spots (or rings) spread symmetrically about the z-axis in both transverse directions (x and y). The number of zero-irradiance points (i.e. gaps between spots) corresponds to the order of the underlying polynomial function and is used to label each mode. This is done with two subscripts, thus: $TEM_{mn}$ – where $m$ and $n$ identify the mode order for each independent transverse coordinate (x and y, retangular; or r and $\theta$, cylindrical). TEM refers to the mode being a free-space transverse electromagnetic wave. Thus the lowest order (Gaussian) mode is labelled $TEM_{00}$. Sometimes a third subscript '$q$' is added – corresponding to the longitudinal mode number; i.e. the number of half-wavelengths within the cavity. This is rarely of any significance in cavities greater than a few $\lambda$ in length, as it is a very large, and usually uncertain, number.

Full descriptions of the different higher-order mode distributions are available in most basic texts on lasers. These details, however, are of little practical value for two reasons. Firstly, non-Gaussian mode outputs from real (stable) lasers are generally complex super-positions of the modes forming these basic sets, with the final distribution actually determined by various minor details of the cavity/gain-medium construction (Chapters 9 and 10 describe ways of handling the propagation of multimode laser beams.) Secondly, the higher-order modes are usually not wanted and in most laser designs, efforts are made to actively suppress them.

This latter objective follows from the usual desire to exploit the basic uniqueness of a laser as a light source with high spatial and temporal coherence. Because they have the same wave-front curvature as the fundamental Gaussian, but with a larger cross-sectional area, the higher-order modes have a lower radiance (W $m^{-2}$ $Sr^{-1}$) at a given power level. This reduced spatial coherence is a result of the multi-phased nature of the higher-order modes – adjacent lobes being $\pi$ out of phase with each other. The axial Gaussian mode, by contrast, is uni-phase. The higher order modes also exhibit different cavity round-trip phase changes as a result of fractionally higher phase velocities. Thus, when more than one transverse mode is oscillating, the laser output is not single frequency (even if it is nominally *single longitudinal mode*) and thus temporal coherence is degraded.

The degree by which the higher-order modes deviate in frequency from the $TEM_{00q}$ axial mode is given by:

$$\Delta\nu_{nm} = \beta(m+n)\,\frac{c}{2nL} \tag{1.28}$$

where

$$\beta = (1/\pi)\,\cos^{-1}\{\pm[\,(1-L/R_1)(1-L/R_2)\,]^{\frac{1}{2}}\} \tag{1.29}$$

Because $\beta$ lies in the range 0 to 1, the higher-order transverse modes can be spaced apart anywhere between the longitudinal mode spacing, $\Delta\nu_q = c/2nL$ (when $R_1+R_2 = L$), down to small fractions of $\Delta\nu_q$ (when $R_1 = R_2 >> L$).

The higher-order transverse modes are usually suppressed by including a restrictive aperture of some sort in the cavity: e.g. a small diameter cavity mirror or an iris. Sometimes the gain medium itself (i.e. the laser rod or discharge tube) will restrict the higher order modes by its small size - but this may also lead to unwanted waveguide effects if not designed properly. The purpose of this aperture is to introduce significant losses for the non-axial modes, while leaving the round-trip loss of the axial $TEM_{00}$ mode little changed. This is possible because the higher-order modes extend further in the transverse direction than the $TEM_{00}$ mode.

The simplest experimental approach to finding the optimum aperture size which just prevents the higher-modes from oscillating (i.e. while not significantly attenuating the axial mode), is to vary an adjustable iris by trial and error. (Indeed many commercial lasers include such a variable aperture.) Alternatively, detailed diffraction calculations can be carried out numerically. The latter show that the optimum mode selection aperture, $a_A$, is such as to give the typical resonator a Fresnel number around unity:

$$a_A \sim (L\lambda)^{\frac{1}{2}} \tag{1.30}$$

For a cylindrically symmetric cavity with $(1-L/R_1)(1-L/R_2) = 0.5$ and $N = 1$, calculations by H. Kogelnik and T. Li show that the $TEM_{00}$ mode has a round trip loss of only ~ 2% while the $TEM_{01}$ mode experiences ~ 20% diffraction loss. For a cavity near oscillation threshold this difference can be sufficient to totally suppress the higher modes.

## 1.5 PROPERTIES OF GAUSSIAN BEAMS

### 1.5.1 The Paraxial Wave Equation and the Complex Beam Parameter.

Any electromagnetic wave propagating in free-space must satisfy the scalar equation:

$$\nabla^2 E + k^2 E = 0 \tag{1.31}$$

where the electric field $E$ is an oscillatory function of time and also varying in space and where $k=2\pi/\lambda$ is the wavevector. There are an infinite number of possible solutions to this equation, dependent upon the boundary conditions that are imposed. The open resonators discussed in the previous section require light to oscillate with minimal divergence along the axis connecting two, generally spherical mirrors. Such a light beam can be described as a plane wave, propagating in the $z$ direction, modified by a complex amplitude, $u$, that varies across the $x/y$ plane in such a manner as to account for the spatial variation of irradiance and the deviations in phase corresponding to the non-planar wave-fronts. Using this paraxial approximation the spatial dependence of the electric field, that must satisfy equation (1.30), can be written:

$$E(x,y,z) = u(x,y,z)\ e^{-jkz} \tag{1.32}$$

Substituting into equation (1.30) gives:

$$\frac{\partial^2 u}{\partial x^2} + \frac{\partial^2 u}{\partial y^2} - 2jk\frac{\partial u}{\partial z} = 0 \tag{1.33}$$

Here the usual slowly varying amplitude approximation has been made, permitting the $\partial^2 u/\partial z^2$ term to be dropped. Equation (1.33) is known as the paraxial wave equation and remains valid for light beams with half cone-angles of up to about 30°. Given the boundary conditions that (i) the waves resonating within a laser cavity must be spherical (i.e. matching the mirror curvatures on reflection) and that (ii) the lowest order low-divergence mode is favoured, then the solution of equation (1.33) is the spherical Gaussian wave assumed in section 1.4.1. (The higher order transverse modes referred to in section 1.4.4, correspond to other, higher-divergence, spherical wave solutions.)

The solution for the complex amplitude, $u(x,y,z)$, corresponding to the Gaussian beam takes the form:

$$u = \exp\left\{-j\left[P(z) + \frac{k(x^2 + y^2)}{2q(z)}\right]\right\} \tag{1.34}$$

where $P(z)$ and $q(z)$ can be determined by substituting into equation (1.33).

This gives:

$$\frac{k(x^2 + y^2)}{q^2} + \frac{2j}{q} + 2\frac{\partial P}{\partial z} - \frac{k(x^2 + y^2)}{q^2}\frac{\partial q}{\partial z} = 0 \tag{1.35}$$

The requirement that equation (1.35) must be valid for any values of $x$ and $y$ leads to two first-order differential equations:

$$\frac{\partial q}{\partial z} = 1 \quad ; \quad \frac{\partial P}{\partial z} = -\frac{j}{q} \tag{1.36}$$

This immediately shows that the $z$-dependence of the parameter $q$ – known as the *complex beam parameter* – is given by the simple equation:

$$q(z) = q(0) + z \tag{1.37}$$

In the solution for $u(x,y,z)$ – equation (1.34) – $q(z)$ was introduced as a reciprocal within the complex exponent. It is useful to expand $1/q(z)$ into its real and imaginary components in the following way:

$$\frac{1}{q(z)} = \frac{1}{R(z)} - \frac{j\lambda}{\pi w^2(z)} \tag{1.38}$$

where $R(z)$ and $w(z)$ are purely real parameters. The reason for doing this becomes apparent when equation (1.38) is used to substitute for $q(z)$ in equation (1.34):

$$u = \exp\left\{-j\left[P(z) + \frac{k(x^2+y^2)}{R(z)}\right]\right\} \cdot \exp\left[-\frac{(x^2+y^2)}{w^2(z)}\right] \tag{1.39}$$

If $x^2 + y^2$ is written $r^2$, $r$ being the radial distance from the axis in the $x/y$ plane, then the last (non–oscillatory) term in equation (1.39) is identical to the Gaussian distribution of equation (1.17) in which $w$ represents the spot radius (see Figure 1.6). The first, spatially oscillatory, term has two components, both describing how the phase varies from the ideal plane–wave solution. The transverse deviation corresponds to a spherical wavefront of radius $R(z)$; confirming the spherical wave nature of this solution. The longitudinal phase term, $P(z)$, can also be written in terms of $w(z)$ and $R(z)$ and corresponds to a small on–axis phase–shift between this Gaussian beam solution and the infinite plane–wave.

If the origin $z$=0 is defined as the position at which the wavefronts are plane ($R=\infty$) then, from equation (1.38):

$$q(0) = \frac{j\pi w_0^2}{\lambda} \tag{1.40}$$

where $w_0 = w(0)$ is the waist radius. With this definition of $q(0)$, equation (1.37) and (1.38) provide two simultaneous complex relations for the spot radius $w$(z) and wavefront radius $R$(z). Equating real and imaginary parts leads to the two equations (1.18) and (1.19) that were used to describe $w$(z) and $R$(z) in section 1.4 and which are discussed further in the next section.

### 1.5.2 Propagation of Gaussian Beams

The Gaussian beam, described in equation (1.17) in terms of the field amplitude variation (Figure 1.6), has an *irradiance* profile:

$$I(r) = I(0)\exp(-2r^2/w^2) \tag{1.41}$$

where the central (peak) irradiance is related to the total power $P$ in the beam by:

$$I(0) = 2P/\pi w^2 \tag{1.42}$$

It can be seen that defining the spot–size $w$ (the radius at which $I$ drops to $I(0)/e^2$) as an effective radius of the beam, that $I(0)$ is exactly twice the "average irradiance" obtained by dividing $P$ by the spot area, $\pi w^2$.

All Gaussian beams have a position along their axis at which the wavefronts become plane and the spot radius goes through a minimum. As indicated previously, it is from this ***waist*** position, where $w = w_0$, that distances are conventionally measured along the beam path (See Figure 1.8).

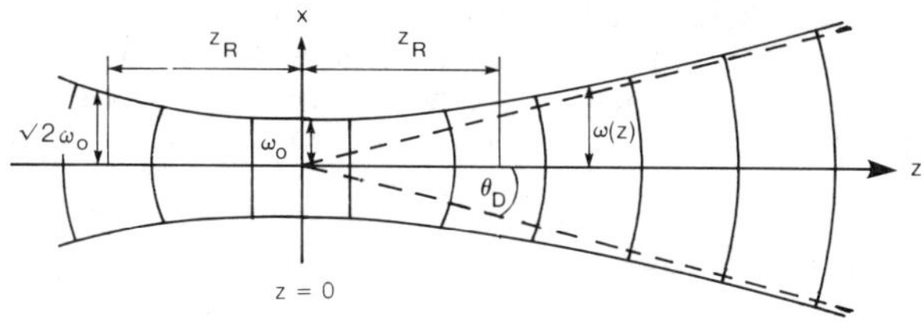


Fig 1.8 The geometry of a Gaussian beam. Showing the focal (near-field) region, i.e. within one Rayleigh range $Z_R$ of the waist position, and the asymptotic approach to the far-field divergence angle, $\theta_D$.

The variation of the radius, $w$, along the propagation axis, $z$, can be written:

$$w(z) = w_0\,[1+(z/z_R)^2]^{\frac{1}{2}} \qquad (1.43)$$

where $z_R = \pi w_0^2/\lambda$, and is known as the Rayleigh range; (Equation (1.43) is, of course, identical to equation (1.18) for the Gaussian resonator mode.) The significance of the Rayleigh range (the waist area divided by $\lambda$) is that at this distance from the waist the beam area has just doubled. The region $-z_R$ to $+z_R$ can therefore be identified as the focal zone, or the region over which the beam is approximately collimated. (An alternative term to the Rayleigh range, which will also be found in the literature, is the *confocal parameter*: $b = 2z_R$).

The wave-front radius of curvature $R(z)$ can also be written in terms of the Rayleigh range:

$$R(z) = z[1+(z_R/z)^2] \qquad (1.44)$$

Differentiation of equation (1.44) readily shows that $R(z)$ goes through a minimum value of $R = 2z_R$ (maximum curvature), when $z = z_R$. It can also be seen that in addition to the wavefronts being plane ($R = \infty$) at $z = 0$, they approach plane wave configuration as $z \rightarrow \infty$. The $z \gg z_R$ regime, as indicated in Figure 1.8, corresponds to a spherical wave diverging from a point source at $z = 0$. In the paraxial approximation, the half-cone-angle of this divergence ($z \gg z_R$) is given by:

$$\theta_D = w(z)/z = \lambda/\pi w_0 = w_0/z_R \qquad (1.45)$$

The two regimes $z \ll z_R$ and $z \gg z_R$ are generally known as the near-field region and far-field region respectively.

The relationship between $R$, $w$ and $z$ is neatly summarised in terms of the complex beam parameter, $q(z)$, as defined in equation (1.38). Given $q$ at some position $z$, equation (1.37) gives the complex beam parameter at a distance $L$ further along the beam, as simply:

$$q(z+L) = q(z)+L \tag{1.46}$$

It can be seen that $q$ may be viewed as a complex curvature, by comparison with the equivalent relation for a spherical–wave diverging from an ideal point source: $R_s(z + L) = R_s(z) + L$. Given the wavefront radius of curvature, $R$, and spot radius, $w$, at some arbitrary position along the beam, then both the size and position of the waist can be deduced by combining the complex beam parameter equations (1.37), (1.38) and (1.40) to obtain:

$$\left[ \frac{1}{R} - j\frac{\lambda}{\pi w^2} \right]^{-1} = z + \frac{j\pi w_o^2}{\lambda} \tag{1.47}$$

Equating real and imaginary parts gives the distance to the waist:

$$z = R\left[1 + \left[ \frac{\lambda R}{\pi w^2}\right]^2 \right]^{-1} \tag{1.48}$$

and the waist radius:
$$w_o = w\left[1 + \left[ \frac{\pi w^2}{\lambda R} \right]^2\right]^{-\frac{1}{2}} \tag{1.49}$$

### 1.5.3 Power Distribution in Gaussian Beams

The total power within radius '$a$' from the axis of a Gaussian beam can be obtained by integrating equation (1.41):

$$P_a = \frac{2P}{\pi w^2} \int_0^a 2\pi r \exp(-2r^2/w^2)dr \tag{1.50}$$

to obtain the result:
$$P_a/P = 1\text{-}\exp(-2r^2/w^2) \tag{1.51}$$

This reveals the interesting result that at the radius at which the irradiance drops to a level $xI(0)$, the fraction of beam power enclosed within that radius is simply $1-x$. Thus, for example, at the spot radius ($r = w$), when $I = I(0)/e^2 = 0.135 \cdot I(0)$, the power enclosed is 0.865 of the total power in the beam.

The above relationship also demonstrates that the use of the central part of a Gaussian beam to provide uniform irradiance illumination of a surface is extremely inefficient. For example, if the Gaussian is expanded to cover the relevant area such that the irradiance drops by only, say, 5% at the edges, then only 5% of the total power is being used.

Equation (1.51) also shows that, although the Gaussian profile theoretically extends to infinity, ~99% of the total power lies within radius $1.5w$. Thus in designing an optical system through which a Gaussian beam is to be relayed, it is usually sufficient to make all apertures, lenses, etc., $\geqslant 3w$ in diameter, provided the beam remains accurately centred. (NB. Somewhat larger apertures may be required in a long–path optical system in which Fresnel diffraction fringes need to be strongly suppressed).

### 1.5.4 Gaussian Beam Focussing

The effect of a thin lens on a diverging, or converging, spherical wave is to transform the radius of curvature from $R$ to $R'$, according to:

$$1/R' = 1/R - 1/f \tag{1.52}$$

where $f$ is the lens focal length. This applies equally to a Gaussian beam, although for completeness it should be written in terms of the complex beam parameter (curvature):

$$1/q' = 1/q - 1/f \tag{1.53}$$

In fact, because the spot radius is left unaltered by the lens, equations (1.52) and (1.53) are identical relations.

Knowing the spot radius at the lens, and given the transformed wave–front radius of curvature, equations (1.48) and (1.49) can be used to deduce the position and size of the new waist (focal spot). If the beam entering the lens is characterised in terms of its far–field divergence, $\theta_D$, and spot–radius at the lens, $w_L$, then the new waist radius is given by:

$$w_0' = \frac{\lambda}{\pi} \left\{ \theta_D^2 + \left[\frac{w_L}{f}\right]^2 - \frac{2\lambda}{f} \left[ \left[\frac{\pi w_L \theta_D}{\lambda}\right]^2 - 1\right]^{\frac{1}{2}} \right\}^{-\frac{1}{2}} \tag{1.54}$$

This simplifies considerably under certain particular conditions. For example, for a lens placed in the near field region of the initial beam (i.e. $w_L \approx w_0$), equation (1.54) becomes

$$w_0' = f\theta_D\left[1 + \left[\frac{f\theta_D}{w_L}\right]^2\right]^{\frac{1}{2}} \tag{1.55}$$

Assuming the lens f–number, $F^{\#} = f/2w_L$ , is significantly less than $1/\theta_D$ then this reduces to $w_0 = f\theta_D$ and the new waist position is distance $f$ from the lens (as expected for a collimated beam).

When the lens is placed one focal–length away from the initial waist the new waist size is again $w_0' = f\theta_D$ (with no approximations) and the position is, once more, $f$ from the lens. **This is an unexpected result.** In geometric optics a point source at the initial waist position would be collimated, i.e. focus at infinity.

A third case is where the lens is placed in the far field, $z \gg z_R$. In this case equation (1.54) reduces to

$$w_0' = \frac{\lambda}{\pi} \left[ \frac{w_L}{f} - \theta_D \right] \tag{1.56}$$

and the distance from the lens to the new waist position, $z'$, is given by

$$1/z' = 1/f - 1/z_L \tag{1.57}$$

which is the usual geometric optics imaging formula (thin lens, paraxial approximation). Equation (1.57) follows directly from equation (1.52), when it is noted that in the far–field the Gaussian beam looks like a spherical wave radiating from a point source at the waist, i.e. with $R = z$. Assuming, again, that the lens f–number is small compared to $1/\theta_D$, then equation (1.56) reduces to $w_0' = \lambda f/\pi w_L$

It can be seen from the above examples that, provided the focussing power of the lens dominates the initial divergence, the new waist will be near the focal point of the lens and its radius will be given by:

$$w_0' = \frac{\lambda f}{\pi w_L} = 0.64 \; F^{\#}\lambda \tag{1.58}$$

where $F^{\#}$ is the $f$-number at the lens ($f/2w_L$). This can be compared to focussing of a circular uniform plane-wave, where the Airy-disc radius is given by:

$$r_A = 1.22 \; F^{\#}\lambda \tag{1.59}$$

It is interesting to note that the power falling within the Airy-disc is ~85% of the power passing through the lens, compared to ~86% of a Gaussian into radius $w_0$, and thus $r_A$ and $w_0$ are comparable measures of the degree by which an optical beam is being focussed. Although equations (1.58) and (1.59) imply that a Gaussian beam forms a better focus than a uniform plane-wave assuming comparable focussing $f$-numbers, it must be remembered that the lens diameter must be significantly larger than the Gaussian beam diameter ($2w$) if significant power loss is to be avoided. Thus to ensure this focussing advantage, the *lens* $f$-number must be greater when handling the Gaussian beam. Conversely, if the lens dimensions are fixed, and it is assumed that its diameter is twice that of the incident Gaussian beam diameter (to ensure negligible truncation), then the effective $f$-number of the focussing Gaussian will be half that of a uniform plane-wave passing through the lens and consequently, from equations (1.58) and (1.59), the two focal spots will be equal in size.

Finally, to conclude with, *three* points of warning. *Firstly*, the properties of Gaussian beams described above should not be confused with the term Gaussian Optics which refers to the study of geometric optics under the paraxial ray approximation. *Secondly*, the difference between using a lens to relay a Gaussian waist, and imaging of some plane, can sometimes be considerable. For example, a positive lens placed $2f$ from the waist will form a 1:1 image of the waist plane at a distance of $2f$. However the new waist position will vary, depending on the actual values of $f$, $\theta_D$ and $\lambda$, anywhere between $2f$ and just $f$ from the lens. *Thirdly*, when considering the focusing of Gaussian laser beams, diffraction based calculations are rarely sufficient (see Section 10.5). Geometric aberrations (particularly spherical aberration) are usually significant contributors to actual focal spot-sizes. Only when large $f$-numbers are being used (e.g. $F^{\#} > 10$, for visible wavelengths) will the focus produced by a single (optimum shape) thin lens be the size predicted by Gaussian beam formulae.

## 1.6 REFERENCES

1. For the basics of stable resonator design and Gaussian beam propagation, see: H Kogelnik and T Li 1966 Appl. Opt. 5 1550
2. For more general discussion of all aspects of resonator/laser design see: A E Siegman Lasers, (Mill Valley California: University Science Books) 1986

# 2 Unstable Resonators

*P E DYER*

## 2.1 INTRODUCTION

Unstable resonators have proved to be of great value for many laser systems, particularly those with large mode volumes and high gains which are capable of operating at high powers on a pulsed or cw (continuous wave) basis. In the early days of the laser it was only necessary to be aware of the fact that such resonators were definitely to be avoided, their high *diffraction–loss* being anathema to the laser designer! The subsequent development of volume scalable systems, notably the $CO_2$ and HF gas discharge lasers and the pioneering theoretical work by Siegman and co–workers (Siegman 1965, Siegman and Arrathoon 1967, Siegman and Miller 1970) altered this situation. It was rapidly realised that unstable resonators provided a *complementary* rather than competitive alternative to stable–resonators (reviewed in the previous chapter), each class providing a valuable solution to laser power extraction depending on the system involved.

This chapter discusses the advantages unstable resonators offer the laser designer, how the basic mode properties can be analysed using a simple geometrical model, and to what extent agreement with experimental results is obtained. Information gained from wave calculations of mode properties is then examined and evaluated, leading to the formulation of a set of design steps for a resonator. A model for mode evolution in short–pulse lasers is then presented which illustrates the useful role that unstable resonators can play for such systems. The chapter concludes with a brief consideration of injection controlled, novel geometry, and variable–reflectivity–mirror unstable resonators (see also Chapter 6).

For a more detailed and comprehensive treatment of unstable resonators the reader is directed to the recent book by Siegman (1986); excellent physical insight into the behaviour of these resonators has also been provided by Anan'ev and co–workers in a series of publications (Anan'ev *et al* 1968, 1969, 1972, Anan'ev 1975).

## 2.2 ADVANTAGES

The following, intertwined, list of advantages for choosing an unstable resonator design can be put forward:

- Single transverse–mode, near diffraction limited output beam at high Fresnel numbers.

- High radiance emission; that is the power radiated per unit area of output per unit solid-angle.
- Controllable mode-volume.
- All reflective optics capability.
- Rapid mode evolution in short-pulse lasers.
- Novel geometries.
- Well suited to injection locking.

Of these the second is probably the most important, and encapsulates the major advantage of the unstable-resonator for large mode-volume devices. As radiance is a conserved quantity in a lossless optical system, it means that high irradiance ($Wcm^{-2}$) beams can be achieved by focussing, or low divergence beam propagation attained with suitable collimating optics. For the diffraction-limited case it can be shown (Dyer 1989) that the radiance, $I$ is:

$$I \simeq 0.3 \times P_L/\lambda^2$$

where $P_L$ is the laser power and $\lambda$ the wavelength. Thus, noting the $\lambda^2$ dividing factor, if a large-volume laser with concomitantly high power output is operated near 'diffraction limited', the radiance will be extremely high. A close approach to this condition is often possible with a correctly designed unstable resonator.

## 2.3 BASIC CONCEPTS

A simple two-mirror optical resonator is classed as unstable if (Chapter 1, section 3.1):

$$g_1g_2 > 1 \quad \text{or} \quad g_1g_2 < 0$$

where $g_1 = 1 - L/R_1$ and $g_2 = 1 - L/R_2$, $R_1$ and $R_2$ being the mirror radius of curvatures and L the mirror spacing.

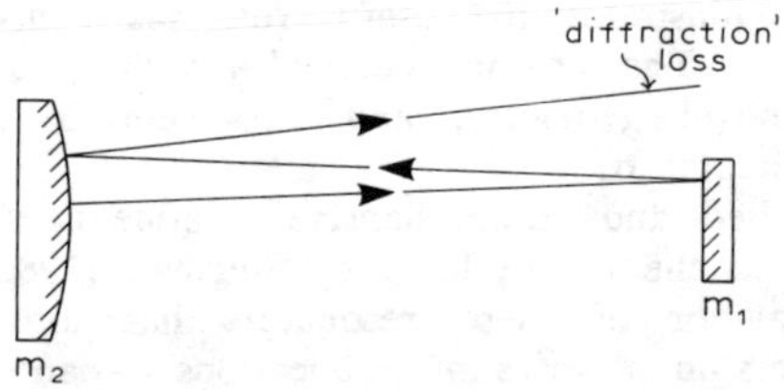


Fig 2.1 In an unstable resonator rays 'walk' away from the resonator axis.

Rays in such a resonator tend to walk away from the axis rather than being confined as illustrated in Figure 2.1. Alternatively, using a wavefront description, the mode continually spreads out across the resonator on each round-trip. The effect of this is that the so-called 'diffraction loss', that is radiation that goes around the edge of the mirror(s), is large. (Since in the case of unstable resonators, the mirror geometry enhances this, it is perhaps more appropriately called geometrical loss). This outward spreading means that if a gaussian distribution 'test' beam is launched into the structure then the central flat portion of the beam fills the resonator after a few round trips. A gaussian beam propagation analysis is then no longer suitable (or possible), as this theory is valid only for small (diffraction) losses at the edge of the beam. Instead, as the radiation distribution tends to become uniform, analysis falls in the domain of conventional optics which is founded on the behaviour of point sources providing uniform irradiance over the aperture of the imaging element.

The tendency of the central beam region to spread-out gives a clue as to why this type of resonator can provide good spatial coherence over large apertures. An initially small patch of coherence near the axis is rapidly communicated across the aperture by the repeated expansion process. In contrast, in a stable resonator a similar coherence zone is confined to a small size on the resonator axis limiting the size of the single transverse mode it can support.

(a)

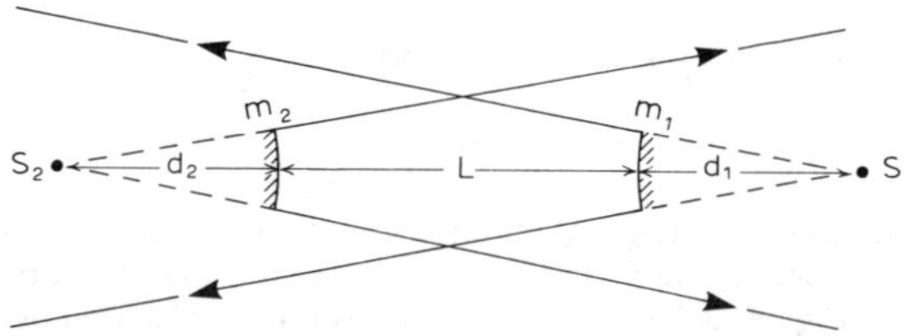


(b)

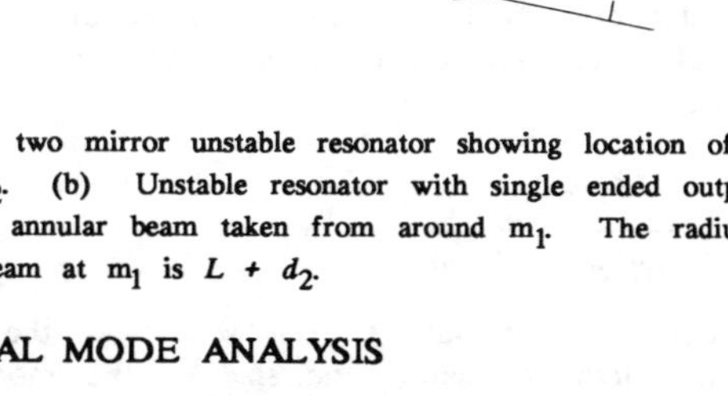


Fig 2.2 (a) Generalized two mirror unstable resonator showing location of virtual source points $S_1$ and $S_2$. (b) Unstable resonator with single ended output coupling in the form of an annular beam taken from around $m_1$. The radius of curvature of the output beam at $m_1$ is $L + d_2$.

## 2.4 GEOMETRICAL MODE ANALYSIS

Figure 2.2a shows a generalised unstable resonator consisting of two fully reflecting mirrors $m_1$ and $m_2$ and with output coupling taken from around the mirrors. On the basis of the above discussion a self-reproducing field distribution for the resonator can be sought using geometrical optics (Siegman

1965). It is asssumed radiation originates from a virtual point source, $S_1$, located as shown a distance of $d_1$ behind $m_1$. This is then imaged by $m_2$ to produce a second virtual source, $S_2$, which, for self-reproducing conditions, must form $S_1$ when imaged in $m_1$. Using the expression for image formation by spherical mirrors the locations of $S_1$ and $S_2$ which satisfy this requirement are (Steier 1979):

$$d_1 = L\{g_1 - 1 \pm (g_1^2 - g_1/g_2)^{\frac{1}{2}}\}^{-1}$$
$$d_2 = L\{g_2 - 1 \pm (g_2^2 - g_2/g_1)^{\frac{1}{2}}\}^{-1} \quad (2.1)$$

The appropriate solutions are those yielding $d_1/L,\ d_2/L > -1$.

A beam originating from $S_2$ travelling to the right is magnified such that its dimension transverse to the axis at $m_1$ is $M_{21}$ times its size at mirror $m_2$, where by geometry:

$$M_{21} = (L + d_2)/d_2$$

Likewise, for a leftward directed beam:

$$M_{12} = (L + d_1)/d_1$$

If output coupling is required from only one end of the resonator, as is usual in practice, one mirror is made sufficiently large to intercept all the radiation (Figure 2.2b). The round trip magnification is:

$$M_{12} \times M_{21} = M_{\pm} = 2g_1g_2 \pm 2[g_1g_2(g_1g_2 - 1)]^{\frac{1}{2}} - 1 \quad (2.2)$$

For a given resonator there are thus two possible M values, one of these describing a demagnifying beam $|M| < 1$, the other a magnifying beam, $|M| > 1$; these magnifications satisfy $M_+M_- = 1$. Both have self-reproducing sets of virtual centres but only the magnifying beam reproduces its field (irradiance) distribution on each round-trip and thus has the correct characteristics of a mode. The demagnifying beam, however, can have significant influence on the resonator properties as will be discussed later.

The output beam for a spherical mirror resonator is an annulus (if mirror $m_1$ has a circular cross-section), with an inner diameter equal to that of $m_1$, that is $2a_1$, and outer diameter of $2Ma_1$. The fraction of the beam returned by mirror $m_1$ is:

$$\frac{\pi a_1^2}{\pi M^2 a_1^2} = \frac{1}{M^2}$$

and the fraction coupled out, $1 - 1/M^2$. As the magnification does not depend on the mirror diameter, the size of the mode can be controlled by choice of $a_1$; it can then be arranged that:

$$2Ma_1 = D$$

where $D$ is the characteristic diameter of the laser gain medium.

For a resonator with cylindrical mirrors the output appears as two coherent strips, one from each side of the resonator, and the output coupling fraction is $1 - 1/M$. In both cylindrical and spherical mirror geometries the radius of curvature of the output wavefront at $m_1$ is $(L + d_2)$. This is controlled by the resonator g parameters (equation 2.1) and for certain geometries a plane output wavefront $(d_2 \rightarrow \infty)$ can be generated as will be seen below.

As with stable resonators there is a constraint placed on the lasing frequencies of the unstable resonator by the requirement that the round-trip phase shift in the resonator is some integer multiple of $2\pi$. This requires (Chapter 1):

$$\varphi + \frac{2\pi nL}{\lambda} = 2q\pi$$

where $q$ is an integer, $nL$ the optical path length of the resonator and $\varphi$ a constant. The *axial* modes defined by different $q$ values are spaced in frequency by $\Delta\nu = c/2nL$, and how many of these appear in the output is governed by the gain-linewidth of the lasing medium.

In summary, the geometrical model predicts that the fundamental mode of an unstable resonator is a uniform-irradiance, magnifying beam with a well-defined spherical (cylindrical) or plane wavefront. The large geometrical or 'diffraction' loss is employed as useful output coupling. The resonator exhibits axial-mode characteristics similar to the stable resonator.

(a)

(b)

(c)

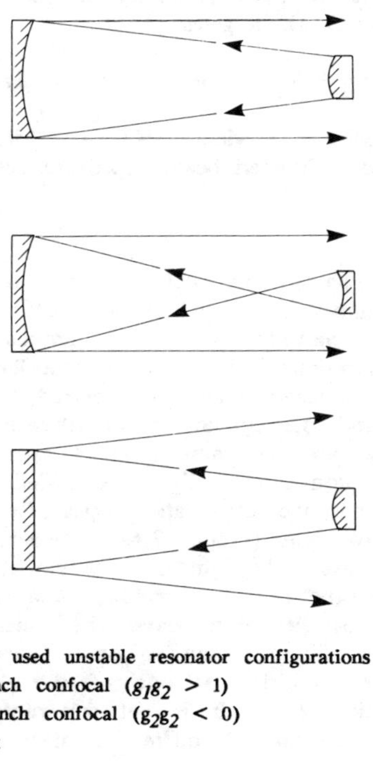

Fig 2.3 Three commonly used unstable resonator configurations

(a) Positive branch confocal ($g_1g_2 > 1$)

(b) Negative branch confocal ($g_2g_2 < 0$)

(c) Plano-convex

## 2.5 PRACTICAL RESONATORS

Many unstable resonator configurations can be devised; for practical purposes, however, three major types are used as shown in Figure 2.3. Of these the positive (Figure 2.3a) and negative (Figure 2.3b) branch confocal unstable resonators are the most popular (Krupke and Sooy 1969) as they produce a collimated (plane wavefront) output beam. The negative branch resonator is the less attractive of the two for high–power lasers because of its internal focus. The plano–convex resonator (Figure 2.3c) has the virtue of being easily fabricated as it requires only one customised spherical mirror. It is useful for preliminary tests of unstable resonator laser performance and for diffraction–grating tuned lasers but does not have the benefit of a collimated output beam.

The confocal resonators satisfy:

$$g_1 + g_2 = 2g_1g_2$$

the positive branch having $g_1g_2 > 1$, the negative branch $g_1g_2 < 0$. The resonator mirrors share a common focal point so equivalently they can be described by:

$$R_1 + R_2 = 2L$$

The magnification follows from the geometry of the resonators in Figure 2.3a and *b* or from equation (2) which gives:

$$M = 2g_1 - 1 \qquad \text{or} \qquad M = 1/(2g_1 - 1)$$

The first solution is chosen as it yields $|M| > 1$ for a resonator designed to give a rightward directed collimated beam. This reduces to:

$$M = -R_2/R_1$$

giving positive magnification for the positive branch resonator ($R_1$ is negative), and negative for the negative branch ($R_1$ is positive).

A valuable way of viewing the positive branch confocal resonator (or indeed any other configuration) is by an equivalent lens sequence as shown in Figure 2.4. Here the concave mirror is replaced by a convex lens of the same focal length $f_2$ and the convex mirror likewise by a negative lens of focal length $f_1$. These are the same distance apart as the mirrors. The self–reproducing configuration can readily be identified as a plane wave passing from the convex to concave lens and repeatedly being intercepted and expanded by the negative lens (Figure 2.4a). Aperture losses represent the output coupling in this case. As light is reversible along the same path it is also possible to consider the reverse situation; that of a demagnifying beam (Figure 2.4b). Here a parallel beam leaves the concave lens and is focussed by the positive lens, recollimated and so on. This demagnifying beam or *backward* wave can be identified as one of the sets of the solutions of equation 2.1 and equation 2.2. It is not self–reproducing, however, as it continues to contract in size until its diffraction divergence offsets the focussing effect of the convex lens and the beam starts to expand. The effect of this is that after a finite number of round trips the backward wave converts to the self reproducing forward wave in Figure 2.4a (Zemskov *et al* 1974, Anan'ev 1975).

(a)

(b)

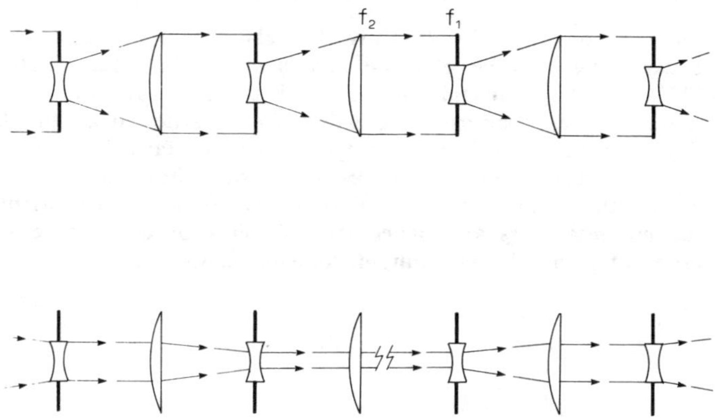


Fig 2.4 Lens sequence used to represent an unstable resonator, in this case the positive branch confocal configuration with $f_2 - |f_1| = L$. (a) Forward wave or magnifying beam. (b) Backward wave or demagnifying beam.

## 2.6 NEAR AND FAR-FIELD PATTERN

In the near-field, that is in the region to the output mirror, a conventional unstable resonator ideally produces a uniformly illuminated annulus with a plane or spherical wavefront (Figure 2.5a). Although at first sight the missing centre would appear to limit its usefulness (and sometimes does!), the beam converts in the far-field to have an on-axis maximum with a series of surrounding rings of progressively lower irradiance. The far-field diffraction pattern is obtained at sufficiently far distances that the phase across the output (diffracting) aperture as seen from the viewing point, varies linearly with position in the aperture. This is most conveniently achieved by viewing the distribution in the focal plane of a lens or mirror.

For a uniformly illuminated, uniphase, annulus the far-field irradiance (or fluence) pattern is (Sinclair and Cottrell 1967):

$$I(x) = A \left[ \frac{J_1(x)}{x} - \frac{J_1(x/M)}{Mx} \right]^2$$

where $J_1$ is the first order Bessel function, $A$ is a constant and $x = 2\pi\theta M a_1/\lambda$, with $\theta$ being the angular co-ordinate measured from the axis of the output aperture. This pattern takes the form of an on-axis maximum with surrounding rings (Figure 2.5a) that become larger in relative height as $M$ decreases. As the pattern has circular symmetry the energy or power contained within a circle of angular radius $\theta$ can be found from:

$$P = B \int_0^x I(x')x'dx'$$

where $B$ is a constant. In this way the relative output power encircled by the various dark rings can be obtained and is shown in Figure 2.5b as a function of $M^{-1}$. For large magnifications, $M^{-1}$ tends to zero and the distribution approaches the Airy pattern of a uniformly irradiated disc, the first dark ring encircling ~ 84% of the total output. As $M$ falls it is evident that this fraction drops steadily so that at $M = 2$ about 50% and at $M = 1.3$ only ~ 20% is in the central spot. This tends to limit the usefulness of low magnification resonators and hence the suitability of unstable resonators for low gain lasers which need small output coupling factors.

(a)

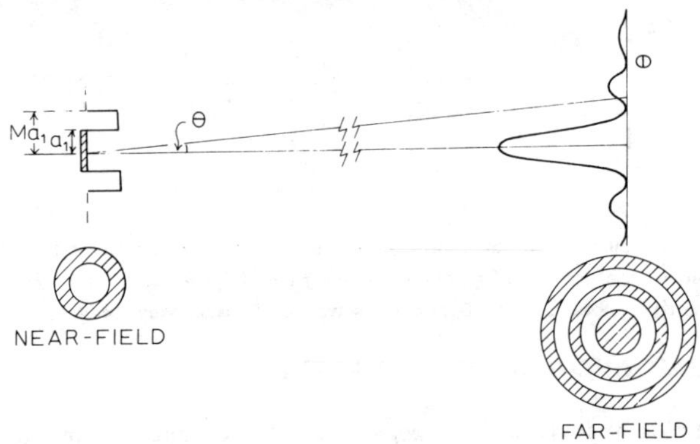


(b)

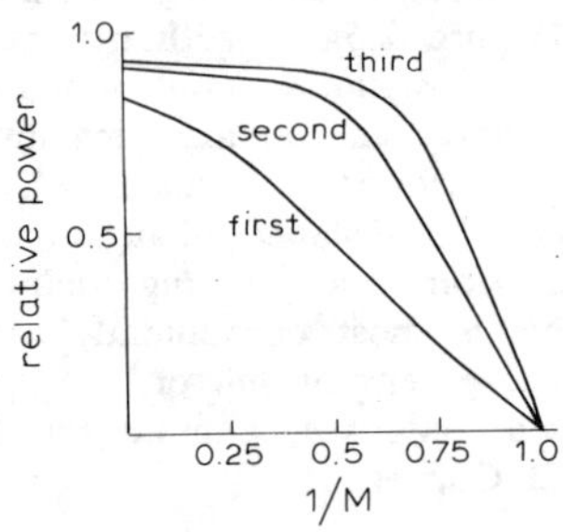


Fig 2.5 (a) Near-field and far-field patterns for a uniform annular output unstable resonator. (b) Relative power (or energy) encircled by the first, second and third dark rings of the far-field pattern as a function of the inverse resonator magnification $M^{-1}$ (Dyer et al 1972).

Roughly, as a guide to *diffraction limited* performance, the full-width at half -maximum angular divergence of the central spot of the far-field pattern is:

$$\theta_d \simeq \lambda/2Ma_1 \simeq \lambda/D$$

where $D$ is the outer diameter of the output beam.

The angular extent of the ring system will be approximately:

$$\theta_R \simeq \frac{\lambda}{(M-1)a_1}$$

## 2.7 EXPERIMENTAL STUDIES

There have been many experimental studies of unstable resonator beam properties on a variety of pulsed and cw lasers. The results show that the predictions of the geometrical model are largely borne out and, in particular, the attainment of single transverse mode operation appears to be the rule rather than exception in most cases. In carefully designed systems, with the appropriate quality gain medium, the far-field pattern approaches 'diffraction-limited' with discrepancies that can be accounted for by non-uniform irradiance in the annulus rather than significant phase-front errors. As an example, Figure 2.6 shows early results for a pulsed TEA $CO_2$

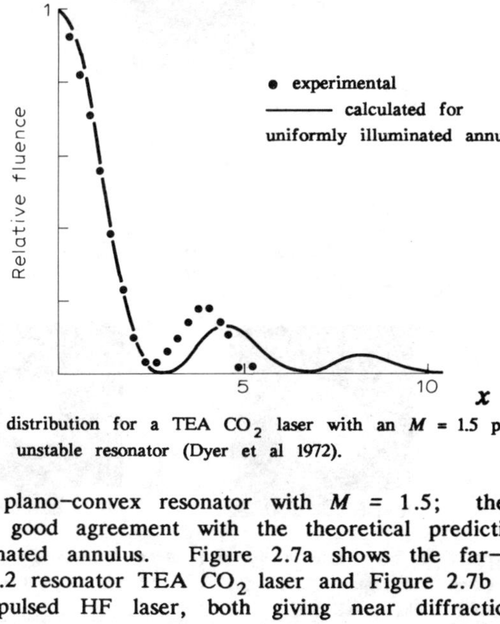


Fig 2.6 Far-field fluence distribution for a TEA $CO_2$ laser with an $M$ = 1.5 plano-convex unstable resonator (Dyer et al 1972).

laser operated with a plano-convex resonator with $M$ = 1.5; the far-field pattern is in relatively good agreement with the theoretical predictions based on a uniformly illuminated annulus. Figure 2.7a shows the far-field burn patterns for an $M$ = 2.2 resonator TEA $CO_2$ laser and Figure 2.7b for an $M$ = 1.5 and $M$ = 3 pulsed HF laser, both giving near diffraction limited performance.

Many similar results have been reported for $CO_2$ (Krupke and Sooy 1969, Freidberg *et al* 1972), chemical (Chodzko and Chester 1976, Deka *et al* 1976, Greiner *et al* 1973), solid-state (Anan'ev *et al* 1968, 1969), visible (Sinclair and Cottrell 1967, Ewanizky 1974, Zemskov *et al* 1974) and uv (Ishchenko *et al* 1975, McKee *et al* 1977) lasers with spherical mirror resonators. In certain cases cylindrical mirror resonators have also been studied (Anan'ev *et al* 1972, Dyer and James 1975a).

The near-field irradiance or fluence distribution from an experimental unstable resonator laser deviates significantly from the uniform annulus model in that both large scale non-uniformities and small-scale 'diffraction' effects appear (Figure 2.8). The tapering-off of the the irradiance towards the edge of the beam (Figure 2.8b) tends to reduce the output loss below that expected from geometrical coupling and, as noted above, slightly alters the far-field although as the phase is preserved and this effect is quite small (Chodzko and Chester 1976). Figure 2.8c shows the near-field burn pattern for a high energy TEA $CO_2$ laser with a square cross-section output mirror (Salvetti 1979). The interference pattern associated with edge diffraction at the mirror is clearly evident on the output beam.

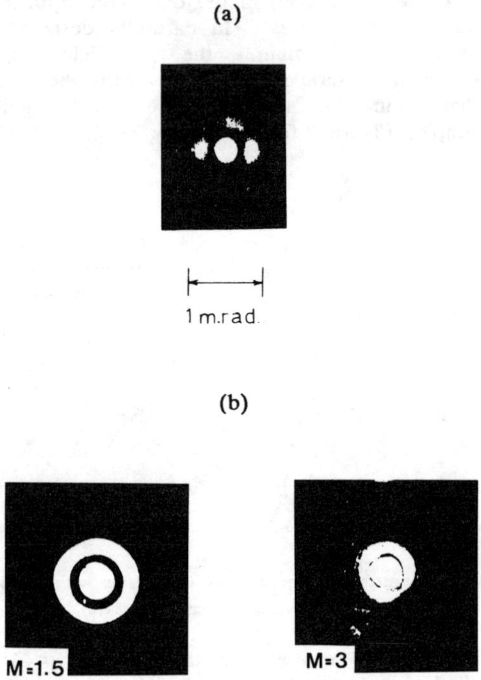


Fig 2.7 Far-field burn patterns for (a) $M$ = 2.2 unstable resonator TEA $CO_2$ laser (Dyer and James 1975b) (b) $M$ = 1.5 and $M$ = 3 unstable resonator pulsed HF laser (Deka et al 1976).

## 2.8 WAVE ANALYSIS

Unstable resonator mode properties have been studied extensively using both analytical and computational solutions of the Fresnel-Huygens integral (Siegman 1986). Such modelling is able to predict the non-uniformity and diffraction structure that appears in the near-field output beam. It is found that the calculated wavefront (phase surface) agrees well with that expected from the geometrical model.

One particularly interesting feature of the wave calculations is an apparent periodic variation of resonator loss with the Fresnel number of the resonator which is not expected from the geometrical model. This behaviour was first identified by Fox and Li (1964), and rationalized by Seigman and Arrathoon

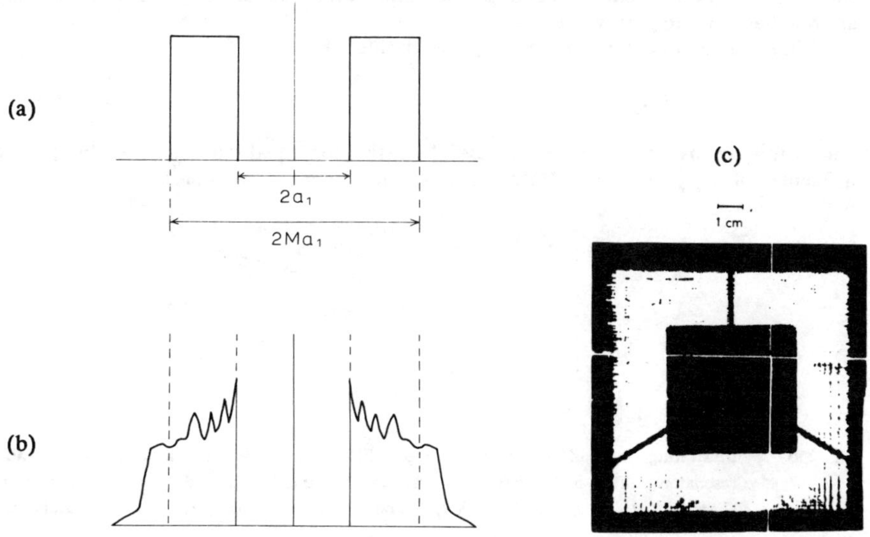


Fig 2.8 Near-field irradiance: (a) Geometrical model. (b) Typical observed distribution show large scale non-uniformity and diffraction effects. (c) Experimental near-field output from an $M = 1.95$, square mirror unstable resonator TEA $CO_2$ laser delivering 106J (Salvetti 1979).

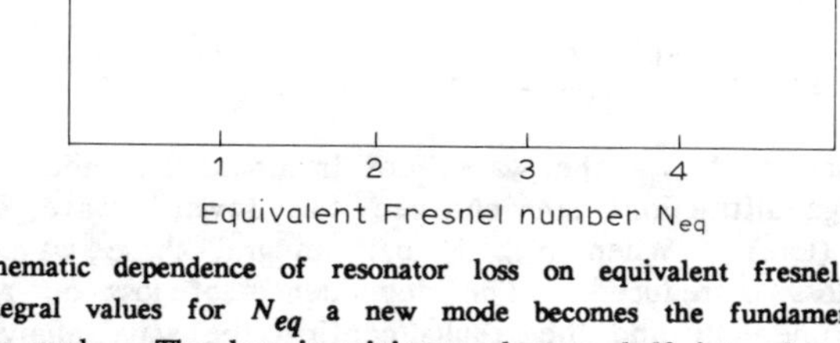


Fig 2.9 Schematic dependence of resonator loss on equivalent fresnel number, $N_{eq}$. At integral values for $N_{eq}$ a new mode becomes the fundamental, that is, lowest loss mode. The loss is minimum close to half integral values for $N_{eq}$ (after Siegman 1986).

(1967) in terms of an equivalent fresnel number, $N_{eq}$. The form of the loss as a function of $N_{eq}$ is shown schematically in Figure 2.9 for a circularly symmetric resonator with sharp-edged mirrors. The loss is minimum when $N_{eq}$ is half integral and maximum when it is an integer. This behaviour is not simply due to there being an oscillatory loss for one mode but rather the result of a new mode becoming the one with lowest loss each time $N_{eq}$ approaches an integral value.

For the confocal resonator, $N_{eq}$ is defined by:

$$N_{eq} = a_1^2(M-1)/2L\lambda \tag{2.3}$$

and this resonator provides a useful basis for qualitatively explaining the influence of $N_{eq}$ on loss. With reference to Figure 2.10, consider the

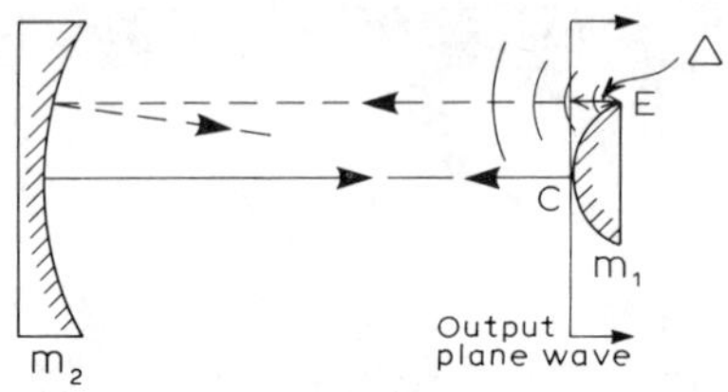


Fig 2.10 Illustrating the effect of the edge diffracted wave for a confocal unstable resonator. The path difference between the axially directed wave returned from C and that at *E* is $2\Delta = \lambda N_{eq}$ where $N_{eq} = a_1^2(M-1)/2L\lambda$ is the equivalent fresnel number.

out-going plane-wave (forward wave) that is the fundamental mode of this resonator. For a sharp-edged output mirror there will be a narrow zone near the periphery of $m_1$ which is 'diffracted' as it returns (Anan'ev and Sherstobitov 1971); thus while the majority of the wavefront is returned as a divergent beam to be recollimated by $m_2$, a fraction of the edge scattered wave can excite the backward wave in the resonator (ie. that shown in Figure 2.4b). This radiation, shown as a dotted line parallel to the axis in Figure 2.10, converges towards the axis and subsequently converts back to the forward wave as discussed earlier. The extra distance travelled by the edge diffracted wave compared with the plane wave reflected from the centre of $m_1$ is $2\Delta$ (Figure 2.10). From the resonator geometry, $2\Delta$ can be written as

$$2\Delta = \frac{a_1{}^2(M-1)}{2L} = \lambda \times \frac{a_1{}^2(M-1)}{2L\lambda} = \lambda\ N_{eq}$$

For integral values of $N_{eq}$, the waves are in phase and the backward wave produced by edge-diffraction can enhance the foward wave increasing the output coupling (loss). When $N_{eq}$ is half-integral these waves are out of phase and the loss is reduced. The dependence of loss on $N_{eq}$ has been investigated experimentally and the results confirm the wave analysis predictions (Freiberg *et al* 1972).

To avoid the possibility of two transverse modes co-existing in the unstable resonator it is quite common to employ a design with $N_{eq}$ equal to a half integer value. Whether or not this is really necessary is something of a moot point as wave calculations show that at the $N_{eq}$ integral mode-crossing points, the complex amplitudes of each mode become essentially identical (Siegman 1986). This mode crossing phenomenon is eliminated in poorly aligned resonators, those with a 'soft' edged output mirror or those lacking mirror cross-section symmetry (Siegman 1986).

## 2.9 DESIGN PROCEDURE

The starting point for designing an unstable resonator will normally be the selection of a magnification appropriate to the available laser gain-length product. If possible, $M$ is chosen to provide an effective reflectivity, $M^{-2}$, ($M^{-1}$ for cylindrical geometries), which optimises the extraction efficiency of the laser. A high $M$ value is desirable to obtain a large fraction of the output in the central lobe of the far-field pattern. Once $M$ and the type of resonator (confocal, plano-convex, symmetric etc.) are chosen, the $g$ parameters are automatically fixed as $M = f(g_1g_2)$ (Equation 2.2). The size, $a_1$, of the output coupling mirror is then determined to provide the best coverage of the available gain cross-section by requiring $2Ma_1$ is equal to the transverse dimensions of the gain medium; a circular cross section mirror is usual, although this could be square, rectangular or indeed any shape that is most suitable. Finally, if it is decided to operate with $N_{eq}$ half integral, then the resonator length, $L$, can be selected to satisfy this constraint (Equation 2.3). As $g_1$ and $g_2$ are fixed by the earlier design steps this means $R_1$ and $R_2$, defining the mirror curvatures, can be calculated and the practical implementation of the resonator can commence.

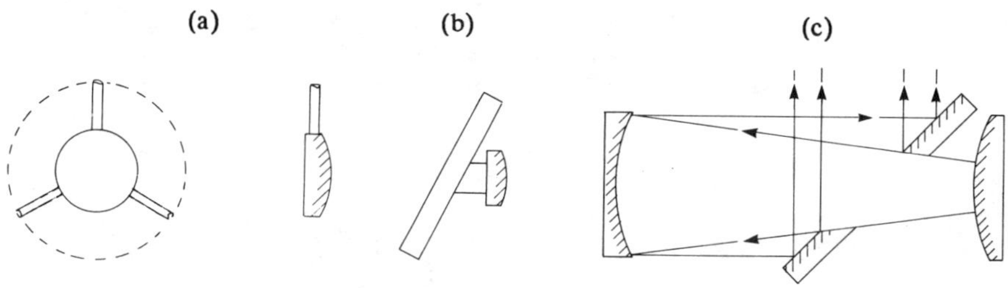


Fig 2.11 Methods for coupling radiation out of an unstable resonator (a) Mirror on stalks (b) Mounting on transmissive flat (c) 45° scraper mirror

To allow output coupling, the small mirror is, in some cases, simply mounted on a thin stalk or set of stalks (Figure 2.11a) or on a tilted (and preferably antireflection coated) transmissive flat (Figure 2.11b). A convenient alternative, however, which avoids obscuring the annular beam or the need for a mounting window, is to use a 45° *scraper* plate as shown in Figure 2.11c (Sinclair and Cottrell 1967). Thus, apart from the final output window, all-reflective metal optics can be used. This is an advantage in certain parts of the spectrum because of limited availability of good transmissive elements, and also eases the mirror cooling problem in high power devices. In very high-power lasers damage limitations at the final output window can be eliminated by use of a gaseous 'aerodynamic' window (Anderson 1976).

It is important to stress that whilst the theoretical mode quality of a correctly designed unstable resonator can be very high, maintaining near diffraction–limited performance requires excellent quality optics and a high–quality gain medium. This can be particularly demanding for large aperture lasers ($\gtrsim$ 10 cm) operating at shorter wavelengths ($\lambda \lesssim 1 \mu m$), where the theoretical diffraction limited beam–spread corresponds to $\lesssim$ 10 $\mu$radians. Even small optical imperfections or medium inhomogeneities can significantly degrade the far–field beam quality.

## 2.10 MODE CONTROL IN SHORT GAIN–LIFE LASERS

One particularly valuable property of the unstable resonator is its capability for attaining a diffraction–limited output beam after only a small number of round–trips are made by photons (waves) in the structure (Zemskov *et al* 1974). This allows high–quality output beams to be obtained from a range of lasers that have high optical gains that last for only a short time, usually less than 100ns. As each round–trip can take several to ten or so nanoseconds, only a few passes of the resonator are thus possible for such devices.

(a)

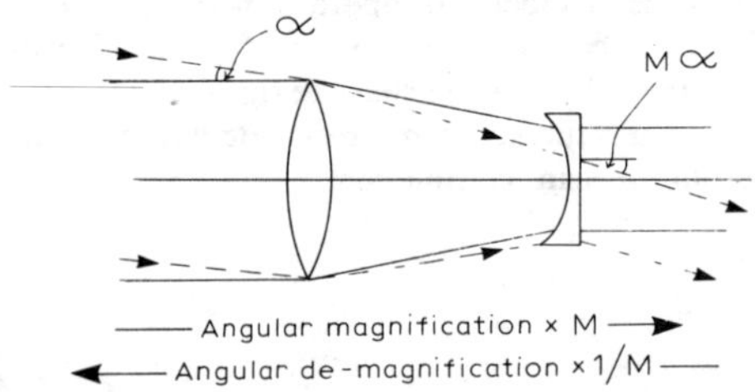


(b)

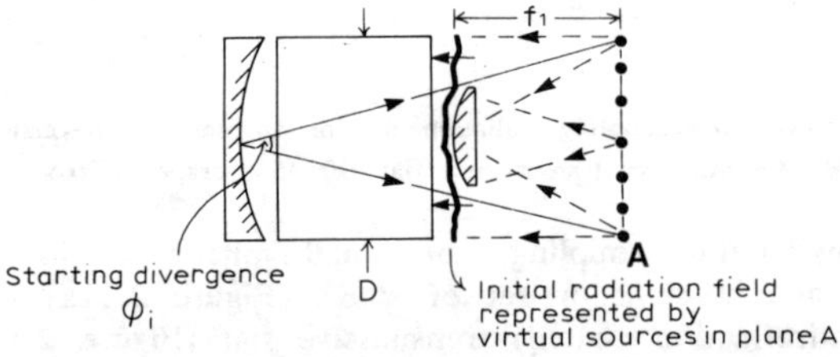


Fig 2.12 **Mode evolution in an unstable resonator (a) Angular magnification and demagnification property of the confocal resonator. (b) Initial noise field in resonator (thick line) and virtual source representation.**

To analyse this behaviour, the lens equivalent of a confocal unstable resonator is used as shown in Figure 2.12a. This, in addition to the property of producing spatial magnification of the forward wave (Figure 2.4a), gives angular demagnification by a factor $1/M$. (The backward wave receives angular magnification, a feature capitalized upon in telescopes – confocal resonators are thus often referred to as telescopic resonators.) Referring to Figure 2.12b, the initial radiation field at threshold that originates from spontaneous emission (noise), is represented by an incoherent set of point sources at the focal point of the convex mirror as shown (Anan'ev 1975). The starting divergence $\varphi_i$ of radiation from the resonator is then (Figure 2.12b):

$$\varphi_i = D/Mf_1$$

with $D$ being the transverse dimension of the gain medium and $f_1 = |R_1|/2$. After $n$ round-trips of the resonator this is reduced by the angular demagnification produced by the mirrors to:

$$\varphi_n = D/M^n f_1$$

If the diffraction-limited case is taken to be $\varphi_n = 2\theta_d = 2\lambda/D$ then this condition is reached after:

$$n = \frac{1}{\ln M} \ln \frac{D^2}{2\lambda f_1} = \frac{1}{\ln M} \ln 2N_F(M-1)$$

round trips, where $N_F = D^2/4\lambda L$ is the conventional Fresnel number of the laser. The number of round-trips needed to reach the diffraction limit is shown in Figure 2.13 as a function of $M$ for $N_F = 10$ and 100. As can be seen for $M > 3$, five or less round-trips are adequate, which would normally mean a mode evolution time of $\lesssim$ 30ns. Note that as $M$ tends to unity (the stable-unstable resonator boundary) $n$ increases rapidly suggesting that the development of a diffraction-limited mode is slow in such resonators.

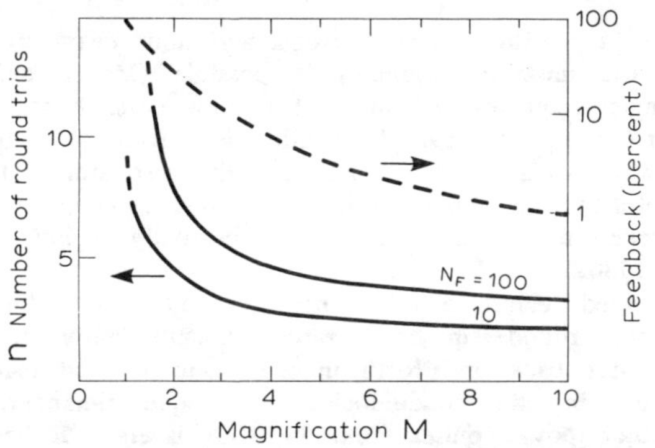


Fig 2.13 Number of round trips, $n$, to reach diffraction limited output as a function of resonator mangification $M$. The feedback factor (effective output mirror reflectivity) is also plotted. $N_F$ is the conventional fresnel number.

The advantages of high magnification unstable resonators for beam control have been demonstrated experimentally for a number of important high-gain, short gain duration lasers, notably: copper vapour (Zemskov *et al* 1974), dye

(Ewanizky 1974), excimer (McKee *et al* 1977), nitrogen (Ischenko *et al* 1975) and HF/DF chemical (Deka and Dyer 1978) devices. In designing such systems it should be borne in mind, however, that the feedback factor (effective reflectivity, $M^{-2}$) falls dramatically with increasing $M$ (Figure 2.13). 'Parasitic' lasing from spurious reflections may then dominate if care is not taken. In addition the output mirror may be difficult to fabricate because of its small size.

## 2.11 SPECIAL TECHNIQUES

### 2.11.1 Injection locking

The basic idea behind injection locking is to use a small, low-power, laser with refined beam properties to control the performance of a larger device. The unstable resonator laser is well suited to this as radiation injected in through, say, a small hole in one of the mirrors (Figure 2.14) rapidly spreads out to fill the resonator volume. A weak injected signal is often sufficient to *capture* the oscillation of the larger *slave* device as this signal has only to compete with the spontaneous emission noise.

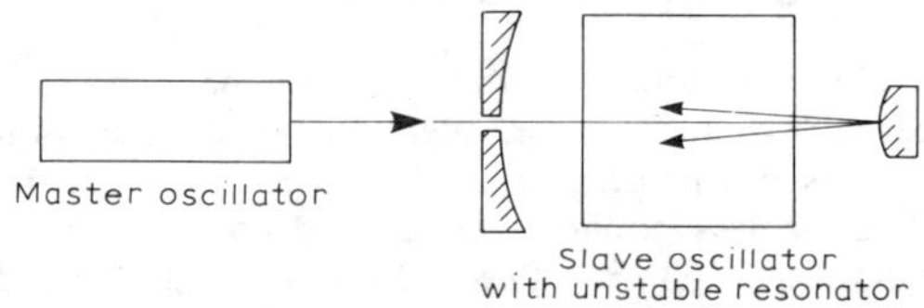


Fig 2.14 Injection locked unstable resonator.

Injection locking with a narrow frequency input beam to excite only one axial-mode of the unstable resonator is possible (Meyer 1976), leading the high-power, single-frequency, output. It is also possible to control the laser wavelength (if adequate gain-linewidth is available) by injecting a line-narrowed wavelength tunable signal into the resonator. This technique is particularly useful for excimer lasers, allowing excellent spatial and good temporal coherence to be attained from a normally highly incoherent device (Goldhar *et al* 1980).

Another related technique is to inject a single short duration pulse (less than the resonator round-trip time) which, if sufficiently strong to dominate the noise, circulates back and forth in the resonator and leads to an output pulse train. This 'injection mode-locking' technique (Belanger *et al* 1978) has produced very high power pulses from TE $CO_2$ lasers. It has been found in such experiments that the unstable resonator laser is very sensitive to low levels of radiation fed back by external scattering into the 'backward' wave (Baldis and Corkum 1979). The reason for this is that this wave (Figure 2.4b) suffers no output loss while it is demagnifying and thus experiences a substantially higher net-gain than the forward wave. Satellite pulses can then show-up in the mode-locked output, even with extremely weak levels of back scattering.

### 2.11.2 Novel geometries

A great variety of unstable resonators has been devised which have novel properties and allow improvements over the conventional confocal system for specific applications. Off-axis output mirrors, dual stable-unstable resonators, geometries incorporating axicon mirrors to convert the annulus into a circular beam, ring-geometries and multiple-mirror resonators have been reported (Siegman 1986); large mode-volume cylindrical mirror resonators have also been studied. One particularly interesting system that will be covered briefly here, and in more detail in Chapter 6, is the variable-reflectivity-mirror resonator.

(a)

(b)

(c)

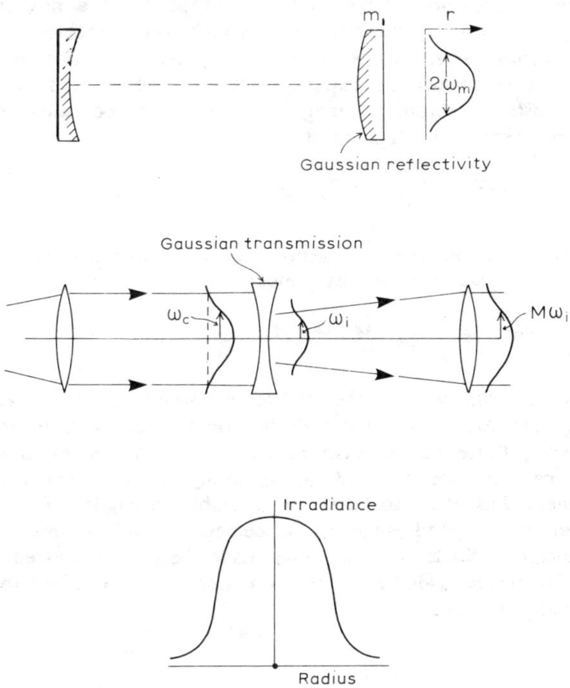


Fig 2.15 **(a) Variable reflectivity mirror unstable resonator. (b) Lens-sequence equivalent, with the concave lens having a gaussian transmission function. (c) Schematic near-field output profile.**

### 2.11.3 Variable reflectivity mirror resonator

Two undesirable features of the unstable resonator are its annular output beam and the possibility of two modes co-existing if operated near the integer $N_{eq}$ point. One method for overcoming these limitations is to use a soft-edged output mirror, that is with a reflectivity that tapers to zero towards the mirror edge. Theoretical calculations show this type of mirror leads to clear mode loss separation (Anan'ev and Sherstobitov 1971), and it is also possible to obtain a *filled-in*, smooth, near-field irradiance pattern (McCarthy and Lavigne 1985).

For simplicity of analysis consider a confocal unstable resonator with an output mirror having a radially varying amplitude reflectivity that is described by a gaussian function with an $e^{-1}$ full-width of $2w_m$ as shown in Figure 2.15a. This is represented by its lens equivalent in Figure 2.15b, with the diverging element having the same form for its transmission as the mirror reflectivity. It is assumed that the mode of this resonator is a gaussian field distribution, which is collimated and has a gaussian half-width of $w_c$ when it passes from the convex to concave lens. This produces a new gaussian profile of half-width $w_i$ following transmission through the concave lens. Assuming now that geometrical optics apply (ie neglecting the subtleties of gaussian beam propagation), this beam will be expanded by the convex lens to $Mw_i$ where $M$ is the usual unstable resonator magnification. To be self-reproducing, the transmission and expansion steps must satisfy:

$$e^{-r^2/w_m^2} \times e^{-r^2/w_c^2} = e^{-r^2/w_i^2}$$

where the first term on the left represents transmission through the gaussian lens. As, however, $w_c = Mw_i$ this yields:

$$w_c = w_m (M^2 - 1)^{\frac{1}{2}}$$

This indicates that the size of the collimated beam can be controlled through choice of $w_m$ and $M$. By correct design of the gaussian mirror transmission function (allowing finite transmission at its centre) the radiation coupled out of the resonator can provide a filled-in, smoothly varying, beam (Figure 2.15c). This arrangement has the attributes of a stable resonator $TEM_{00}$ type mode whilst maintaining the good single-mode selectivity and volume coverage of the unstable resonator. Such configurations have been investigated experimentally giving successful results (McCarthy and Lavigne 1985) and suitable mirrors are now commercially available.

## 2.12 CONCLUSIONS

It is now well established that unstable resonators provide the designer with a very valuable means of obtaining high quality output beams from lasers. The application of these resonators now ranges from their use on ultra-small diode lasers to the largest high-average-power lasers used in materials processing. The field is by no means static, however, and important developments continue to occur both on the theoretical and experimental front. Likely future trends include the further exploitation of novel geometries such as the stable-unstable resonator, and extension of the variable-reflectivity-mirror concept.

## 2.13 REFERENCES


Anan'ev Yu A, Sventsitskaya N A and Sherstobitov V E 1968 Sov. Phys. - Doklady 13 351

Anan'ev Yu A, Sventsitskaya N A and Sherstobitov V E 1969 Sov. Phys. -JETP 28 69

Anan'ev Yu A and Sherstobitov VE 1971 Sov. J. Quantum Electron. 1 263

Anan'ev Yu A, Chernov V N and Sherstobitov V E 1972 Sov. J. Quantum Electron. 1 403

Anan'ev Yu A 1975 Sov. J. Quantum Electron. 5 615

Anderson J D 1976 Gas Dynamics Lasers (London: Academic Press) Chapter XI

Baldis H K and Corkum P B 1979 Appl. Opt. 18 1346

Belanger P A, Tremblay R and Lapierre P 1978 Opt. Commun. 26 256

Chodzko R A and Chester A N 1976 Handbook of Chemical Lasers, ed. R W F Gross and J F Bott (Toronto: Wiley) Chapter 3

Deka B K, Dyer P E and James D J 1976 Opt. Commun. 18 462

Deka B K and Dyer P E 1978 IEEE J. Quantum Electron. QE-14 661

Dyer P E 1989 Lasers in Physical Chemistry ed. D.K. Evans (New York: Marcel Dekker) Chapter 1

Dyer P E, James D J and Ramsden S A 1972 Opt. Commun. 5 236

Dyer P E and James D J 1975a Opt. Commun. 15 20

Dyer P E and James D J 1976b Appl. Phys. Lett. 26 331

Ewanizky T F 1974 Appl. Phys. Lett. 25 295

Fox A G and Li T 1964 Quantum Electronics III ed. P Grivet and N Bloembergen (New York: Columbia Univ. Press) 2 pp 1263-70

Freiberg R J, Chenausky P P and Buczek C J 1972 IEEE J. Quantum Electron. QE-8 882

Goldhar J, Rapoport W R and Murray J R 1980 IEEE J. Quantum Electron. QE-16 235

Greiner N R, Arnold G P and Wenzel R G 1973 J. Appl. Phys. 44 3202

Ishchenko V N, Lisitsyn V N and Starinskii V N 1975 Sov. J. Quantum Electron. 5 851

Krupke W F and Sooy W R 1969 IEEE J. Quantum Electron. QE-5 575

McCarthy N and Lavigne P 1985 Opt. Lett. 10 553

McKee T J, Stoicheff B P and Wallace S C 1977 Appl. Phys. Lett. 30 278

Meyer J 1976 Phys. Lett. 58A 167

Salvetti G, PhD Thesis, University of Hull, 1978.

Siegman A E 1965 Proc. IEEE 53 277

Siegman A E 1986 Lasers (Oxford: Oxford Univ. Press) Chapters 22 and 23

Siegman A E and Arrathoon R 1967 IEEE J. Quantum Electron. QE-3 156

Siegman A E and Miller H Y 1970 Appl. Opt. 9 2729

Sinclair D C and Cottrell T H E 1967 Appl. Opt. 6 845

Steier W H 1979 Laser Handbook Vol. 3, Ed. M.L. Stitch (Amsterdam: North Holland) pp3-39

Zemskov K I, Isaev A A, Kazaryan M A, Petrash G G and Rautain S G 1974 Sov. J. Quantum Electron. 4 474

# 3 Theory of Waveguide Laser Resonators

*C A HILL*

## 3.1 INTRODUCTION

The first two chapters of this book ignore any interaction of light with the laser cavity side walls. Apertures and mirror edges may be used for transverse mode control, but otherwise the light is assumed to propagate freely between the resonator mirrors. This chapter describes a class of resonator with intentional wall effects: the resonator path includes a "tube" or *waveguide*, in which light is guided by a series of Fresnel reflections from the tube walls. We will see how some resonator properties of waveguide lasers may be modelled theoretically, and how such models compare with actual laser devices.

A general waveguide resonator is shown in Figure 3.1. The total influence of all the optical elements, including the guide, will determine what self-consistent field patterns may exist. Our modelling problem is to find these *resonator modes*, with their round-trip losses and resonant frequencies. The important point here is that, inside a sufficiently narrow guide, the Gaussian-beam modes and propagation equations of Chapter 1 are inappropriate. Instead, we must recognise guide wall effects by using another set of functions (modes) which obey boundary conditions at those walls. We must also treat the coupling of optical radiation at the guide apertures and the free-space propagation to and from the mirrors.

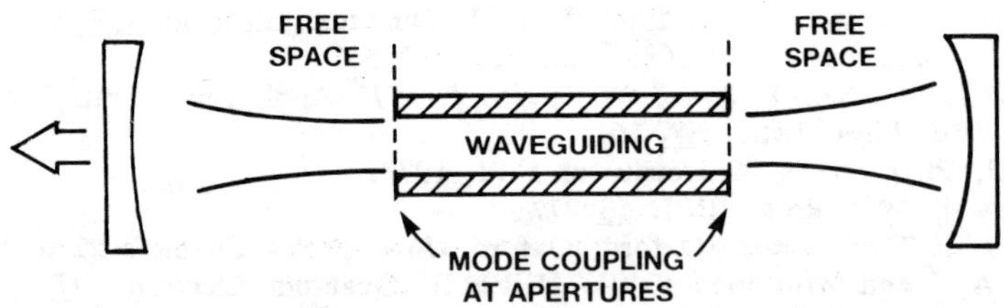


Fig 3.1 Sketch of a general linear waveguide resonator.

Early workers on waveguide lasers sought definite *gain/bandwidth* advantages. In particular, because of favourable wall interactions, narrow-bore waveguide lasers with $CO_2$ as the active medium could run at much higher pressures than conventional Gaussian-resonator lasers, and thus offered increased frequency tunability. Chapter 17 says more about $CO_2$ waveguide laser device characteristics and technology. Here we concentrate on the optical properties of a "waveguide resonator", where *the presence of the waveguide significantly affects the mode structure of the optical resonator.*

We restrict our discussions to infrared hollow dielectric guides. These are hollow guides as opposed to solid or clad optical fibre guides. They are assumed to be uniform, straight, and of rectangular (usually square) or circular cross-section, with half-widths or radii much greater than the radiation wavelength ($a \gg \lambda$). Such structures possess sets of *waveguide modes*. These are field patterns which can propagate along the guides, with tolerably low loss, maintaining their transverse shape. Formally, they are solutions of Maxwell's equations obeying the appropriate boundary conditions at the guide walls. The lowest-order linearly polarised mode of a square-bore guide, for instance, is called $EH_{11}$ and looks rather like the lowest-order linearly polarised free-space mode $TEM_{00}$. As we shall see later, and in Chapter 17, the fact that guide attenuation losses are strongly mode-dependent may be used to our advantage in controlling the transverse mode pattern actually emitted by the laser. When talking of mode fields, coupling losses and resonators, we will not use specific guide dimensions or wavelengths, since the theory is fairly general. The most common real devices, and most of our examples in this chapter, involve square guides and the familiar $CO_2$ laser medium ($\lambda \simeq 10$ $\mu$m). There is no direct treatment here of laser media, or the effects of guide walls on discharges, or their consequences for laser design and engineering. Some of this material is covered in Chapter 17.

This chapter has four main sections. In section 3.2 we briefly discuss light propagation and modes in hollow dielectric waveguide structures. Section 3.3 is a rather general account of *waveguide resonator analysis*. Section 3.4 presents some detail of first-order (single mode) waveguide resonator theory. Finally, section 3.5 is concerned with the resonator properties of some real waveguide lasers, and how we interpret them in terms of available resonator theory.

## 3.2 PROPAGATION IN HOLLOW DIELECTRIC WAVEGUIDES

In 1964, Marcatili and Schmeltzer proposed the use of hollow dielectric waveguides as low-loss components for laser amplifiers. They calculated the allowed *modes of propagation* of hollow dielectric waveguides by solving the standard wave equation in free space but with dielectric boundary conditions; field components were matched at the guide walls instead of falling away to zero at infinity. These modes are field patterns which keep their shape, but experience a mode-dependent phase shift, as they propagate along the guide. Hollow dielectric waveguides used in typical gas waveguide lasers have half-widths or radii $a \simeq 100\lambda$, so the assumption $a \gg \lambda$ is reasonable. They cannot be made very much smaller without severe attenuation ($\propto a^{-3}$), and are inherently *multimode* or *overmoded*. When terms of first or higher order in $\lambda/a$ are removed, the rather complex expressions for the field amplitudes of the waveguide modes reduce to simple "first-order" ones. But the corresponding simplified propagation constants include essential terms of order $\lambda/a$ and $(\lambda/a)^2$. For each waveguide mode the propagation constant can be written in the form $k = \beta + i\alpha$, where $\beta$ is the phase constant and $\alpha$ the amplitude attenuation constant. Propagation along the long axis of the waveguide is described by a term $\exp(ikz)$.

### 3.2.1 Waveguide Mode Expressions

In this section we describe the simplified fields and propagation constants for the square and circular waveguide geometries. The lack of consistent notation for waveguide modes is usually the first thing that hampers workers in this subject. A brief study of Degnan (1976) or other reviews may help.

For circular cross section dielectric guides (radius $a$, polar coordinates $r$ and $\varphi$) as in Figure 3.2(a), we usually have a single dielectric constant $\epsilon = (n_a - ik_a)^2$ for the wall material. Circular guides may support transverse circular electric modes ($TE_{0m}$), transverse circular magnetic modes ($TM_{0m}$), and sets of higher order hybrid modes ($EH_{nm}$) which have both radial and tangential electric and magnetic fields. Figure 3.2(b) shows the first-order field amplitudes and transverse patterns for the lowest-order mode of each type. In practice, linear polarisation with a fixed axis is usually desired, and often enforced with a Brewster window, diffraction grating or some grid impressed on the mirror surface. We can express an arbitrary linearly polarised transverse field pattern within a circular guide as a combination of $LP_{nm}$ modes, where $n$ is the azimuthal or angular mode number and $m$ is the radial mode number. The $LP_{0m}$ modes are identical with the $EH_{1m}$ modes of Marcatili and Schmeltzer (1964) or the $HE_{1m}$ modes of Snitzer (1961). This important set of zero-angular-order modes is sufficient to describe any circularly symmetric field pattern, and to model many experiments. Figure 3.2(c) shows the amplitude distribution of the fundamental hybrid mode $EH_{11}$ (or $HE_{11}$ or $LP_{01}$). The higher-order modes must be included for a full description of even well-aligned lasers, and are essential for any account of misalignment effects.

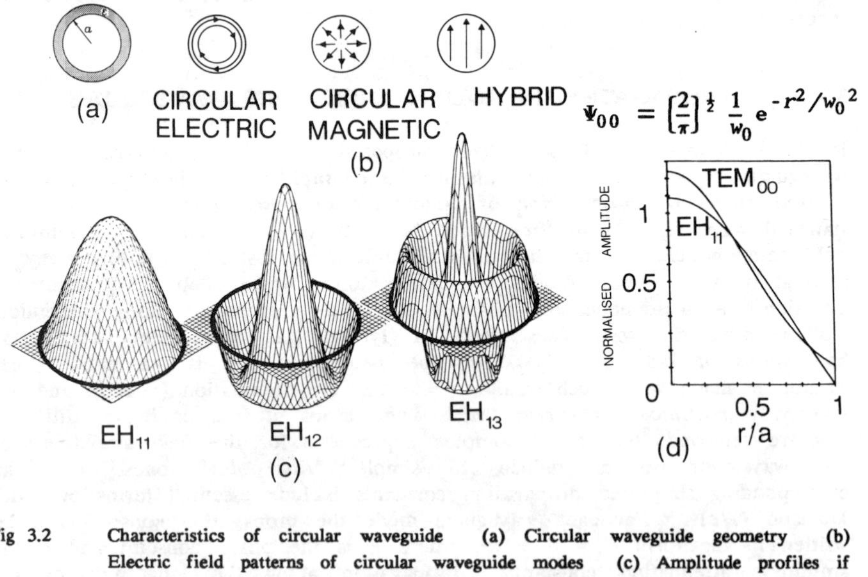


Fig 3.2 Characteristics of circular waveguide (a) Circular waveguide geometry (b) Electric field patterns of circular waveguide modes (c) Amplitude profiles if low-order $EH_{1m}$ modes (d) Amplitude profile of $EH_{11}$ and its best fit Gaussian $\Psi_{00}$ ($w_0$= 0.64$a$). $EH_{11}$ is normalised according to Eq. (3.2) and (3.3) with a=1.

The first-order fields are given by:

$$E_{nm} = f_{nm}\, J_n(\rho_{nm} r/a)\, \cos(n\varphi) \qquad 0 \leq r \leq a \qquad (3.1)$$

where $\rho_{nm}$ is the $m^{th}$ root of the $n^{th}$ order Bessel function, i.e. $J_n(\rho_{nm}) = 0$, and:

$$\left.\begin{aligned} f_{nm} &= \surd 2\ [\surd\pi\ a\ J_{n+1}(\rho_{nm})]^{-1} && if\ n = 2,3,4,\ldots \\ &= [\surd\pi\ a\ J_1(\rho_{0m})]^{-1} && if\ n = 0 \end{aligned}\right\} \qquad (3.2)$$

for the even $LP_{nm}$ ($n \neq 1$) modes. The axis of polarisation is arbitrary: neither the field strengths nor the propagation constants depend on it. The field strengths for the even linearly polarised ($n$ = 1) modes obey the same form:

$$E_{1m} = \surd 2 [\surd\pi\ a\ J_2(\rho_{1m})]^{-1}\ J_1(\rho_{1m} r/a)\ \cos\varphi \qquad (3.3)$$

but the axis of polarisation is not arbitrary. This choice of even (cosine) modes forces the choice of circular magnetic $TM_{0m}$ modes (plus $HE_{2m}$) for $x$-polarised light, and circular electric $TE_{0m}$ modes (plus $HE_{2m}$) for $y$-polarised light. The normalisation and the propagation constants are given by:

$$\int_0^{2\pi}\int_0^{a} E_{nm}(r,\varphi)\ E_{n'm'}(r,\varphi)\ r dr\ d\varphi = \delta_{nn'}\ \delta_{mm'} \qquad (3.4)$$

$$\beta_{nm} \simeq \frac{2\pi}{\lambda}\left[1 - \frac{1}{2}\left[\frac{\lambda\rho_{nm}}{2\pi a}\right]^2\right] \qquad (3.5)$$

$$\alpha_{nm} \simeq \frac{1}{a}\left[\frac{\lambda\rho_{nm}}{2\pi a}\right]^2 \mathrm{Re}[\tfrac{1}{2}(\epsilon+1)(\epsilon-1)^{-\frac{1}{2}}] \qquad (3.6)$$

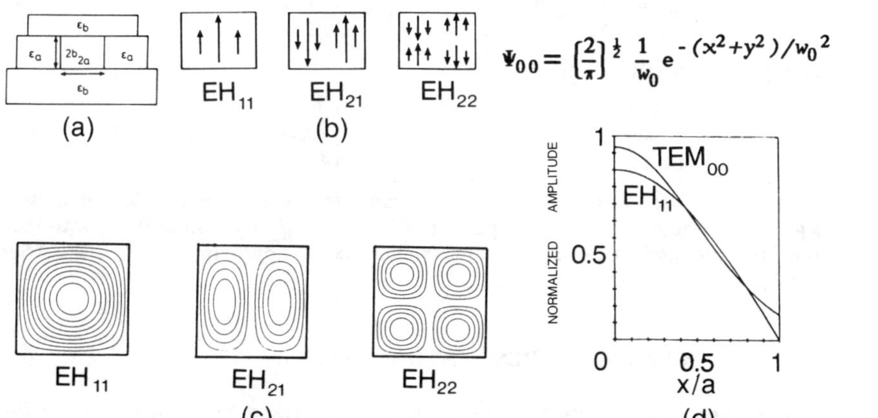


Fig 3.3 (a) Rectangular waveguide geometry (b) Electric field lines of low-order $EH_{mn}$ modes (c) Low order amplitude contours for a square bore guide (d) Amplitude profile of $EH_{11}$ and its best fit Gaussian $\Psi_{00}$ ($w_0$ = 0.70$a$). $EH_{11}$ is normalized according to equation 3.7 with $a$=$b$=1.

Figure 3.3 shows a general rectangular cross section dielectric guide (width $2a$ and height $2b$) with two different dielectric materials. In practice, many guides use a single material, but many others are "hybrid" with one or two metal walls (which may be electrodes for transverse RF excitation). Common guide materials are aluminium, alumina ($Al_2O_3$) and beryllia (BeO). Figure 3.3 shows two complex relative dielectric constants:

$$\epsilon_a = (n_a - ik_a)^2 \qquad \epsilon_b = (n_b - ik_b)^2$$

where $n_a$ and $n_b$ are refractive indices, and $k_a$ and $k_b$ are extinction coefficients; all four are positive real numbers. These rectangular guides support two sets of linearly polarised $EH_{mn}$ hybrid modes, but do not support circularly polarised modes. In a ceramic–metal hybrid guide with one or two horizontal metal walls, the horizontally polarised $E^xH_{mn}$ modes are usually favoured over the vertically polarised $E^yH_{mn}$ modes, chiefly because of the loss factors associated with the metal.

The first-order fields for the linearly polarised $EH_{mn}$ modes are

$$E_{mn}(x,y) = (ab)^{-\frac{1}{2}} \begin{bmatrix}\cos\\ \sin\end{bmatrix}\left[\frac{m\pi x}{2\ a}\right] \begin{bmatrix}\cos\\ \sin\end{bmatrix}\left[\frac{n\pi y}{2\ b}\right] \ ; \ m,n = \begin{matrix}odd\\ even\end{matrix} \tag{3.7}$$

with normalisation

$$\int_{-a}^{a}\int_{-b}^{b} E_{mn}(x,y)\ E_{m'n'}(x,y)\ dx\ dy = \delta_{mm'}\ \delta_{nn'} \tag{3.8}$$

The propagation constants $k_{mn} = \beta_{mn} + i\alpha_{mn}$ are given by

$$\beta_{mn} \simeq \frac{2\pi}{\lambda}\left[1 - \frac{1}{2}\left[\frac{m\lambda}{4a}\right]^2 - \frac{1}{2}\left[\frac{n\lambda}{4b}\right]^2\right] \tag{3.9}$$

and (for $x$–polarised modes)

$$\alpha_{mn} \simeq \frac{m^2}{a}\left[\frac{\lambda}{4\ a}\right]^2 \mathrm{Re}(\epsilon_a(\epsilon_a-1)^{-\frac{1}{2}}) + \frac{n^2}{b}\left[\frac{\lambda}{4\ b}\right]^2 \mathrm{Re}((\epsilon_b-1)^{-\frac{1}{2}}) \tag{3.10}$$

Figure 3.3 also has some contour maps for the lowest-order modes in a square waveguide, and a comparison of the $EH_{11}$ field amplitude with that of the fundamental Gaussian beam which best approximates it (see Section 3.4.1).

## 3.3 WAVEGUIDE RESONATOR ANALYSIS

This section gives a rather general account of resonator analysis as applied to waveguide lasers. A waveguide resonator (Figure 3.1) may be viewed as a free–space Gaussian resonator perturbed by an aperturing waveguide tube, or as a tube resonator perturbed by the addition of free–space sections. We summarise the usual methods of calculating the properties of waveguide resonators, and later examine some of the geometric constraints and principles in the design of a resonator. Having read Chapters 1 and 2, readers should be familiar with the main features of *optical resonators* and free-space Gaussian beams, so that concepts such as round-trip eigenvalues and their associated *resonator modes* may be introduced and examined without difficulty. Even so, large amounts of time and effort can be saved if such terms as "mode", "resonator mode", "multimode", and "mode loss" are clearly understood. At the risk of some duplication, we discuss these key concepts as they arise.

### 3.3.1 The Concept of Resonator Modes

The real waveguide lasers used in the factory or on the bench are more or less complicated assemblies with electrical, mechanical and optical elements including active media. We wish to model important laser properties such as output power, beam shape, frequency tunability and frequency purity. We discard all specific real-life elements and consider the very simple, and fairly abstract, idea of an *optical resonator* which possesses some *resonator modes*. The link with real life devices will reappear shortly. A resonator mode is a field distribution which repeats itself in shape and in phase after one round trip of the resonator.

Let us start with a well-aligned Gaussian resonator with two large-aperture spherically curved mirrors, and recall parts of Chapter 1. We know that a standard free-space scalar wave equation, when we require the light to stay near the $z$-axis with the dominant variation being simply the axial propagation term $\exp(ikz)$, yields Gaussian beam functions. These are the $\mathrm{TEM}_{pq}$ free-space *modes of propagation*; their properties change very slowly with wavelength, and their transverse shape is given by a simple Gaussian $\exp(-r^2/w^2)$, multiplied by Laguerre polynomials (in cylindrical coordinates) or Hermite polynomials (in Cartesian coordinates). We can define a common beam waist position $z_0$, and a common beam waist radius $w_0$, for the whole orthonormal set of beams. Then, formally, we can express the free-space field $E_{f-s}$ anywhere along the resonator axis as a linear combination of Gaussian beams:

$$E_{f-s} = \Sigma\, b_{pq}\, \Psi_{pq}(z-z_0, w_0) \tag{3.11}$$

$$b_{pq} = \int E_{f-s}\, \Psi^{*}_{pq}\, dA \tag{3.12}$$

where the $\Psi_{pq}$ are the $\mathrm{TEM}_{pq}$ Hermite–Gaussian or Laguerre–Gaussian functions *including* the $z$-dependent amplitude and phase factors given in Chapter 1. The '*' indicates complex conjugation, and the integral is performed over the infinite cross section (though only the area near the $z$-axis contributes significantly). The $b_{pq}$ are complex coefficients which, during lossless free-space propagation, do not depend on axial position $z$.

These two equations show a "decomposition" of some arbitrary function $E_{f-s}$ into an orthonormal set of functions $\Psi_{pq}$ representing various spatial frequencies, just as in Fourier analysis we split some function $f(x)$ into sines and cosines. If the original field $E_{f-s}$ is associated with a precise temporal frequency, then so is each term of the sum: the precise frequency information is extra, and not contained in the spatial form of the modes of propagation.

By contrast, resonator modes are *modes of oscillation*. This concept has two essentials: the mode shape must be one of the self-repeating *transverse modes* of the cavity, and the mode frequency must yield a precise *axial* resonance. Conventionally, we assign two transverse mode integers $r$ and $s$, and one longitudinal mode integer $j$, where $j$ is the number of full $2\pi$ propagation phase shifts per round trip. Thus, a general resonator mode frequency is "$\nu_{j,rs}$". For most real lasers, $j$ >10,000 and $r,s$ <10, with $r$ and $s$ referring to the transverse mode order in the $x$ and $y$ (or radial and azimuthal) directions. Usually a change of 1 part in $10^4$ in $\lambda$ will make no important difference to the transverse part (the shapes and amplitudes of the Gaussian beams), whereas the change $j \rightarrow j+1$ (or a single-pass phase change $\simeq \pi$) is crucial in laser frequency studies. This means that laser axial modes and transverse modes are physically decoupled and can be treated separately. They are both present in the idea of an oscillating or resonating mode. It often helps to imagine a laser emitting some beam of arbitrary transverse shape with a single well-defined frequency; this shape, and this frequency $\nu_{j,rs}$, constitute a resonator mode by definition.

Formally, the self-consistency condition translates into the complex eigenvalue equation:

$$ME_{f-s} = \gamma E_{f-s} \tag{3.13}$$

where each self-consistent field $E$ has its complex eigenvalue $\gamma$. We can define in advance a *round-trip loss* $1-|\gamma|^2$ and *relative phase shift* $\arg(\gamma)$. Since the round-trip matrix $M$ is complex and may be large, we may be no further forward. However, we see that if enough of $M$ can be specified with reasonable accuracy, Equation (3.13) can be solved in principle. For stable open resonators, the resonator transverse modes are accurately given (see Chapter 1) by pure Gaussian beams with appropriate phase shifts: that is, by choosing $w_0$ and $z_0$ correctly we can make $M$ diagonal. Each eigenvector can then be written as a single term:

$$E_{f-s} = \Psi_{pq}(z-z_0,w_0) \tag{3.14}$$

The $\Psi_{pq}$ relative phases are given by $2(p+q+1)\cos^{-1}[\pm(g_1g_2)^{\frac{1}{2}}]$, where the 2 refers to a *round* trip or double pass between the mirrors, and the parameters $g_{1,2} = 1-L/R_{1,2}$ were introduced in Chapter 1. But in general, with misalignments, obstructions or perturbations, $M$ is not diagonal and each eigenvector is a mixture or linear combination as in our Equation (3.11); the $b_{pq}$ may now depend on $z$. We are still free to choose any $w_0$ and $z_0$, but no choice will yield pure $\Psi_{pq}$ as the self-consistent fields, and Equation (3.13) will generally be very tedious to solve without a computer. Applying this empty resonator theory to active lasers, we *assume* that the various $\gamma$ and $E$ are unaffected by the laser medium – except that the overall cavity loss $1-|\gamma|^2$ is exactly balanced by the round-trip gain. This neglect of the active medium is convenient but (not surprisingly) may lead to inaccurate results.

If we understand all this, then the "free-space" part of our problem is solved. To analyze a given resonator we break one round trip into a sequence of well-defined segments, such as free-space propagation between the mirrors or reflection from a mirror. We choose a base set of free-space Gaussian beams $\Psi_{pq}$. In practice, with limited computing resources, $p$ and $q$ cannot take infinitely many values, so our base set is ***judiciously truncated***. By the methods of Chapter 1, or by explicit Huygens–Fresnel diffraction calculations, we mathematically describe the effect of each segment on our arbitrary field (= linear combination of $\Psi_{pq}$) by a matrix. The ordered product $M = M_n \ldots M_3 M_2 M_1$ of all these matrices is the grand round-trip matrix, whose eigenvectors are self-consistent linear combinations and whose eigenvalues define the round-trip losses and phase shifts.

We often hope that the largest eigenvalue (lowest loss) refers to some field:

$$E = b_{00}\Psi_{00} + b_{01}\Psi_{01} + b_{10}\Psi_{10} + \cdots \qquad (3.15)$$

(truncated at some $p_{max}$, $q_{max}$) which is nearly pure $TEM_{00}$; that is, if we normalise to $\Sigma |b_{pq}|^2 = 1$, then we want $1 - |b_{00}|^2 \ll 1$. This seems less likely as larger perturbations are introduced, and is usually untrue of unstable resonators. The perturbation that most concerns us here is a hollow dielectric waveguide, lying along the $z$-axis and acting as a "three-dimensional aperture". From our present point of view, some obvious questions are: How do we form a matrix to describe the effect of this guide on our Gaussian beams? That is, how does radiation couple into the guide, propagate along it, and couple out again? Is it easy to decide whether a given waveguide will significantly perturb the first few resonator modes? (If it will not, we may ignore it for most purposes.) And, importantly, does the eventual waveguide resonator matrix theory show any agreement with experiment?

### 3.3.2 Waveguide Modes

Suppose, by analogy with Equation (3.11), we have an orthonormal set of guide modes, and can express the field $E_{wg}$ anywhere inside the guide as

$$E_{wg} = \Sigma\, a_{mn}\, E_{mn} \qquad (3.16)$$

$$a_{mn} = \int E_{wg}\, E^*_{mn}\, dA \qquad (3.17)$$

where we integrate over the guide cross section. These functions $E_{mn}$ will usually stand for the linearly polarised $EH_{mn}$ modes, just as the $\Psi_{pq}$ functions stand for the $TEM_{pq}$ modes. Because waveguides are lossy, the complex coefficients $a_{mn}$ depend on $z$. There is no analogous mode waist radius $w_0$ or waist plane at some $z_0$; the guide mode has the same spatial shape at every $z$ within the guide.

For Gaussian beams, an "ideal" reflector is a large-aperture mirror whose curvature exactly matches that of the incident beams. Each beam is reflected back on itself, with no coupling to any other beam: that is, the Gaussian beam coupling matrix of the mirror is diagonal. For waveguide modes, the corresponding reflector is a plane mirror aligned perpendicular to the guide axis and placed immediately against the guide end. Here again the modes are reflected back along the $z$-axis without cross-coupling, so we have amplitude self-consistency. It is then fairly clear that, with two such mirrors, we have a simple waveguide resonator whose *modes of oscillation*, or resonator modes, are pure waveguide modes – with, again, the important extra condition of phase self-consistency. The resonator matrix M can be written diagonally; its nonzero elements are the propagation constants $\exp(i2k_{mn}L)$. This diagonal feature is unique to this (in principle) simple tube resonator, called *dual Case-I* (see Section 3.4.1).

Later, we will meet other designs which offer *nearly* pure modes. We will also see that in practice even this simplest resonator does not always follow 'first-order' theory. In general, with perturbations, $M$ must be nondiagonal and the eigenvectors will be linear combinations of the $E_{mn}$ with several significant nonzero terms $a_{mn}$. Note that we can order the values of $m$ and $n$ so that $\Sigma\ a_{mn}\ E_{mn}$ is a column vector and $M$ is a square matrix. If we understand this, the waveguide part of our problem will be solved also.

### 3.3.3 Mode Coupling, Coupling Losses, and Mode Losses

The missing step so far is the coupling between free space and the waveguide. In mathematical terms this is a change of basis; given the two orthonormal sets of field patterns, we express the coupling coefficients as overlap integrals across the guide aperture, and form these coefficients into a coupling matrix. When describing *mode coupling loss*, we must specify the modes from which and into which radiation is coupled. For example, let us consider the fundamental waveguide mode $EH_{11}$, together with a general curved-mirror reflector (Figure 3.4).

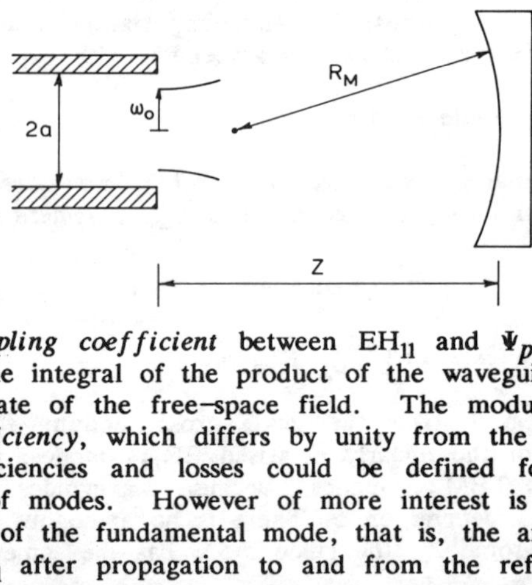


Fig 3.4 Sketch of general curved mirror -waveguide combination.

The *amplitude coupling coefficient* between $EH_{11}$ and $\Psi_{pq}$ is given by $\int E_{11}\ \Psi^*_{pq}\ dA$, that is, the integral of the product of the waveguide mode field and the complex conjugate of the free-space field. The modulus squared of this is the *coupling efficiency*, which differs by unity from the *coupling loss*. Similar coefficients, efficiencies and losses could be defined for any $EH_{mn}$ mode or for any sum of modes. However of more interest is the amplitude *self*-coupling coefficient of the fundamental mode, that is, the amount of $EH_{11}$ which reappears as $EH_{11}$ after propagation to and from the reflector. This is defined as the overlap integral of $E_{11}$ with the returned field due to a "launched" $E_{11}$. The modulus squared will be the $EH_{11}$ self-coupling efficiency $|c_{11}|^2$, and the self-coupling *loss* will be:

$$\Gamma_{11} = 1 - |c_{11}|^2 \tag{3.18}$$

$\Gamma_{11}$ is commonly called the "$EH_{11}$ coupling loss". We can find these fields and integrals by diffraction theory or by assembling and tracking our linear combination of $\Psi_{pq}$, but the task may not be easy. There is much more on this subject in the literature, and Section 3.4 summarises some results.

A particular combination of waveguide modes at the guide exit ($\Sigma\ a_{mn}$ $E_{mn}$), associated with a given resonator mode, will experience its own self–coupling efficiency and loss. Because this is a coherent assembly of guide modes, the resonator mode coupling efficiency is not generally the weighted sum ($\Sigma |a_{mn}|^2 |c_{mn}|^2$) of the individual $EH_{mn}$ coupling efficiencies, but instead depends crucially on the relative phases of the waveguide modes. This is the simple but vital reason why a *multimode* approach is needed for all but the simplest resonators, or all but the crudest estimates of laser loss and transverse mode quality. Unless great pains are taken to force single–mode operation, significant higher–order mode amplitudes ($a_{mn}$) may creep in and cause unpredictable spatial interference; and small changes of only 1–2% in round–trip loss can significantly alter the available laser power. The relative higher–order mode powers ($|a_{mn}|^2$) may seem very small, but interference involves complex amplitudes.

Similarly, the power lost per unit length through guide attenuation of waveguide mode combinations is not generally $\Sigma\ 2\alpha_{mn} |a_{mn}|^2$. In first–order theory (Section 3.4), where terms of order $\lambda/a$ are neglected, the $EH_{mn}$ mode fields form a complete orthonormal set across the guide aperture and all the fields vanish at the walls. But to describe attenuation we must treat the small *nonzero* fields at the walls; and, when two or more modes are present, these fields include interference terms. Existing resonator models tend to neglect this fact, either ignoring it totally or assuming that, with several modes travelling up and down the guide, the interference terms will rapidly "wash out" to leave a smooth average loss. We should be aware of such neat but inaccurate assumptions, and realize that the distribution of loss over the round–trip path in real lasers is seldom easy to calculate.

### 3.3.4 Single–mode, Few–mode, and Multimode Theory

Once our method of resonator analysis is settled, and we can describe each resonator segment, the question usually arises: How many modes should, or must, we include? Broadly speaking, there have been three answers: one, a few, and many: that is, single–mode theory, few–mode theory, and multimode theory.

In *single–mode* theory we assume that the lasing resonator mode is pure $EH_{11}$ (or, rarely, another pure $EH_{mn}$ mode); therefore the resonator loss is formed by pure $EH_{11}$ attenuation and coupling losses, and no interference effects are considered. This may be a good approximation for some lasers with well–aligned near–field plane mirrors, or strong inbuilt mode selection; it is usually hopeless for lasers with gratings or misaligned mirrors or almost any perturbation. It requires all higher–order modes to be either very faintly excited or very strongly damped. But higher–order modes tend not to be suppressed very effectively by attenuation in typical guides: if we begin with equal amounts of $EH_{11}$ and $EH_{12}$, it may take many metres of guide propagation before the ratio is even 2:1 in favour of $EH_{11}$. (To check this we evaluate $L \simeq 0.5(\alpha_{12}-\alpha_{11})^{-1}\ln 2$.) Also, by making guides with low $EH_{11}$ losses, we inevitably reduce any mode discrimination due to guide loss: the inference is that an "ideal" laser of minimum $EH_{11}$ loss, with an excellent straight smooth guide and plane mirrors near the guide ends, has negligible inbuilt transverse mode discrimination and may emit beams whose transverse shapes are quite unlike $E_{11}$ or $\Psi_{00}$.

To model real lasers in any detail, we must consider at least the *first few modes*. It is generally considered that about five or ten is adequate: the lowest-loss resonator mode and its loss will be modelled with fair accuracy, so that adding extra modes does not greatly change the results; beyond five or ten modes the first-order theory must be suspect; and, since with $N$ modes we generally have an $N \times N$ complex matrix, the computing load for $N > 10$ may be too heavy.

*Multimode* resonator models, when compared with real lasers (see Section 3.5), tend to give encouraging but not *very* accurate results. They are far better than single-mode models, and considerably better than three- or five-mode models, but yield diminishing returns as our initial approximations become strained and then broken.

In fact, these approximations are likely to collapse before we achieve numerical convergence to "final" resonator predictions. This is, again, simply because typical waveguides are short and fat. Although the capillary waveguides used in lasers appear relatively long and slender, in terms of transverse mode discrimination they are not. The influence of interfering high-order modes can be complicated, and resonator losses can seem to depend (in theory) very sharply on precise guide geometry (see Section 3.5). Some caution is needed when we interpret rapidly varying loss curves, derived from idealised matrix equations, in terms of real laser behaviour. Real lasers have imperfections – roughnesses, random mirror defects, scattering centres – which may be very difficult to model accurately but which, on the whole, tend to smooth out such "ideal" sharp variations or spikes. In practice we must draw a line at some reasonable level of theoretical complexity; it becomes unproductive to consider hundreds of modes in search of an extra small fraction of numerical accuracy. Section 3.4 offers some rough limits to common waveguide laser assumptions. The only general lesson is that we should keep clearly in mind the type of results and the degree of accuracy that will satisfy us; use the necessary number of modes to that end; and *lower our expectations* if we find that number too high.

## 3.4 FIRST ORDER THEORY AND ITS LIMITS

This section discusses the *single-mode* theory of waveguide resonators. This theory, built on the waveguide mode expressions in Section 3.2 and the coupling coefficients in Section 3.3, gives our first view of the impact of a waveguiding region within the resonator. We quote some important results from the literature on coupling theory, the dual-Case I resonator, and Rigrod modelling, but we stress that the resonator properties of most waveguide lasers are poorly described by single-mode theory.

### 3.4.1 Coupling Loss Theory of Single-Mode Waveguide Resonators

This topic introduces the essential concepts in resonator analysis, and remains interesting for three main reasons. First, if we trust that our resonator transverse mode is almost a pure waveguide mode, we want to estimate coupling losses and judge their effect on laser power (see Section 3.3.5 where the Rigrod equation is discussed). Secondly, in resonator design, we want to exploit mode-dependent coupling losses and *force* near-single-mode operation; here the relative losses of several modes are important. The third reason is that waveguide transmission coupling losses are also important.

We consider the general reflector (waveguide plus spherical mirror) of Figure 3.4. A guide mode is launched from the aperture and, after reflection, couples to itself and to other modes. Researchers have found the returned field distribution in two main ways. One is to express the initial field as a linear combination of Gaussian modes and propagate it to and from the mirror according to the Gaussian beam equations of Kogelnik and Li (1966), assuming that the mirror aperture is effectively infinite. The other is to use scalar diffraction integrals. The two approaches give identical results (Hill and Hall 1985). The early work (Abrams 1973, Abrams and Chester 1974, Degnan and Hall 1973) predicted the existence of three reflector configurations which result in low $EH_{11}$ coupling loss. For a circular guide of radius $a$, with

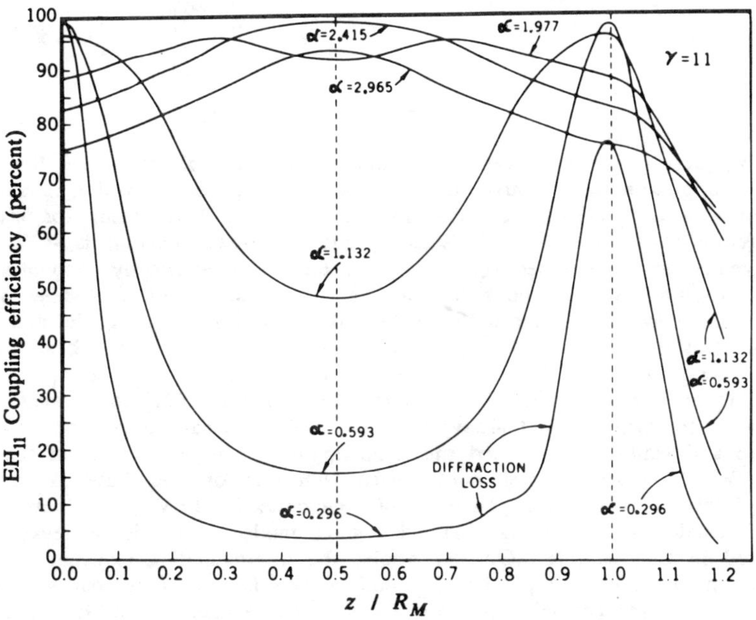


Fig 3.5 $EH_{11}$ coupling losses for circular guides. Curves are shown for various values of $\alpha = (ka^2/R_M)$ and a constant mirror half width of $11a$.

E($r$) the launched field and E'($r$) the returned field, equation (3.18) may be written:

$$\Gamma_{11} = 1 - \left| \int_0^a E(r)\ E'(r)\ 2\pi r\ dr \right|^2 \qquad (3.19)$$

Evaluating this for a range of mirror–guide distances and mirror curvatures yields the set of curves in Figure 3.5 (Degnan and Hall 1973). The three low–loss reflector geometries (Figure 3.6) were named by Degnan and Hall:

| | |
|---|---|
| *Case I*: | Large $R$ mirrors very near the guide ($z \simeq 0$, $z/R \simeq 0$) |
| *Case II*: | Large $R$ mirrors centred near the guide entrance ($z \simeq R$) |
| *Case III*: | Mirrors with $R \simeq 2b$ and $z \simeq b$, where $b \equiv \pi w_0^2/\lambda$. |

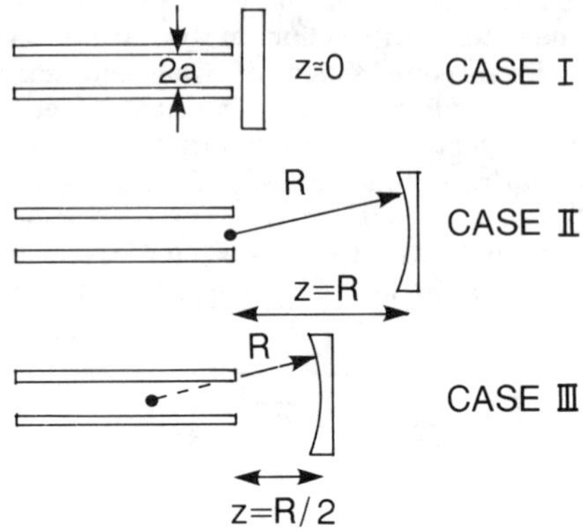


Fig 3.6 Low–loss coupling configurations for fundamental waveguide mode.

Here $w_0$ and $b$ are the beam waist radius and Rayleigh range of the $EH_{11}$ mode's *approximating Gaussian*, i.e. the $TEM_{00}$ beam having maximum overlap with $EH_{11}$ across the guide aperture ($z = 0$). This occurs for $w_0/a \simeq$ 0.64 (circular) or $w_0/a \simeq 0.70$ (square), and the power overlap is $\simeq 98\%$ in both geometries. We expect that any reflector which efficiently recouples this $TEM_{00}$ to itself will do the same for $EH_{11}$. Thus the simplest possible model of a general waveguide resonator, with only $EH_{11}$ inside the guide and only $TEM_{00}$ outside, suggests the choice of *phase–matched mirrors,* whose curved surfaces coincide with the phasefronts of this special $TEM_{00}$ at $z_1$ and $z_2$, according to $R = z + b^2/z$. For such mirrors, Abrams (1972) found that $\Gamma_{11}$ is $\simeq 1.48\ \%$ for Case III and always less than 7 % (circular bore). For Case III, Degnan and Hall (1973) revised this estimate to $\simeq 1.38\ \%$ and found that $\Gamma_{12}$ $\simeq 78\ \%$. A reasonably well–built device with one or two Case III mirrors usually offers the best chance of guaranteed $TEM_{00}$–like operation. Experimental results showing that very good mode quality is achievable are discussed in Chapter 17. On the whole, the approximating Gaussian concept is neat and intuitive and predicts $\Gamma_{11}$ minima with fair accuracy, but with short fat guides its practical value has not always been realised. *Note*: this definition $b \equiv \pi w_0^2/\lambda = Z_R$ is common in the waveguide laser literature. It is half the "*b*" of Chapter 1 and Kogelnik and Li (1966).

In the far field, any launched field pattern will revert to a roughly spherical wave centred near the guide aperture. Case II is a far–field phase–matched reflector, for which *any* guide mode has low coupling loss. The poor mode discrimination and large $z$ make it unpopular, unless the extra space is needed for intracavity elements.

Case I is used in the great bulk of today's commercial and scientific gas waveguide lasers: just a plane mirror at each end of the guide, placed against, or within a couple of millimetres of, the exit. There has been much study of the $EH_{11}$ coupling loss for small (realistic) departures from perfect Case I ($z = 0$, $R = \infty$, zero tilt ), with surprising disagreements. Some simple and useful results are:

***EH*$_{mn}$ *Coupling Losses*** $\Gamma_{11}(z \ll b)$

Degnan and Hall give $\simeq 38.4\ N^{-1.5}$ % for circular guides, where $N = a^2/\lambda z$ is the reflector Fresnel number (numerical approximation to diffraction integral results). Boulnois and Agrawal (1982) give $\simeq 33N^{-1.5}$ % for square guides (asymptotic approximation); the same answer is obtained from a Hermite–Gaussian expansion (Hill 1988).

***EH*$_{11}$ *Self-coupling Efficiencies for Square Guides*** $(N \gg 1)$

Calculations by Boulnois and Agrawal yield:

$$|c_{mm}|^2 \simeq \left[ 1 - \frac{m^2}{6\ N^{1.5}} - \frac{\pi\ m^4}{240\ N^{2.5}} + \frac{m^4}{72\ N^3} \right]^2 \tag{3.20}$$

from which (keeping two terms) $\Gamma_{mn} \simeq (1/6)\ (m^2 + n^2)\ N^{-1.5}$. Their analytical expressions, not given here, greatly reduce the computer time needed for square-bore coupling loss calculations, if misalignment is not involved, and coupling amplitudes and cross-coupling coefficients are not required.

### 3.4.2 Dual-Case I Waveguide Lasers

We recall that in a dual-Case I resonator a single round trip is very nearly equivalent to propagation along an undistorted guide of length $2L$, and the resonator transverse modes are essentially the transverse modes of the waveguide itself. The mode frequencies are found from a standard phase equation:

$$2\beta_{mn} n(\nu_{j,mn}) L = 2j\pi \tag{3.21}$$

where the laser medium refractive index $n(\nu)$ includes a small anomalous dispersion term. By taking the difference between two mode frequencies, we can remove the axial mode integer $j \simeq 10^5$. In the absence of mode pulling ($n(\nu) = 1$) we have:

$$\nu_{j,mn} \simeq \frac{jc}{2L} + \frac{c\lambda}{32} \left[ \frac{m^2}{a^2} + \frac{n^2}{b^2} \right] \qquad \text{(rectangular)} \tag{3.22}$$

$$\nu_{j,nm} \simeq \frac{jc}{2L} + \frac{c\lambda \rho_{nm}^2}{8\pi^2 a^2} \qquad \text{(circular)} \tag{3.23}$$

Equations (3.22) and (3.23) show roughly how the mode frequencies depend on waveguide geometry, although in practice the exact frequencies will depend on many other small corrections. In particular, the difference in frequency between two different transverse modes depends strongly on waveguide bore but weakly on waveguide length; in terms of typical manufacturing tolerances the derivatives $\partial\nu/\partial a$ vary as $a^{-3}$.

By contrast with stable open resonators (Chapter 1), these waveguide laser transverse mode spacings vary as $m^2$ and $n^2$ or $\rho_{nm}^2$ and can exceed the axial mode spacing $c/2L$. Homogeneously broadened lasers tend, with some mostly unwelcome exceptions, to operate on the one mode (of all the available modes) with the highest ratio of small-signal gain to threshold loss. Now, if we fix $|\nu_{j,11}-\nu_{j,12}|$ at an integer multiple of $c/2L$, then the $EH_{12}$ mode always coincides with a lower-loss $EH_{11}$ mode and, *other things being equal*, $EH_{11}$ must be preferred. This "coincidence" of transverse mode frequencies, if it can be achieved, will hold very well for different axial modes under the same line, and fairly well for different $CO_2$ lines. Formally, this implies:

$$L = 16sa^2/\lambda\Delta \qquad s = 1, 2, 3, 4, \ldots \text{(rectangular)} \tag{3.24}$$

where $\Delta = |m^2 - m'^2 + n^2 - n'^2|$. This is for a square-bore dual-Case I laser and any two modes $EH_{mn}$ and $EH_{m'n'}$. For the circular-bore dual-Case I device we obtain:

$$L = 4\pi^2 sa^2/\lambda(|\rho_{nm}^2 - \rho_{n'm'}^2|) \qquad \text{(circular)} \tag{3.25}$$

for any two modes $LP_{nm}$ and $LP_{n'm'}$. There is some evidence that such choices of guide geometry can improve the average laser mode quality by increasing the proportion of the active signature over which the fundamental mode is preferred.

It may be asked, "When is a Case I ?" We can show from Equations (3.10) and (3.20) that at $\lambda$=10$\mu$m the plane-mirror coupling loss $\Gamma_{mn}$ ($N \gg 1$) becomes comparable with the first-order single-pass alumina guide loss $2\alpha_{mn}$ when $z \simeq 0.18L^{0.67}$ ($z$ and $L$ in mm), independent of guide width and nearly independent of mode number (Hill and Colley 1988). It seems reasonable to propose, as a working definition of a Case I mirror, that $z$ should be considerably less than this value, so that the coupling loss can be neglected. However, this typically requires $z \leq 1$ mm, and rules out many existing Case I mirrors. Another reasonable definition is that a dual-Case I laser, where the mirror-guide distances $z$ are gradually increased, ceases to be dual-Case I when the fundamental resonator mode has appreciable non-$EH_{11}$ content.

There are no universally accepted definitions of mode quality. Formally, we would wish to know all the significant terms in the decompositions $\Sigma\, a_{mn}E_{mn}$ or $\Sigma\, b_{pq}\Psi_{pq}$, over the full laser signature and for a range of cavity perturbations. This is a most unrealistic goal. Many users are satisfied if a laser, after a brief warm-up, yields a "reasonable" single-dot transverse intensity pattern with long-term power variations of a few per cent. In this case, extensive but simple beam measurements are adequate (see Chapter 17). Other users may set fierce limits on short-term frequency fluctuations, unwanted mode beating, and departure from pure $EH_{11}$ or pure $TEM_{00}$. They may also demand wide tunability, and no line-hopping or multilining.

### 3.4.3 Rigrod Analysis for Waveguide Lasers

A useful and deliberately simple model for the output power of a homogeneously broadened gas laser has been widely adopted in the guise of one or other form of the Rigrod equation. It allows comparisons of waveguide resonator theory with experiment by expressing a measurable quantity, the laser output power, in terms of resonator losses. The gain is assumed constant along the length $L$ of the laser and does not depend on direction; when homogeneous broadening is dominant it is fully described by:

$$g(z) = \frac{g_0(\nu_0)}{1 + I_+ + I_-} - 2\alpha = \frac{1}{I_+}\frac{dI_+}{dz} = -\frac{1}{I_-}\frac{dI_-}{dz} \qquad (3.26)$$

This $g(z)$ is the saturated gain in intensity per unit length experienced by both the *forward* and *backward* travelling plane waves in the amplifying section of the resonator. These plane wave intensities are normalised to the line–centre saturation intensity $I_s(\nu_0)$, the intensity in whose presence the available gain drops to $g_0(\nu_0)/2$ (half the small–signal value). Thus:

$$I_+ \equiv I(\text{forward})/I_s(\nu_0) \qquad\qquad I_- \equiv I(\text{backward})/I_s(\nu_0)$$

There is a uniform loss $2\alpha$ per unit length. By considering the boundary conditions defined by the mirror reflectances $R_1 \equiv 1 - A_1 - T_1$ and $R_2 \equiv 1 - A_2 - T_2$ (where $A$ is the mirror loss and $T$ the transmittance), and manipulating this gain equation, Rigrod obtained essentially this expression for line–centre output power when $\alpha = 0$:

$$P = \frac{I_s(\nu_0)\ A_b\ \sqrt{R_1}\ T_2}{(\sqrt{R_1}+\sqrt{R_2})\ \ (1-\sqrt{R_1 R_2})}\left[\ g_0(\nu_0)L - \ln(R_1 R_2)^{-\frac{1}{2}}\ \right] \qquad (3.27)$$

This represents the normalised intensity incident on the outcoupler, multiplied by (i) the normalisation factor $I_s(\nu_0)$ (ii) the area of the incident beam $A_b$ (iii) the outcoupler transmittance $T_2$. Often there is only one outcoupler ($T_1$ = 0) and all the dissipative loss is lumped into one term $A$; then $R_1 = 1-A$ and $R_2 = 1-T$. To account for a small uniform guide loss we can put $R_1 = 1-A-4\alpha L$. The full Equation (3.26) cannot be integrated explicitly for a resonator, but Rigrod (1978) derived a useful approximate form. For a homogeneously broadened laser with a Lorentzian linewidth $\Delta\nu$ and $f \equiv (\nu-\nu_0)/\Delta\nu$ :

$$P(\nu) = \frac{I_s(\nu_0)\ A_b\ \sqrt{R_1}\ T_2\ \ [(g_0(\nu_0)-2\alpha(1+4f^2))L+(1+4f^2)\ln(R_1 R_2)^{\frac{1}{2}}]}{(\sqrt{R_1}+\sqrt{R_2})\ (1-\sqrt{R_1 R_2})\ (1-2\alpha L/\ln(R_1 R_2)^{\frac{1}{2}})} \qquad (3.28)$$

The waveguide beam area is usually defined as $A_b \equiv \pi w_e^2 = \frac{1}{2}\pi w_0^2$, where $w_e \equiv w_0/\sqrt{2}$ is the 1/e–power radius of the $EH_{11}$ approximating Gaussian (see Section 3.4.1). For a square guide $A_b \simeq \pi\ (0.70/\sqrt{2})^2 \simeq 0.78a^2$. If the resonator mode *is* pure $EH_{11}$, the assumption of uniform $A_b$ (and hence uniform active medium "filling factor") will hold well – better than for open resonators with Gaussian beams of varying $w(z)$.

The waveguide distributed loss is $2\alpha$, not $\alpha$ as sometimes assumed in the literature. Unluckily, if the guide is narrow enough to force $EH_{11}$-like operation, then $2\alpha_{11}$ is often not greatly less than the *saturated* gain; but if it is wide enough to make $2\alpha_{11}$ very small, perturbations and lack of discrimination may introduce higher-order modes whose losses are not very small. Common dual-Case I $CO_2$ waveguide lasers tend to have lumped losses $A$ of 1–3%; higher values naturally occur in more complex resonators. The best obtainable alumina waveguides seem to have losses of $2\alpha_{11} \gtrsim 1\%\ m^{-1}$ for $2a = 1.5$ mm, and the actual laser mode is never pure $EH_{11}$. Thus, it is usually unwise to ignore guide losses. Equation (3.28) is simple and useful; it pays at least lip service to distributed losses; and in many cases (see Chapter 17) it can be fitted fairly well to measured power.

In a single-mode model, the coupling loss is considered as part of the lumped loss $A$. In a multimode model, or in nontrivial resonators with several sources of loss, the easiest solution is often to insert an average distributed loss $-(1/L)\ln|\gamma|$ in Equation (3.28). As mentioned in section 3.3.3, we hope that loss variations will "wash out" along the resonator path, but few or no real lasers have had their dissipative loss distributions examined in detail.

Rigrod analysis is often used to extract best-fit values of laser parameters, such as the $g_0$, $I_s$ and $A$. One parameter, such as $T_2$ or active gain length, is varied while the others are kept constant; and output power is recorded for various pressures (or gas mixes, input powers, and so on). Even after much curve-fitting the error bars are almost always quite large ($\simeq \pm 10$ % or worse). The equations are so simple that this is not surprising. What use is the idea of constant $A_b$, when we know that changing the pressure or overpumping the discharge will seriously alter the output mode (and hence $A_b$, and hence $2\alpha$, and probably the effective $g_0$ and $I_s$ too) ? Why suppose that the loss is uniform in a waveguide laser, when we know that injecting an $EH_{11}$-matched beam into a typical guide gives clear periodic variations in loss? The answer is that Rigrod analysis is *simple* and, though obviously deficient, *accurate enough to be useful* (see Section 3.5) in understanding and designing waveguide laser resonators.

## 3.5 REAL WAVEGUIDE RESONATORS: EXPERIMENT AND THEORY

*C'est magnifique, mais ce n'est pas Laguerre*

### 3.5.1 Distant Mirrors

From time to time, the versatile toolbox of waveguide resonator theory has been refined on real-life lasers. Some problems, such as a spatially resolved treatment of gain and saturation, or accurate determination of waveguide $E_{mn}$ and $k_{mn}$, are very hard. A few are both interesting and not too difficult, at least initially, and are summarized here: plane mirrors, then tilted plane mirrors; intraline tunability and line selection; finally, unpleasant behaviour such as line-hopping and resonator mode "hooting". Multimode predictions and detailed power/mode/frequency studies have rarely been made for real devices. Some well-known and some recent results are included here.

If we begin with a square-bore dual-Case I resonator, in theory the fundamental transverse mode is pure $EH_{11}$ and the others are pure higher-order $EH_{mn}$ (which can be ranked in order of increasing loss). As one mirror is moved away, the round-trip losses for pure $EH_{mn}$ modes rise roughly as $4\alpha_{mn}L + (1/6)(m^2 + n^2)N^{-1.5}$, and these modes begin to couple among themselves. The fundamental *resonator* mode at a fixed wavelength is now whatever field minimizes the sum of guide loss and coupling loss. Thus, the resonator may do better than $4\alpha_{11}L + \Gamma_{11}$ by mixing in (say) some $EH_{13}$, $EH_{15}$, and so on. But it may also do worse, and the fundamental mode loss may be higher or lower than the pure $EH_{11}$ loss. The resonator cannot choose pure $EH_{11}$, which is no longer a *resonator* mode.

If we fix $L$ and vary $z$, the fundamental loss curve often shows non-monotonic behaviour (wiggles): This is confirmed by experiments where laser power fades when $z$ is increased initially, but revives when the plane mirror is pushed yet further away. If we fix $z$ and vary $L$ (easier to model than do), the fundamental loss curve often shows rather sharp periodic "spiky" behaviour. The periods correspond with the "beat wavelengths" over which pairs of important modes (notably $EH_{11}$ and $EH_{13}$) change their relative guide propagation phase shifts by $2\pi$. The published predictions of loss spikes (Gerlach *et al* 1984, Hill 1988) are no doubt unrealistically sharp (see section 3.3.4), but experiments confirm that phase-period effects are important in real lasers – often more important than attenuation effects. This is further proof of shortness and fatness; if guides were very long and thin, high-order modes would be damped out efficiently.

### 3.5.2 Tilted Mirrors and Folded Lasers

Some people, presented with a "well-aligned" waveguide laser and an $EH_{11}$-like output, cannot resist "tweaking" a mirror to be sure that it *is* well-aligned. The immediate result is familiar: the power drops sharply, the mode becomes asymmetric or multidot, and things are never quite as good again. Tilt experiments are interesting for several reasons. First, of course, they are relatively easy, often easier than varying $z$ or $L$. Figure 3.7(a) shows a typical arrangement. A tilted plane mirror is a fair first approximation to a laser diffraction grating (see 3.5.3). Interesting things tend to happen with tilts of order $\lambda/a$, which is typically several milliradians for $CO_2$ guides: microradian precision is not needed, and the important dispersive tilts in $CO_2$ lasers happen to be 2–8 mrad. A curved guide can readily be modelled as a sequence of short segments with small tilts between them. Moreover, tilting is a simple and controllable way to introduce and study non-$EH_{11}$ guide modes.

Suppose we tilt an ideal Case I plane mirror ($z \simeq 0$) around its vertical diameter: call this a pure $x$-tilt $\varphi_x$. To first order, we get a linear phase shift $\exp(i2\pi x\varphi_x/\lambda)$ across the aperture width. The $EH_{mn}$ fields are sine and cosine functions, and their tilt-coupling coefficients emerge as sums of sinc functions (Hill and Hall 1986). There is no "$y$-coupling" between modes of different $n$. With circular guides (Bessel functions and polar coordinates), any tilt couples together modes with different $n$ and different $m$.

For nonzero $z$, the tilt causes both a linear phase shift and a displacement, $\Delta x$ of the recoupling field, and numerical solution is generally required. As $\varphi_x$ and $\Delta x$ grow to (typically) several mrad and a significant fraction of the guide width, the number of modes needed for a given accuracy tends to increase sharply. The theoretical dependence of resonator loss on $z$ or $L$, for modest and relevant tilt values such as 6–8 mrad, can be surprisingly acute, and "Case I" assumptions may be improper even for $z \simeq$ 1–2 mm (Hill and Colley 1988). Recent results for a small square-bore rf-excited laser are shown in Figure 3.7(b). The measured and predicted powers $P(\varphi_x)$ agree very well, and the Rigrod values for $g_0$ and $I_s$ fall within the usual (rather wide) limits.

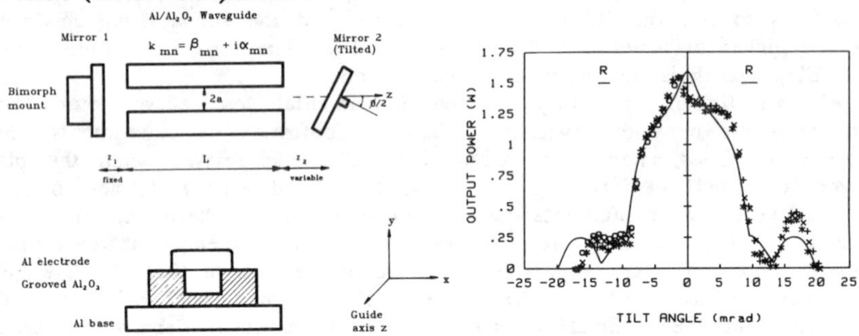


Fig 3.7 Square-bore rf-excited $CO_2$ waveguide laser with adjustable mirror mounts (Hill and Colley 1988).

Folded-waveguide $CO_2$ designs are popular because they squeeze long active lengths into compact devices and are well suited to slablike rf excitation structures (Newman and Hart 1987). A V-fold or U-fold with plane mirrors is no harder to model than a distant plane mirror in a linear resonator. Curved folding mirrors can improve output mode quality, but they cause astigmatism and the mode coupling coefficients need heavy computation (Banerji *et al* 1989). Folds with "partial waveguiding" or Brewster-cut prisms (Chapter 17) reduce the losses and increase the theoretical difficulty.

### 3.5.3 Tunability and Line Selection

For each available laser transition, a resonator has a set (perhaps small, perhaps very large) of potentially oscillating resonator modes. Many $CO_2$ devices can choose from several tens of distinct lines, each with $\simeq$ 1–10 potential resonator modes. Well-behaved lasers ought to select the one mode/line combination offering the highest ratio of small-signal gain to threshold loss; the competitive processes characteristic of homogeneous broadening should suppress all other laser frequencies. This is often a reasonable description of real lasers. The main problem of $CO_2$ tunability theory is then to calculate the relevant gains and losses and find how far away from line centre we may move our chosen mode (usually the fundamental mode) before it slips below threshold or another mode takes over.

Setting $P(\nu) = 0$ in Equation (3.28) gives the maximum possible continuous tuning range:

$$2\left|\nu - \nu_0\right|_{zero\ power} = \Delta\nu \left[ \frac{g_0(\nu_0)\ L}{2\alpha L + \ln(R_1 R_2)^{-\frac{1}{2}}} - 1 \right]^{\frac{1}{2}} \qquad (3.29)$$

In these simple terms, the modelling problem is just to find the average loss $2\alpha \equiv -(1/L)\ln|\gamma|$. However, no systematic check of laser tuning ranges against multimode theory seems to have been performed. To model "hopping" to other resonator modes under the same $CO_2$ line, we need to know their losses and relative frequencies. These depend on guide material constants (poorly known) and precise guide geometry (hard to measure accurately and, if distorted, hard to model). To model hopping to another line, we need to know the relative gains of the two lines and the precise positions of the resonator modes within the lineshapes. Thus the total problem of mode selection and tunability looks very difficult.

Nevertheless it is important to see, at least qualitatively, why waveguide $CO_2$ lasers are especially prone to line-hopping. Their ability to run cw at high pressures (200–400 torr) offers a linewidth $\Delta\nu$ of 1–2 GHz. It is possible, but rather difficult, to obtain reliable dual-Case I single-mode $CO_2$ tuning ranges of $\gtrsim$ 1 GHz (Abrams 1974, Gonchukov *et al* 1978). Complex cavities can raise this to $\gtrsim$ 1.5 GHz. Usually, unwanted lines or modes break in after a few hundred MHz. With narrow guides only 100–300 grating lines are typically illuminated, but "loss of resolution" is an incomplete explanation.

Lasing on a desired line, on the fundamental mode in a dual-Case I grating-tuned laser, will cease for one of three main reasons. We cannot exceed $|\nu-\nu_0| = c/4L$ because an adjacent axial mode will take over. We may fall below threshold according to Equation (3.29). Or another mode may achieve a higher gain/loss ratio. If this is a mode under the same line, we can infer the mode frequency separation from the laser signature or from heterodyne measurements, estimate the mode losses, and hope that the experimental mode-hopping point coincides with the theoretical point where the gain/loss ratios are equal. This seems a straightforward test of theory. It is not, and is seldom if ever tried. Real lasers do not obey ideal theory, and unwanted modes tend to be seen as nuisances to be eliminated, not as resonator features to be studied.

### 3.5.4 Resonator Mode Degeneracies: Hopping and Hooting

So far we have accepted that a cw $CO_2$ waveguide laser chooses, from instant to instant, the one resonator mode with the highest ratio of small-signal gain to threshold loss. All other potential modes are frozen out by rapid collisional-broadening effects in the high-pressure active medium. At many times and for many devices this is not true. Real gas discharge waveguide lasers are imperfect, and cannot suppress other modes unless the gain/loss ratios are *clearly* larger than that of the fundamental mode; but what "clearly" means for a specific laser depends on molecular excitation rates and transverse mode spatial overlaps, and would be remarkably hard to quantify. The rate equations can be readily solved only for cases well removed from real life. In very simple terms, the laser cannot be expected to run reliably single-frequency when two equal-loss modes lie within a few MHz; or when mode-hopping is imminent and the gain-loss ratios are very nearly equal; or, similarly, when line-hopping is imminent. Thus we may see more than one frequency emitted on one line, or more than one line emitted, or both; and this is highly irritating to many users.

If, with a multimode resonator model, we plot the losses for the first few resonator modes as functions of $z$ or $\varphi$, comparing curves for several known $CO_2$ wavelengths, it is tempting to identify regions of curve-crossing or near-coincidence with observations of multifrequency output. This cannot be pushed too far, because even the multimode approaches outlined above are highly idealized and restricted to empty resonators, whereas multifrequency waveguide $CO_2$ lasers are (by our definitions) nonideal and badly behaved. For instance, we have not tried to model any effect that the two or more resonating modes have on each other; that is a nonlinear problem needing much more work. Despite these reservations, recent work on cw mode beating ("hooting") and line-hopping is encouraging and provides impressive support for multimode methods.

## 3.6 SUMMARY

The general aim of this chapter has been to offer readers a brief and not mathematically detailed review of waveguide laser resonators, so that the literature can be approached with a knowledge of the theory's present limitations, and without a few common misconceptions. We began by introducing *modes of propagation* (both free-space and waveguide). Resonator modes are self-consistent field patterns, with certain definite round-trip phase shifts, which are potentially available to the laser cavity as *modes of oscillation*; and each is associated with a *single* frequency. The presence in the laser field of several guide modes does not imply several different oscillation frequencies.

From our point of view there is nothing special about *waveguide* resonators as such. It is possible, and may sometimes help, to see them as conventional open resonators with more or less strongly perturbing 3-D apertures. The transition from free-space propagation to waveguiding, as the perturbation strengthens, is not necessarily sharp. And we stress that many existing commercial and scientific "waveguide" devices are not safely clear of this transition region; still less are they safely within the domain of single-mode theory. They are ***Short and Fat***. This forces us towards a multimode treatment if we want reasonably accurate performance comparisons and predictions.

Results so far, some of them presented in this chapter, show that current theory cannot explain all our experimental observations. Nevertheless, the current steady expansion of waveguide laser technology in industrial, medical and remote-sensing applications should prompt continuing improvements in the depth and accuracy of waveguide resonator theory.

## 3.7 REFERENCES

***Reviews***

Abrams R L 1979 Laser Handbook Vol 3 ed. M.L.Stitch (Amsterdam: North-Holland) pp 41-88

Degnan J J 1976 Appl. Phys. 11 1

Hall D R and Hill C A 1987 Handbook of Molecular Lasers ed. P K Cheo (New York and London: Marcel Dekker) pp 165-258

Smith P W, Wood O R (II), Maloney P J and Adams C R 1981 IEEE J. Quantum Electron. 17 1166

*Waveguide resonators*

Abrams R L and Chester A N 1974 Appl. Opt. 13 2117
Bel'tyugov V N *et al* 1988 Sov. J. Quantum Electron. 18 599
Chester A N and Abrams R L 1972 Appl. Phys. Lett. 21 576
Degnan J J and Hall D R 1973 IEEE J. Quantum Electron. QE–9 901
Gerlach R, Wei D and Amer N M 1984 IEEE J. Quantum Electron. QE–20 948
Hill C A and Hall D R 1986 IEEE J. Quantum Electron. QE–22 1078

*Hollow Dielectric Waveguides*

Degnan J J 1973 Appl. Opt. 12 1026
Hill C A 1988 IEEE J. Quantum Electron. 24 1936
Laakmann K D and Steier W H 1976 Appl. Opt. 15 1334
Marcatili E A J and Schmeltzer R A 1964 Bell Syst. Tech. J. 43 1783

*Coupling Losses*

Abrams R L 1972 IEEE J. Quantum Electron. 8 838
Banerji J, Davies A R, Jackson P E and Jenkins R M 1989 Appl. Opt. 28 in press
Boulnois J–L and Agrawal G P 1982 J. Opt. Soc. Am. 72 853
Henderson D M 1976 Appl. Opt. 15 1066
Hill C A and Hall D R 1985 Appl. Opt. 24 1283

*Tunability and Resonator Design*

Abrams R L 1974 Appl. Phys. Lett. 25 304
Abrams R L and Bridges W B 1973 IEEE J. Quantum Electron. 9 940
Gonchukov S A, Kornilov S T and Protsenko E D 1978 Sov. Phys. Tech. Phys. 23 1084
Merkle G and Heppner J 1983 IEEE J. Quantum Electron. 19 1663
Rigrod W W 1978 IEEE J. Quantum Electron. 14 377
Roullard F P (III) and M Bass 1977 IEEE J. Quantum Electron. 13 813
Tang F and Henningsen J O 1987 Appl. Phys. B 44 93

# 4 Ring Laser Resonators

*T A KING*

## 4.1 INTRODUCTION

The ring resonator finds widespread application in many different types of laser. The ring laser cavity can oscillate simultaneously or independently in clockwise or counterclockwise propagating directions supporting travelling waves, rather than the standing waves of linear cavities. This eliminates the creation of an inhomogeneous depletion of inversion in the laser medium, known as spatial hole burning. The travelling wave more uniformly saturates the gain medium leading to enhanced extraction of power. A further consequence of the presence of travelling waves is the ability to obtain more stable outputs and an order of magnitude higher power in single frequency operation, compared with performance of the linear laser. The increased competition between adjacent modes means that the laser can be pumped further above threshold while maintaining single frequency operation. The ring cavity gives flexibility in resonator design, such as the ability to incorporate long collimated beam sections with large beam diameters in order to extract power from large volumes of laser media. Additional beam waists can be incorporated into the ring for the placement of intracavity elements, such as modulators and non–linear crystals. The ring structure offers decreased sensitivity to misalignment in the plane of the cavity, insensitivity to feedback from back reflections into the cavity, where the back–reflected beam goes into the non–oscillating direction in the ring. This latter trait is important in several applications such as injection locking.

In this article, the emphasis is on *resonator* aspects of *ring lasers*; the passive ring resonators used in interferometry, beam diagnostics and fibre–optic sensing are not described in any detail. Particular attention is paid to the non–planar ring resonator because of its novel features and importance in recent applications in the ring laser gyroscope and in diode–pumped non–planar solid–state ring lasers.

## 4.2 FEATURES OF RING CAVITIES

### 4.2.1 Cavity Optics

Some illustrations of ring cavities are shown in Figure 4.1. In a linear cavity the beam cross–section is generally circular, characterised by a beam waist, a spot size at the $1/e^2$ intensity point and with a plane phase front at the beam

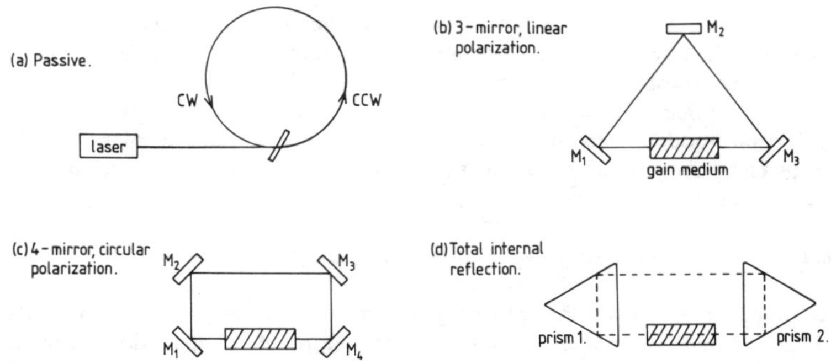


Fig 4.1 Representative ring laser resonators.

waist. The mode structure in ring resonators is similar to that of linear cavities, but with some significant differences. These have been discussed by Collins (1964). The wavefronts of the beam inside the resonator are determined by the geometry of the ring path and the curvature of the confining reflectors, which determines the spatial modes of the resonator (Figure 4.2). For the ring resonator the beam meets the confining mirror at

Fig 4.2 Three-mirror ring cavity with one curved mirror indicating the beam waist.

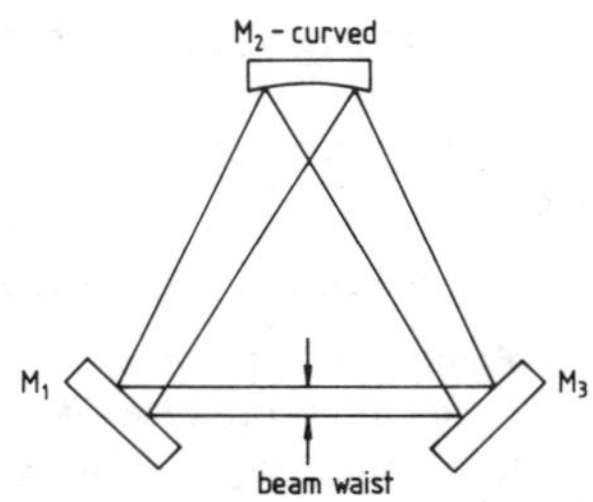


an oblique angle, the radius of the beam in the plane of the ring and normal to it being different. Consequently, the beam profile is elliptical with its major axis normal to the plane of the ring and the ring resonator is astigmatic. Astigmatism of the cavity can be compensated by choice of cavity parameters (Kogelnik *et al* 1972, Cojocaru *et al* 1989)

The ring resonator differs from the linear laser in the resonant conditions for the frequencies of the longitudinal and transverse modes. In the ring resonator, the electric field distribution is reproduced after one complete transit. For a ring of total perimeter length $L$ the resonant frequency of the $n^{\text{th}}$ axial mode is:

$$\nu_n = n \frac{c}{L} \tag{4.1}$$

giving a frequency separation of longitudinal modes:

$$\Delta\nu = \frac{c}{L} \tag{4.2}$$

which is twice that derived in Chapter 1 (Equation 1.2) for linear resonators.

In the ring resonator with an odd number of confining reflectors, an off–axis ray traces a path which appears on the opposite side of the axis after one transit of the ring when viewed in the plane of the ring. Transverse modes which have an odd function distribution across the beam in the plane of the ring must undergo an additional $\pi$ phase shift in each ring transit. Thus the mode is displaced by $c/2L$ from its usual frequency. For a p-polarised mode (with its E-vector perpendicular to the plane of incidence), a $\pi$ phase change occurs upon each reflection in a less dense medium and a similar frequency shift of $c/2L$ occurs.

### 4.2.2 Image Rotation In Non-planar Rings

Non-planar rings have the property of image rotation. On each transit of the ring an image is rotated about the direction of propagation, the degree of rotation is dependent on the geometry of the ring with the entire transverse field distribution and the polarization state being rotated.

The reflection of an ideal mirror may be expressed in a matrix equation as:

$$\underline{E}' = -F\underline{E} \tag{4.3}$$

$\underline{E}$ is the electric field column vector and $F$ is the orthonormal mirror reflection matrix. Successive reflections from $N$ mirrors gives an $\underline{E}$ field distribution:

$$\underline{E}' = F_N \cdots\cdots F_2 F_1 \underline{E} \tag{4.4}$$

For a ring, the net rotation $R$ is given by the ordered product of the successive reflections:

$$R = F_N \cdots\cdots\cdots F_2 F_1 \tag{4.5}$$

As for planar rings, the condition for resonance requires the electromagnetic field to reproduce itself in phase, amplitude and polarization after each transit of the ring. For a ring with an even number of mirrors, the polarization eigenvalue equation for the electric field is:

$$R\underline{E} = (F_N \cdots\cdots\cdots F_2 F_1)\ \underline{E} = g\ \underline{E} \tag{4.6}$$

After each transit the electric field reproduces itself within a constant phase factor $g$. The ideal non-planar cavity can be reduced to a single equivalent ring element which alters the polarization state by a pure rotation.

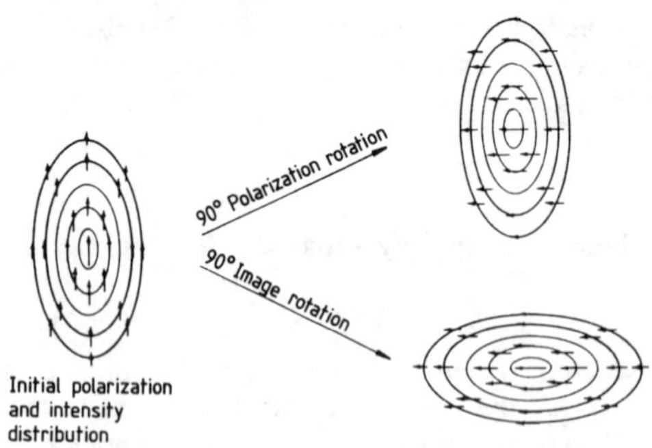


Fig 4.3 Image and polarization rotation.

The eigenvectors of the field are right– and left–handed circularly polarized waves with eigenvalues $\exp(jq)$ and $\exp(-jq)$. The fields are circularly polarized for any value of image rotation other than for multiples of $\pi$ (for which the polarization states are indeterminate). Real mirrors exhibit phase birefringence and loss upon reflection, which can cause the eigenvalues to become elliptically polarized. The Jones matrix with birefringence and loss is then used in the eigenvalue equation.

An image rotator and a polarization rotator are not identical operations on the field. As shown in Figure 4.3 the image rotator acts on the entire field distribution but the polarization rotator acts only in the direction of the field and the complete distribution can be unaffected.

### 4.2.3 Spatial Hole Burning

One limitation to the attainment of a large output power in a single mode in a standing wave cavity is set by spatial hole burning. In a standing wave linear cavity the laser mode has a field distribution with a spatial sequence of nodes and antinodes along the axis of the gain medium. Saturation of gain leads to a spatial modulation of inversion. There are then regions of unsaturated inversion in the position of the nodes of the standing wave. In gas lasers, spatial hole burning is largely smoothed out by the rapid thermal diffusion of the active medium between nodal and antinodal regions. For solid–state and liquid lasers, spatial hole burning can be strong because the thermal displacement of active atoms or molecules is small compared with the wavelength of light, and spatial averaging cannot take place.

The effect of spatial hole burning is eliminated for travelling waves in a ring resonator. The spatial distribution of inversion in the medium remains uniform while mode competition is increased between adjacent axial modes. For homogeneous gain profiles, the ring laser accesses the total population inversion within the active mode volume. A further consequence of spatial hole burning is the difficulty in producing single frequency operation from lasers in which it occurs. Excited atoms or molecules in regions near the nodal field positions of one axial mode will be able to contribute to other modes. This may allow sufficient gain for simultaneous operation on additional longitudinal modes. For inhomogeneous gain profiles, spatial hole burning is eliminated by the unidirectional ring, and the laser can then be pumped further above threshold while remaining in single frequency operation and more power is extracted. In this way higher power in a single frequency can be generated compared with the standing wave cavity.

As well as the use of travelling waves in a ring laser configuration, spatial hole burning can be avoided by several other methods, including the use of circularly polarized light, the use of mechanical movement, and the use of electro–optic modulation. In the CW dye laser for example, it is possible to reduce the effect of spatial hole burning by placing the active medium close to one of the cavity mirrors.

### 4.2.4 Unidirectional Oscillation

The ring resonator allows propagation in the two counterpropagating directions, CW and CCW. It is necessary to suppress one of the travelling waves to prevent backscattered radiation from coupling the two counterpropagating waves and forming a quasi–standing wave. Unidirectional operation can be produced by the use of an optical diode. This creates a loss difference between the CW and CCW waves by producing a polarization difference in the eigenmodes of the two directions of propagation. The optical diode in general is made up of a combination of a reciprocal (direction independent) polarization rotator (such as a c–axis quartz crystal), a nonreciprocal (direction dependent) rotation (such as a Faraday rotator) and a polarizer (Figure 4.4). It is arranged that the reciprocal and nonreciprocal polarization rotations cancel for one direction of propagation and add for the other. For the direction in which the rotation cancels, the eigenpolarizations are the low and high loss linear polarization states aligned with the principal axis of the polarizer. For the direction in which the rotations add, the eigenpolarizations are linear or elliptical polarization states with losses intermediate between the maximum and minimum values.

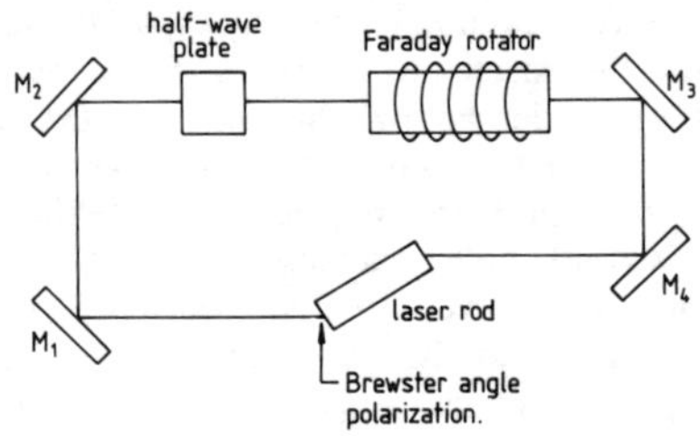


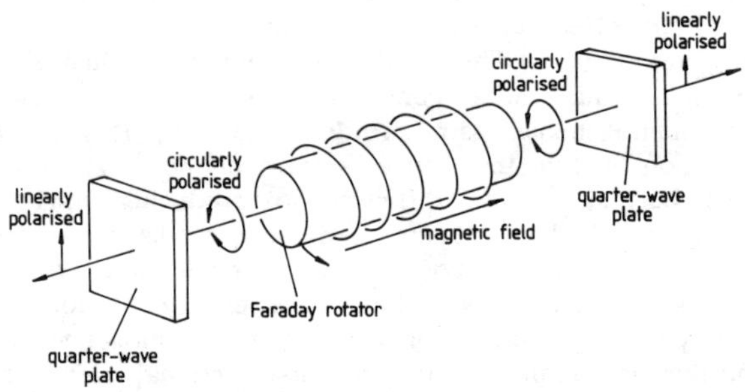


Fig 4.4 Optical diode components (Wilkinson 1987).

The reciprocal polarization quartz crystal is oriented giving a rotation angle $\varphi$ with respect to the polarization of the beams. The magnetic field on the Faraday rotator is set such that the waves travelling in each direction rotate the plane of polarization by an angle $\pm\psi$. In one direction the total rotation of the plane of polarization is $(\varphi + \psi)$. In the other direction the net rotation is $(\varphi - \psi)$. The differential loss, $\Delta\alpha$ between the counterpropagating waves is given by:

$$\Delta\alpha = \sin^2(\varphi + \psi) - \sin^2(\varphi - \psi) \tag{4.7}$$

For maximum power output, the half–wave plate and magnetic field in the Faraday rotator is adjusted such that $\varphi = \psi$. The loss is then only experienced in one direction, while the wave in the alternative direction is unattenuated. One difficulty with optical diodes in general is to maintain good transmission of the Faraday rotator glass, since an increase in Verdet constant is associated with an increased loss from the tail of UV absorption.

### 4.2.5 Unstable Ring Resonators

The unstable ring resonator combines the features of unstable cavities with those of the ring configuration. The unstable cavity enables large mode volumes to provide diffraction-limited output, while the ring with travelling waves provides a single longitudinal mode by elimination of standing waves (Dal Pozzo *et al* 1973). Negative branch resonators and asymmetric confocal rings (Freiberg *et al* 1973, 1974) have been used. The intrinsic geometric magnification of the unstable resonator means that the diameters of the oppositely travelling modes can differ substantially around the ring, depending on the mirror curvatures and positions.

The asymmetric confocal unstable ring resonator has been shown to be well suited for laser application. A long collimated beam can be produced in the active medium and the output coupling can be varied by adjusting the location of the output mirror within the collimated region of the resonator. The asymmetric ring also enables unidirectional operation to be obtained. The negative branch ring unstable resonator has a self-imaging property in that each round trip corresponds to an image relay. In these ways, unstable ring resonators offer a large degree of flexibility.

### 4.2.6 Passive Ring Resonators

We end this section by drawing brief attention to passive ring resonators. These act in either phase sensitive and frequency sensitive configurations. They are used in many applications, in particular both these forms have been used in fibre-optic gyroscopes (Bergh *et al* 1984, Ezekiel and Arditty 1982). In the fibre-optic gyroscope the induced phase shift is:

$$\Delta\varphi = \frac{2\pi LD\Omega}{\lambda P} \tag{4.8}$$

where $L$ is the total length of the fibre of perimeter $P$, $D$ the diameter of each fibre loop and $\Omega$ is the gyroscope ring rotating frequency. The single mode optical fibre which is used allows two orthogonal polarizations in each direction. Temperature fluctuations and mechanical stress cause power to be transferred from one polarization mode to the other. Single mode polarization preserving fibre with high birefringence inhibits the transfer of energies between the two polarizations.

In a description of a frequency sensitive fibre ring (Ezekiel and Balsamo 1977) the CW and CCW beams are frequency shifted by acousto-optic phase shifters so that they are locked to the CW and CCW modes of the cavity. The frequency shift produced by the Sagnac effect is given by:

$$\Delta\omega = \frac{8\pi A\Omega}{\lambda P} \tag{4.9}$$

where $A$ is the area of the ring.

## 4.3 NON-PLANAR RINGS

### 4.3.1 Polarization Rotation

The non-planar ring acts to rotate the plane of polarization. For a wave with propagation vector $\underline{k}$ orthogonal polarizations are defined as the s-component perpendicular to the plane of incidence and the p-component in the plane of incidence. On reflection from a perfect mirror, there is a $\pi$-plane shift in the s-component (Azzam and Boshara 1987, Shurcliff 1966). For a ring cavity with an even number of mirrors, the eigenpolarizations are nearly circularly polarized and no in-cavity polarization selector is required.

Polarization rotation for ring oscillators has been discussed by Statz *et al* (1985) and Nilsson *et al* (1989). A ring with an even number of mirrors gives an even number of space inversions such that in one transit an orthogonal system remains in the same 'handedness' as the original one. The change in orientation of the plane of incidence as the wave travels from one mirror to the next determines the degree of polarization rotation. The change in rotation about the direction of propagation $\underline{k}$ can be determined by

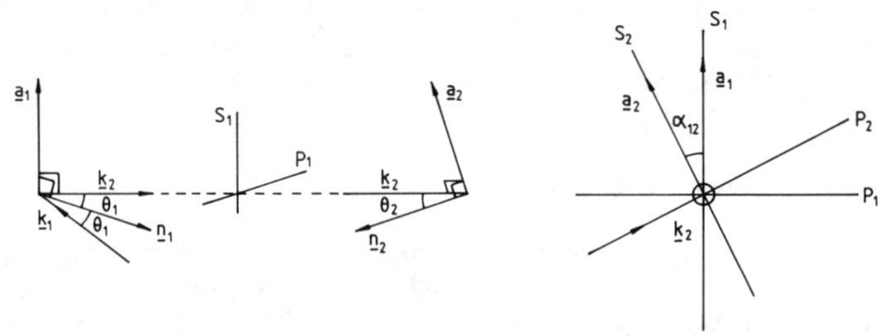


Fig 4.5 Rotation of the plane of incidence on two successive reflections.

considering the rotations of the s and p vectors. In Figure 4.5 $\underline{n}_1$ and $\underline{n}_2$ are the normals to the mirrors $M_1$ and $M_2$ with angles of incidence $\psi_1$ and $\psi_2$. The normals $a_1$ and $a_2$ are perpendicular to the two successive planes of incidence:

$$a_1 = \frac{\underline{n}_1 \cdot \underline{k}_1}{\sin \psi_1} \qquad a_2 = \frac{\underline{n}_2 \cdot \underline{k}_2}{\sin \psi_2} \tag{4.10}$$

The normals to the planes of incidence $\underline{a}_1$ and $\underline{a}_2$ are perpendicular to $\underline{k}_1$ and they both lie in the same plane as $S_1$ and $P_1$.

The new component $S_2$ is the projection of the polarization vector along $\underline{a}_2$:

$$S_2 = S_1 \cos\alpha_{12} - P_1 \sin\alpha_{12} \tag{4.11}$$

where $\alpha_{12}$ is the angle between the planes of incidence of mirrors 1 and 2. The polarization and frequency properties of optical ring resonators are well suited to analysis by the Jones matrix calculus (Gerrard and Burch 1975, Swindell 1975). Polarization may be represented by two-dimensional vectors. The base states, chosen as two orthogonal linearly polarized states with normalised vectors, are:

$$\begin{bmatrix} 1 \\ 0 \end{bmatrix} \quad \text{and} \quad \begin{bmatrix} 0 \\ 1 \end{bmatrix}$$

and for circular modes they are: (4.12)

$$\frac{1}{\sqrt{2}}\begin{bmatrix} 1 \\ j \end{bmatrix} \quad \text{and} \quad \frac{1}{\sqrt{2}}\begin{bmatrix} 1 \\ j \end{bmatrix}$$

The vector length carries information on amplitude, cavity losses and gains.

Rotation of the plane of linearly polarized light by an angle $\theta$ (for example, by a Faraday rotator) is represented by the matrix:

$$\begin{bmatrix} \cos\theta & \sin\theta \\ -\sin\theta & \cos\theta \end{bmatrix} \tag{4.13}$$

For a perfect dielectric mirror, the reflection matrix is:

$$\begin{bmatrix} 1 & 0 \\ 0 & -1 \end{bmatrix} \tag{4.14}$$

The base states are along the directions perpendicular to the plane of incidence (s) and in the plane of incidence (p). An actual mirror requires allowance for the reflectivity coefficient and birefringence. If the amplitude reflection coefficients are $r_s$ and $r_p$, and $g$ is the birefringence ($g = 0$ implies a perfect surface giving $\pi$ phase shift between the s and p polarization) the matrix for mirror $j$ at angle of incidence $\theta_j$ becomes:

$$M_j(\theta_j) = \begin{bmatrix} r_s e^{j\gamma/2} & 0 \\ 0 & -r_p e^{-j\gamma/2} \end{bmatrix} \tag{4.15}$$

The zero off-diagonal elements indicate that circular symmetry is assumed. In the Jones calculus the effect of an optical system on polarized light is represented by a succession of 2 x 2 matrices. The rotation matrix equation is:

$$\begin{bmatrix} P_2 \\ S_2 \end{bmatrix} = R(\alpha_{12}) \begin{bmatrix} P_1 \\ S_1 \end{bmatrix} \tag{4.16}$$

with $$R(\alpha_{12}) = \begin{bmatrix} \cos\alpha_{12} & \sin\alpha_{12} \\ -\sin\alpha_{12} & \cos\alpha_{12} \end{bmatrix} \tag{4.17}$$

where $\alpha$ is the angle through which the electric vector rotates upon reflection.

For a ring cavity with an even number of mirrors, the $\pi$ phase shifts effectively cancel. Where a ring has $N$ mirrors, the total change in the polarization vector for one transit of the ring follows the equation:

$$\begin{bmatrix} P' \\ S' \end{bmatrix} = R(\alpha_{n1})M_n(\theta_n)\ldots.R(\alpha_{23})M_2(\theta_2)R(\alpha_{12})M_1(\theta_1) \begin{bmatrix} P \\ S \end{bmatrix} \tag{4.18}$$

$\alpha_{12}$ is the angle between the planes of incidence of mirrors $i$ and $j$.

The mirror birefringence and cavity anisotropy can be eliminated to produce very pure circular polarization modes.

### 4.3.2 Non-planar Ring and Frequency Modes

To find those modes which will lase in a given cavity it is necessary to construct the cavity matrix; this is the product of all of the individual element matrices taken in the order in which they appear to the wave propagating around the cavity. The polarization modes that lase in the cavity will be those which remain unchanged after one cavity round trip. To find the modes requires a solution for the eigenvectors of the cavity matrix. For the field eigenvector $\underline{E}_n$ giving an eigenvalue $\lambda_n$:

$$M_c \cdot \underline{E}_n = \lambda_n \underline{E}_n \tag{4.19}$$

$\underline{E}_n$ is the field associated with the nth longitudinal mode propagating in one direction. The cavity matrix may depend on the direction of travel in the ring - particularly if nonreciprocal elements (such as a Faraday rotator) are present. An active ring cavity may sustain up to four different polarization modes, two in the CW and two in the CCW directions.

The complex cavity Jones matrix for a ring cavity made up of a phase compensator, rotator and isotropic element (Sanders and Kiehn 1977) is:

$$\mathrm{M}_c = \begin{bmatrix} A & B \\ B^* & A^* \end{bmatrix} \mathrm{e}^{-j\Omega} \tag{4.20}$$

where $\Omega$ is the isotropic phase shift.

The associated eigenvalues and eigenvectors for the CW(+) and CCW(−) modes are:

$$\lambda_\pm = \mathrm{e}^{j[\ \Omega\ \pm\ \cos^{-1} ReA\ ]} \tag{4.21}$$

$$\mathrm{E}_\pm = \begin{bmatrix} e_p B \\ e_s \left\{ -j\ \mathrm{Im}A \pm \left[ \dfrac{ReA^2}{2} - 1 \right]^{\frac{1}{2}} \right\} \end{bmatrix} \tag{4.22}$$

The resonant frequency of the eigenvector, associated with the $n^{th}$ longitudinal mode is:

$$f_n = \frac{c}{L} \left[ n + \frac{\Omega}{2\pi} \pm \frac{\cos^{-1}(ReA)}{2\pi} \right] \tag{4.23}$$

where $L$ is the ring optical path seen by $\mathrm{E}_\pm$. Reflection from a perfect dielectric mirror should yield a $\pi$ phase shift with respect to the s and p components of the electric vector. However real dielectric mirrors are birefringent for reflection at oblique angles.

The condition for resonance in the ring cavity requires that the electromagnetic field reproduces itself in phase, intensity and polarization after each transit of the ring. For an empty non-planar ring, the polarization eigenvalue equation for the electric field is

$$G\ \underline{E} = (F_n \ldots\ldots\ldots F_2 F_1)\underline{E} = g\underline{E} \tag{4.24}$$

where $G$ is the reflection operator, $F_i$ is the operator mirror reflection matrix and $E$ is the column matrix of the electric field. For a ring with an even number of mirrors, the result of operator, $G$ is equivalent to a single reflection operator $R$. Then the ring is represented by a single rotation matrix.

The transformation for the transverse field is given by the matrix:

$$R' = \begin{bmatrix} \cos\rho & \sin\rho \\ -\sin\rho & \cos\rho \end{bmatrix} \tag{4.25}$$

This is the Jones matrix for a polarization rotator. It may also be considered as the Jones matrix for an image rotator. The non-planar cavity is then reduced to a single element giving a pure rotation.

For both phase and polarization, the eigenvalue condition is:

$$(TR)\ \underline{E} = g'\ \underline{E} \tag{4.26}$$

where $T$ is a diagonal matrix with element $e^{jkL}$. This is equivalent to the phase shift for propagation around the perimeter $L$.

$$g' = e^{jkL} \cdot e^{\pm j\rho} \tag{4.27}$$

The ring cavity resonant condition is then

$$f = \left[ n - \frac{\psi \pm \rho}{2\pi} \right] \frac{c}{L} \tag{4.28}$$

where $n$ is the longitudinal mode number and $c/L$ is the ring cavity free spectral range. Here $\psi$ is the geometrical phase advance due to focusing of ring elements: this depends on ring geometries, resonator configuration and the transverse mode number (Arnaud 1969, 1970).

In a ring cavity with right (RCP) and left (LCP) circularly polarized components, their eigenmodes are separated by:

$$\Delta f = \left[ \frac{\rho}{\pi} \right] \frac{c}{L} \tag{4.29}$$

This depends on the degree of image rotation in the ring as shown in Figure 4.6.

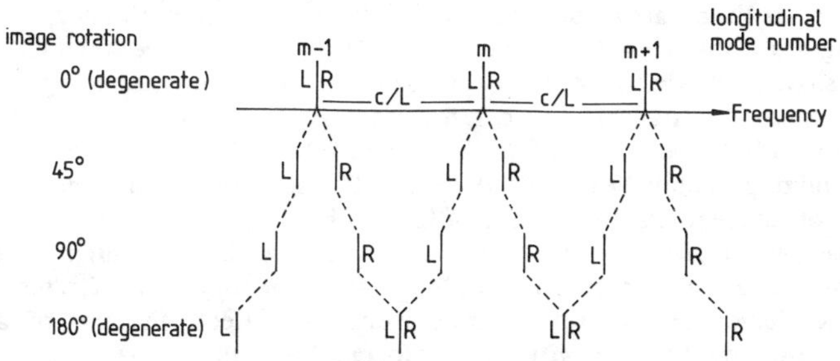


**Fig 4.6 Mode spectrum for a non-planar ring.**

Without a nonreciprocal polarization rotator, the CW and CCW modes of the same polarization (i.e. RCP or LCP) have the same frequency. The nonreciprocal polarization rotator breaks the CW – CCW degeneracy.

The relevant frequency condition with a nonreciprocal polarization rotator is:

$$f_{cw} = \left[ n - \frac{\psi + \theta_F \pm \rho}{2\pi} \right] \left[ \frac{c}{L} \right] \tag{4.30}$$

$$f_{ccw} = \left[ n - \frac{\psi - \theta_F \pm \rho}{2\pi} \right] \left[ \frac{c}{L} \right] \tag{4.31}$$

where $\theta_F$ is the non-reciprocal rotation in radians (referred to as the Faraday rotation). The counterpropagating modes at the two orthogonal polarizations exist at distinct frequencies. In an active laser, the geometry of the cavity of the non-planar ring cavity allows control of the laser frequency splitting between the RCP and LCP modes.

## 4.4 APPLICATIONS

### 4.4.1 Ring Laser Gyroscope

The ring laser gyroscope (RLG) detects rotation following the principle of the Sagnac effect, in which a rotation-dependent phase shift is set up between counterpropagating light waves in a ring interferometer (Aronowitz 1971, Chow *et al* 1985, 1980, Ezekiel and Knausenberger 1978, Jacobs *et al* 1984, Statz *et al* 1985, Wilkinson 1987). Light propagating in the same sense as the rotation of a ring requires slightly greater time to complete each transit of the ring than does light propagating along the same path but against the direction of the ring rotation. In a ring resonator containing a laser medium, the Sagnac phase shift shows itself as a shift in resonant frequency for the two counter-propagating waves. For a clockwise rotation, the resonant frequencies of the CW modes are shifted down in frequency and conversely for the CCW modes.

The difference in frequency of the CW and CCW waves, $\Delta f$, is related to the rotational frequency of the ring $\Omega$ by:

$$\Delta f = \frac{4A}{\lambda P} \Omega \tag{4.32}$$

where $A$ is the area of the ring, $P$ the ring perimeter, $\lambda$ the mean wavelength of the ring laser and S = $4A/\lambda P$ is the scale factor. An idea of the sensitivity of the ring laser gyroscope is obtained assuming that $\Omega$ is the earth's rotation rate of 15 deg/hr, for $P$ = 40cm and $\lambda$ = 632.8nm a frequency difference of 8.85Hz is produced. This is readily measured by optical mixing techniques; indeed, present ring lasers can sense rotation down to $10^{-4}$ of the earth's rate i.e. 0.0015 deg/hr.

The planar ring two frequency resonator exhibits coupling of the CW and CCW waves at low rotation speeds – the phenomenon of "lock-in". Two techniques have been shown to circumvent this effect: the use of a dither bias and the multioscillator cavity. To form the dither bias, the complete ring may be mechanically oscillated at a frequency above the lock-in value; in practical application this mechanical method has several disadvantages. These can be avoided by the use of the non-planar ring cavity.

In the absence of rotation of the ring cavity or polarization elements within the ring, each longitudinal mode of the resonator is four-fold frequency degenerate. Clockwise (CW) and counterclockwise (CCW) waves can exist in each of two possible polarizations. The means to remove this degeneracy is shown in Figure 4.7 by the use of reciprocal (independent of circulating wave direction) (Chow *et al* 1980, 1985, Dorschner 1983, Smith and Dorschner

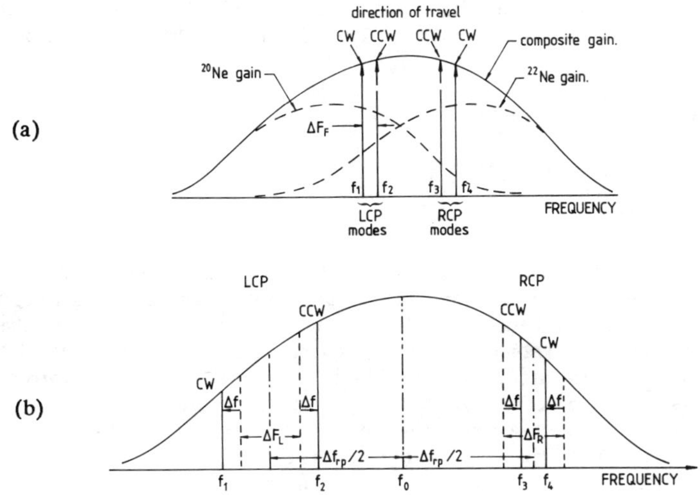


Fig 4.7 Resonant frequencies and shifts of the four lasing modes of a multioscillator ring cavity (Statz *et al* 1985).

1978, Statz *et al* 1985) and nonreciprocal (dependent on circulating wave direction) polarization rotation elements. The reciprocal polarization element introduces an optical path length difference as an apparent difference in refractive index seen by the orthogonal RCP and LCP modes, and the orthogonal modes experience slightly different optical path lengths. Since only circular polarization is unchanged in propagating through the polarization rotator, and is able to reproduce itself upon each transit of the ring, the resonant modes are RCP and LCP. The polarization rotation $\rho$ is equivalent to a differential phase shift of $2\rho$ between the RCP and LCP modes leading to a frequency splitting:

$$\Delta f = \left[ \frac{\rho}{\pi} \right] \left[ \frac{c}{L} \right] \tag{4.33}$$

where $c/L$ is the free spectral range of the empty ring.

The mode spectrum created in the multioscillator cavity is shown in Figure 4.7a. The nonreciprocal polarization rotation gives a differential phase shift between modes of a given polarization travelling in opposite directions. This leads to a splitting of the resonant frequencies: the CW–RCP and CCW–LCP modes are increased in frequency, while the CCW–RCP and CW–LCP modes are shifted down in frequency:

$$\Delta F_L = f_2 - f_1 = \frac{\theta_F}{\pi} \left[ \frac{c}{L} \right] \qquad (4.34)$$

$$\Delta F_R = f_4 - f_3 = \frac{\theta_F}{\pi} \left[ \frac{c}{L} \right] \qquad (4.34)$$

where $\theta_F$ is the nonreciprocal rotation.

Rotation of the ring about its sensitive axis at a rate $\Omega$ shifts the four frequencies as shown in Figure 4.7b. The differential frequency output is then:

$$\begin{aligned}\Delta f &= (f_4 - f_3) - (f_2 - f_1) \\ &= -4\Delta f + (\Delta F_R + \Delta F_L) \\ &= \frac{8A}{\lambda P}\,\Omega \end{aligned} \qquad (4.36)$$

The differential beat frequency is independent of the bias. The sum difference $(\Delta F_R - \Delta F_L)$ is approximately zero since dispersion across the reciprocal polarization frequency splitting is small. The differential oscillator ring resonator gives a rotational sensitive frequency difference twice that for a single two frequency resonator. The earlier differential DILAG laser gyroscope that was based on the combination of the quartz crystal (reciprocal) and Faraday rotation (nonreciprocal) elements, suffered the disadvantage of in–cavity optical elements giving instability and backscatter.

The four–mode multioscillator can be constructed with no additional optical elements placed in the optical path. The non–planar resonator provides a reciprocal splitting and the nonreciprocal splitting is obtained by application of a Zeeman field in an active laser gain medium (Chow *et al* 1980, 1985, Statz *et al* 1985, Dickinson and King 1987) or by the use of a magnetic mirror (Krebs *et al* 1980, Alge and Pfefferkorn 1988). This arrangement, with no additional optical elements in the cavity, is an important improvement in the ring laser gyroscope. Removal of sources of backscattering from optical surfaces, which can lead to coupling of the CW and CCW modes, is one of several advantages.

In the mode spectrum created in the multioscillator ring laser (Figures 4.6 and 4.7) the four frequencies require circular polarization as described in section 4.3.2. The reciprocal frequency shift in the RCP and LCP components is set by the geometry of the non–planar configuration, while the nonreciprocal shift in the CW and CCW waves of each polarization may be achieved by the use of a magnetic mirror or by the application of a Faraday bias via the setting of a Zeeman longitudinal magnetic field on the laser gain medium. The bias can be set such that the two gyros remain far from the lock–in zone (Chow *et al* 1980, 1985, Knipe *et al* 1988).

The magnetic mirror combines the Faraday rotation with one of the cavity mirrors and the resultant Kerr rotation with reflection provides a frequency bias. The circular polarization eigenmodes of the non–planar cavity uses the polar Kerr effect, in which a magnetic field is applied to a magneto–optic material. The magneto–optic Kerr effect causes the reflection coefficients for modes of the same handedness, but propagating in opposite directions, to differ in phase, thus giving a splitting of the resonance modes (similar to the case of the in–cavity Faraday rotator).

For LCP modes, the nonreciprocal splitting is given by:

$$\Delta f_L = (f_2 - f_1) = \left[\frac{\theta_k}{\pi}\right]\frac{c}{L} \qquad (4.37)$$

where $\theta_k$ is the magneto–optic Kerr rotation.

The senses (RCP or LCP) of the polarization of circularly polarized modes change upon reflection; the senses alternate in adjacent legs of the ring. The countertravelling modes undergo a differential phase shift which results in the equivalent of a Faraday splitting.

The elimination of the need for dither gives the multioscillator laser low noise and removes dither–induced errors. The absence of additional in–cavity optical elements also removes the temperature sensitive optical active crystal and Faraday cell. The optical axis of the image rotating non–planar ring is intrinsically stable against small changs of mirror position (Smith 1983); for a planar ring the stability comes from the focusing of one of the cavity mirrors.

### 4.4.2 Non–planar Solid–State Ring Oscillator

The emergence of high power, high efficiency, long lived diode lasers and diode arrays is leading a revolution in the pumping of solid–state lasers (see also Chapter 15). The limitation in these lasers on the power in a single axial mode, set by spatial hole burning in standing wave cavities, is removed by the use of a travelling wave ring laser structure. As discussed earlier spatial hole burning in standing wave solid–state laser systems leads to multi–axial mode operation as the input power is increased to provide output powers greater than the milliwatt level. Multi–longitudinal mode oscillation occurs in the linear Nd:YAG laser since standing waves in the YAG crystal lead to spatially non–uniform depletion of the gain. To suppress multi–longitudinal mode oscillation in Nd:YAG lasers, unidirectional ring structures with discrete elements have been developed, as has more recently, an integral monolithic ring cavity. Single axial mode operation at high power levels can be obtained with the ring cavity. To obtain operation in a single direction the ring needs to contain an optical diode.

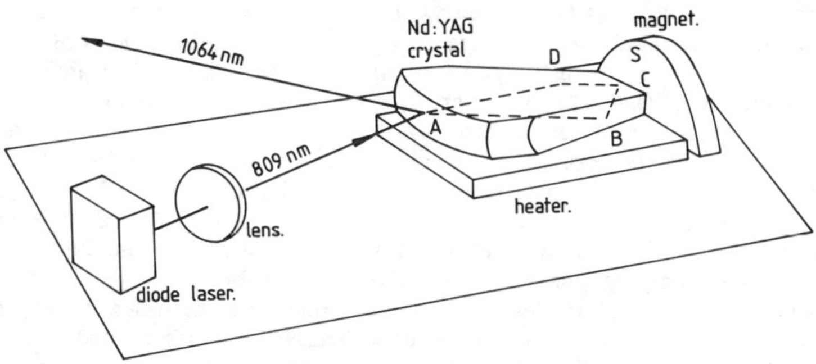


Fig 4.8 Diode–pumped monolithic non–planar ring oscillator (Kane *et al* 1987).

A novel structure has been demonstrated by Kane and Byer (1985) (Nilsson *et al* 1988, 1989) based on a monolithic Nd:YAG crystal in a non–planar ring configuration. The non–planar ring laser cavity with integral optical diode has been used to generate a single unidirectional mode. The laser is illustrated in Figures 4.8 and 4.9 and has been analysed by Nilsson *et al* (1989). Total internal reflection occurs on faces B, C and D, while A is a curved input and output coupler; the curvature of face A determines the

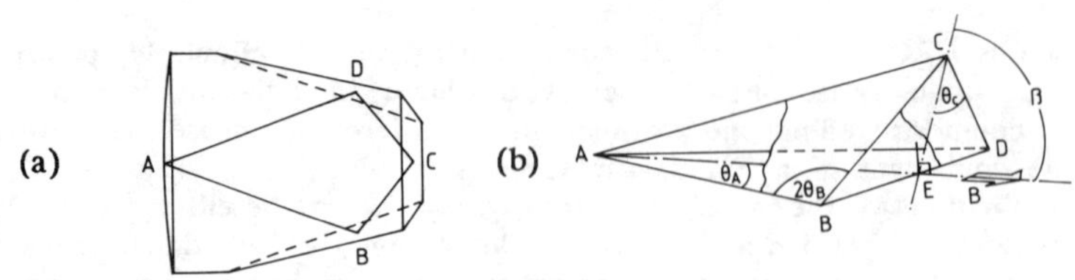


Fig 4.9 Optical path in the monolithic non-planar ring oscillator of Figure 4.8.

spatial mode. For one direction around the ring, the polarization caused by the Faraday rotator and half-wave plate cancel, giving a low loss linear polarization eigenmode. For the opposite direction, the polarization rotations add giving a higher loss, elliptical polarization eigenmode. Unidirectional lasing occurs when the difference between the losses is sufficiently large. The angle of incidence, $\theta_A$ on the output coupler *A*, determines the astigmatism of the resonator, and the difference between the Fresnel reflection coefficients for the s and p polarization components. The distance AE determines the degree of Faraday rotation along the legs AB and DA. The laser is very compact with end-pump excitation. The function of a half-wave plate, Faraday rotator and polarizer are contained in the single monolithic structure. The out-of-plane total internal reflection acts as a λ/2 plate, the magnetic field applied along the crystal acts as a Faraday rotator (Kane and Byer 1985, Trutna *et al* 1987) and the output coupler acts as a partial polarizer. The geometry of the non-planar ring is chosen to give a small rotation angle.

The non-planar ring has to date, operated up to 163mW of single axial mode output when excited by an argon ion laser and up to 25mW single axial mode power when diode laser pumped. The short term free running linewidth of the laser can be < 3kHz, as shown by heterodyne measurements with two Nd:YAG lasers pumped by free running diode laser arrays. The system offers the prospect of substantially improving on this value by the use of stable pump sources and rapid feedback frequency control (Kane *et al* 1987). The monolithic structure is intrinsically stable and can provide high stability single frequency operation (see also Chapter 8). It can also be made immune to feedback from the external pump source. The ring geometry reduces sensitivity to extracavity optical feedback since the back reflections from the output are in a high loss polarization state for its direction of propagation. Also, it is frequency shifted from the resonant frequency for that direction. Thus the reflected light does not strongly couple into the oscillating mode. The diode laser pumping enables short resonators to be used, with consequent large spacing of the axial modes. Also, the diode laser provides a very stable excitation source with very little heat deposited in the laser medium. The most important aberration is astigmatism. This may be reduced by the design of the ring cavity and by compensating astigmatism in the pump beam. While the monolithic device offers mechanical stability advantages, it is not appropriate for applications requiring tuning away from the gain peak or rapid tuning of the wavelength. In such cases, diode-pumped single frequency ring lasers using external mirrors and tuning elements can be used.

### 4.4.3 Other Applications

In addition to application to the ring laser gyroscope using a He-Ne gas laser and diode–pumped Nd–YAG lasers, the ring resonator has been applied to many other types of laser. These include dye lasers, solid–state and tunable solid–state Ti:$Al_2O_3$ lasers (Schulz 1988), colour centre lasers (Beigang 1988), semiconductor lasers and free–electron lasers (Bhowmik *et al* 1988, Tokar *et al*, 1989).

The properties of ring cavities for solid–state lasers have been considered by Magni (1986) and Silvestri *et al* (1988) including resonator stability, mode spot size and misalignment sensitivity. A particular feature of interest is the formation of thermal lensing by variation in the rod refractive index through non–uniform thermal dissipation. The laser rod is represented by a thin lens with focal length:

$$f = \frac{\pi r^2}{kP_{in}} \qquad (4.38)$$

where $r$ is the rod radius, $k$ is a constant and $P_{in}$ is the input pump power. As the pump power varies, the resonator crosses a stability zone. A conclusion of this study is that the pump power–dependent stability range for single transverse mode solid–state lasers is a characteristic of the laser medium, and depends only on the opto–mechanical properties of the rod and not on the resonator configuration. Only one stability zone is predicted for the ring laser compared with two in the linear cavity, while the stability range for the ring laser is twice that for a linear laser. Thus, the pump power stability for $TEM_{00}$ operation is twice as great for the ring compared with the linear resonator.

As described earlier, standing wave resonators show the limitation of spatial hole burning. This is particularly a limiting feature for dye lasers, where it inhibits single mode operation, as well as preventing all molecules within the active mode volume from contributing to the gain and output power of a single mode. In a dye laser, regions of gain outside the antinodes of the main mode can have enough gain to sustain simultaneous oscillation at a different frequency, corresponding to a mode whose antinode matches the unused high gain regions. The ring resonator, with travelling wave unidirectional operation, overcomes this disadvantage. Single mode operation and output powers higher by an order of magnitude may be achieved. This is obtained at some extra complexity in resonator design and alignment requirements than for standing wave resonators.

The titanium–sapphire (Ti:$Al_2O_3$) laser has many attractive features as a tunable solid–state laser. It has a tuning range in the pulsed mode of 660–1180nm, high efficiency, with an output power about three times that of a liquid dye laser and other advantages in stability and convenience over the dye laser. A Ti:$Al_2O_3$ laser with ring resonator has produced stable single frequency output with good beam quality (Schulz 1988). Figure 4.10 illustrates the components of the laser in which the tuning elements are a three–plate birefringent filter and a 0.5mm etalon. The optical diode may be a Faraday rotator and compensator or an acousto–optic modulator. Single frequency operation in this laser is possible, as discussed earlier, by the use of unidirectional travelling wave operation eliminating spatial hole burning. Astigmatism in the cavity was reduced by attention to cavity alignment and the use of astigmatic compensation.

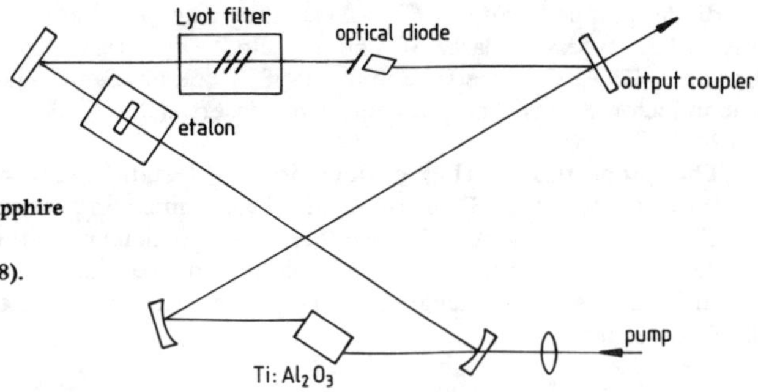


Fig 4.10 Titanium-sapphire ring laser (Schulz 1988).

The ring resonator also offers some advantages for application in non-linear optics for example, in harmonic generation and to produce UV output; by choice of mirror position and curvature, a second beam waist can be formed within the ring where a non-linear crystal or modulator may be placed. The ring resonator dye laser may readily be used with the colliding pulse mode-locker to generate ultra-short pulses. In non-linear optics, ring laser configurations have been used in studies of photorefractive materials (Erbschloe and Solymar 1985, Pauliat *et al*, 1989), phase conjugation, (Yeh *et al* 1983, Yeh and Khoshnevisan 1984) and bistability (Shelby *et al* 1988).

The semiconductor laser with an external composite optical cavity has potential advantages in the ring rather than linear configuration. There is less susceptibility to external reflections, control of cavity power and losses by the use of a variable output coupler, and a convenient means of inserting intracavity elements is possible. Ring cavities formed by both discrete elements or optical fibres have been demonstrated. The semiconductor optical fibre ring laser using birefringent mode selection, produces a single mode with a narrow linewidth at 1.55$\mu$m.

## 4.5 REFERENCES

Alge E and Pfefferkorn R 1988 SPIE 1019

Arnaud J A 1969 Appl. Opt. 8, 189 Arnaud J A and Kogelnik H 1969 Appl. Opt. 1687

Aronowitz F 1971 Laser Applications 1 Ed. M Ross 133 (New York: Academic)

Azzam R M A and Bashara N M 1987 Ellipsometry and Polarized light (Amsterdam: North-Holland)

Beigang R, Klameth K, Becker B, Yoon Z and Welling H 1988 Opt. Comm. 65 383

Bergh R A, Lefevre H C and Shaw H J J 1984 J. Lightwave Tech. LT-2 91

Bhowmik A, Cover R A, Kennedy P K, Bitterly S E and Labbe R H 1988 IEEE J. Quantum Electron. QE-24 802

Chow W W, Gea-Banacloche J, Pedrotti L M, Sanders V E, Schleich W and Scully M O 1985 Rev. Mod. Phys. 57 61

Chow W W, Hambenne J B, Hutchings T J, Sanders V E, Sargent M and

Scully M O 1980 IEEE J. Quantum Electron. QE–16 918
Cojocaru E, Julea T and Herisanu N 1989 Appl. Opt. 28 2577
Collins S A 1964 Appl. Opt. 3 1263
Dickinson M R and King T A 1987 J. Mod. Opt. 34 1045
Dorschner T A 1983 SPIE 412 192
Erbschloe D R and Solymar L 1985 Appl. Phys. Lett 53 1135
Ezekiel S and Balsamo S R 1977 Appl. Phys. Lett. 30 478
Ezekiel S and Arditty H J 1982 Fiber–Optic Rotation Sensors (Berlin: Springer–Verlag)
Ezekiel S and Knausenberger G E 1978 SPIE 157
Freiberg R J, Chenausky P P and Buczek C J 1973 Appl. Opt. 12 1140
Freiberg R J , Chenausky P P and Buczek C J 1974 IEEE J. Quantum Electron. QE–10 279
Gerrard A and Burch J M 1975 Introduction to Matrix Methods in Optics (London: Wiley)
Jacobs S F, Sargent M, Scully M O, Simpson J, Sanders V E and Killpatrick JE 1984 SPIE 487
Kane T J and Byer R L 1985 Opt. Lett. 10 65
Kane T J, Nilsson A C and Byer R L 1987 Opt. Lett. 12 175
Knipe C, Andrews D A and King T A 1988 J. Mod. Opt. 34 441
Kogelnik H W, Ippen E P, Dienes A and Shank C V 1972 IEEE J Quantum Electron. QE–8 373
Krebs J J, Maisch W G Prinz G A and Forester D W 1980 IEEE Trans. Magnetics MAG–16 1179
Magni V 1986 Appl. Opt. 25 107
Nilsson A C, Kane T J and Byer R L 1988 SPIE 912 13
Nilsson A C, Gustafson E K and Byer R L 1989 IEEE J. Quantum Electron. 25 767
Pauliat G, Ingold M and Gunter P 1989 IEEE J. Quantum Electron. 25 201
Dal Pozzo P, Polloni R, Svelto O and Zaraga F 1973 IEEE J. Quantum Electron. QE–9 1061
Sanders V E and Kiehn R M 1977 IEEE J. Quantum Electron. QE–13 739
Schulz P A 1988 IEEE J. Quantum Electron. 24 1039
Shelby R M, Levenson M D and Perlmutter S H 1988 J. Opt. Soc Am B 5 347
Shurcliff W A 1966 Polarized Light (Cambridge: Harvard University Press)
De Silvestri S, Laporta P and Magni V 1988 Opt. Comm. 65 373
Smith I W 1983 SPIE 412 203
Smith I W and Dorschner T A 1978 SPIE 157 21
Statz H, Dorschner T A, Holtz M and Smith I W 1985 in Laser Handbook 4 Eds. M L Smith and M Bass (Amsterdam: North Holland)
Swindell W 1975 Polarized Light (New York: Wiley)
Tokar R L, McVey B D, Thode L E and Gallatin G M 1989 IEEE J. Quantum Electron. QE–25 73
Trutna W R, Donald D K and Nazarathy M 1987 Opt. Lett. 12 248
Wilkinson J R 1987 Prog. in Quantum Electron. 11 1
Yeh P, Tracy J and Khoshnevisan M 1983 SPIE 412 240
Yeh P and Khoshnevisan M 1984 SPIE 487 102

# 5 Multifold Laser Resonators

*H J BAKER*

## 5.1 INTRODUCTION

The output power of most types of laser is fixed predominantly by the volume of active region, since there are physical limitations on the pump power density which can be employed, either through technological problems, or through the inherent properties of the laser gain medium. The emphasis in high power lasers is thus on the design of resonators with a large mode volume, and in this regard, we may seek to increase the transverse dimension of a low order, near–diffraction limited mode using for example unstable resonators (chapter 2) or variable reflectivity mirror cavities (chapter 6). A more straightforward approach (at least conceptually) is to keep to the limited transverse dimensions of stable resonator modes and increase the resonator length. Length scaling of resonators then requires the application of folding techniques to keep the overall dimensions of the laser device within practical limits. In this chapter, the resonator design problems specific to very long cavities are discussed, in terms of the basic properties of Gaussian beams within resonators, and the disturbances of the ideal "empty" resonator by the gain medium.

Long path folded resonators are most commonly applied to diffusion cooled molecular gas lasers, such as the industrial $CO_2$ lasers developed in the early 1970's. In these lasers, the physical limit on the input power density is the conduction of heat through the gas to the cooled walls, resulting in the power output per unit length being fixed, independently of the tube diameter used, typically in the range 50–120 $Wm^{-1}$. For conventional gas discharge tubes of circular cross–section, resonator length scaling is the only route to higher powers. The limitations of heat conduction in static gas lasers have been overcome by fast axial flow lasers, giving typically 5 times improvement in output power density, and some degree of volume scaling through tube diameter. However, folded cavities continue to be useful for these more modern designs, primarily because of the good transverse mode control which is obtainable.

Figure 5.1 illustrates the two basic approaches to obtaining a near $180^0$ fold in a laser resonator. The U–fold, using two $45^0$ mirrors (or perhaps a total internal reflection prism), has the advantage of producing a high packing density, by allowing the arms of the folded resonator to run parallel, but of course involves the absorption losses and alignment requirements of *two* optical surfaces per fold. The V–fold geometry makes use of one mirror, which if required, can be curved, but requires either overlap of beams in adjacent arms of the resonator, or quite large fold angles, reducing the package power density. Both types of fold are used in commercially successful industrial lasers.

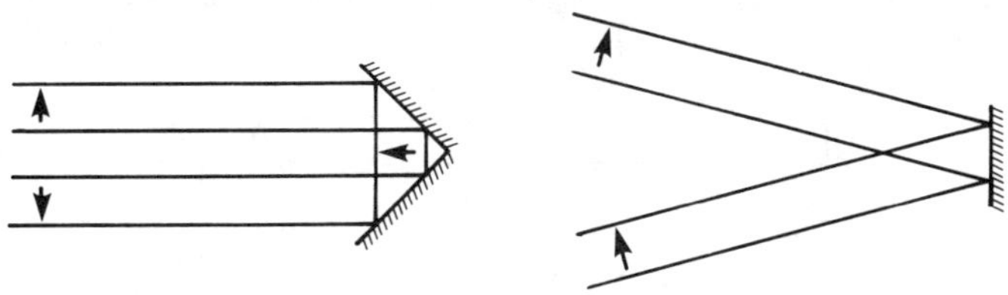

(a) (b)

Fig 5.1 Optical arrangements for folding a cavity resonator mode: a) U-fold with two mirrors. b) V-fold with plane or curved mirror.

The application of cavity folding to long, stable resonators allows the $TEM_{00}$ mode volume to be shaped to match the geometry of non-cylindrical gain media. One example is the use of a planar multi-fold to produce good overlap with the rectangular cross-section of the flow duct of a fast transverse flow laser. Slab-shaped solid-state gain media may be similarly treated. Of recent interest is the extraction of $TEM_{00}$ mode outputs from RF excited $CO_2$ lasers with either planar or annular transverse discharges with large electrode areas. This may be achieved by off-axis, curved mirror folding systems, often referred to as Herriott cells.

It should be noted that the fold mirrors provide an opportunity to incorporate beam correcting elements, which compensate for disturbances to the ideal, passive cavity modes brought about by the gain medium. A simple example is the use of the U-fold geometry in fast transverse flow lasers. A property of the mirror pair in the U-fold is to invert the intensity pattern as illustrated by the arrows in Figure 5.1. Partial compensation of the disturbances in refractive index and gain along the flow direction is obtained by a 2 pass U-folded cavity oriented in the plane of the gas flow. For static gas lasers, or fast axial-flow, where the gain and refractive index variations are cylindrically symmetric, compensating focal power can be conveniently provided by curved mirrors, particularly in the V-fold arrangement. When $45^0$ mirrors are present, as with the U-fold, it is convenient to use the slight polarising effect of off-axis metal mirrors to provide a low loss polariser within the cavity. For dielectrically coated mirrors, this polarisation can be deliberately enhanced by the coating design.

In this chapter, the problems raised by very long folded resonators, operating at high powers are discussed, for both plane and curved mirror folds, and some of these ideas are extended to the design of very compact folded resonators for annular gain media, using the Herriott cell concept.

## 5.2 PLANE MIRROR FOLDED CAVITIES

As a first illustration of the cavity engineering problem of a folded cavity, we will consider a long cavity, terminated by concave spherical mirrors, and folded by plane mirrors as in Figure 5.2. The usual requirement is for a stable resonator mode, ideally $TEM_{00}$, which has not too much variation of

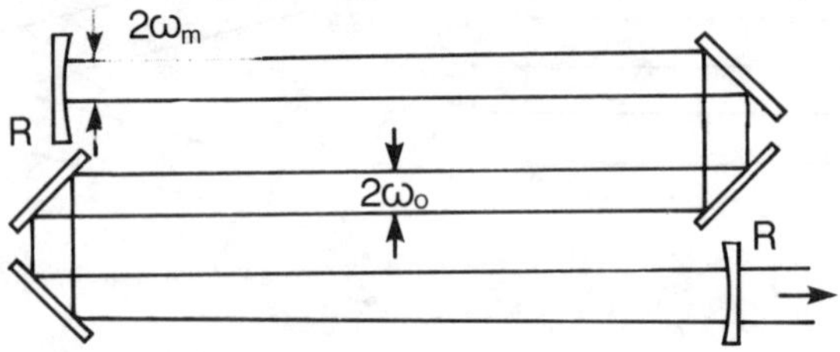


Fig 5.2 Folded resonator using U-fold.

beam diameter through the cavity, whilst still having geometrical parameters which are sufficiently clear of the Gaussian resonator instability boundaries. A reasonable compromise is a cavity which has the spot size at the mirror, $w_m$, which is 10% larger than the beam waist, $w_0$, at the cavity centre. Straightforward application of equations 1.20, 1.24 and 1.25 of chapter 1 lead to the following design equations for a cavity of length $L$.

End mirror radius $\qquad R = 2.88\ L \qquad$ (5.1)

$TEM_{00}$ spot size $\qquad w_0 = 0.588\ \sqrt{\lambda L} \qquad$ (5.2)

We see that as the cavity length is increased, the mirror radius becomes large, corresponding to very weakly diverging, nearly plane wavefronts. For the particular case of diffusion cooled $CO_2$ lasers, folded cavity lengths of 10 metres or more are common to achieve powers approaching 1kW, and here we need mirror radii of 30 meters or so. It should be noted that the strongest wavefront curvature occurs at the mirrors, and most of the cavity has wavefronts of larger radius, which are essentially plane waves.

The problem with long cavities is that these large diameter, weakly focused beams can be significantly disturbed by refractive index gradients within the gain medium, or by poor surface figure in the fold mirrors, either due to manufacturing tolerances or to thermal distortion in use. Generally, the disturbances have an effect which increases proportionally with active length, whilst the ability of the cavity mode to tolerate disturbances decreases with length. The result is that long cavities may have unpredictable beam sizes, varying considerably between nominally identical devices, or beams which are unstable in direction. Since a good knowledge of the beam size of $TEM_{00}$ and higher order modes is essential for the choice of the limiting aperture for transverse mode selection, beam size predictability is essential. It is common practice to seek to achieve mode selection by using the discharge tube diameter as the limiting aperture so as to get a good fill factor and hence efficiency. In such cases, unpredictability of mode diameter causes unwanted high order mode content, reducing the beam brightness of the laser.

At this point it is useful to review the kind of processes which affect the mode stability of a long folded cavity. These may be conveniently grouped into three general classifications, associated with the resonator support structure, mode–mirror interactions, and mode–medium interactions.

## 5.3 RESONATOR SUPPORT STRUCTURE

Misalignment of a cavity mirror produces a lateral movement of the mode axis relative to a limiting aperture, which in turn increases cavity losses through truncation of the desired transverse mode. Assessment of the alignment tolerance requires identification of the optic axis, which for an unfolded cavity is the line joining the centres of curvature of the mirrors. Figure 5.3

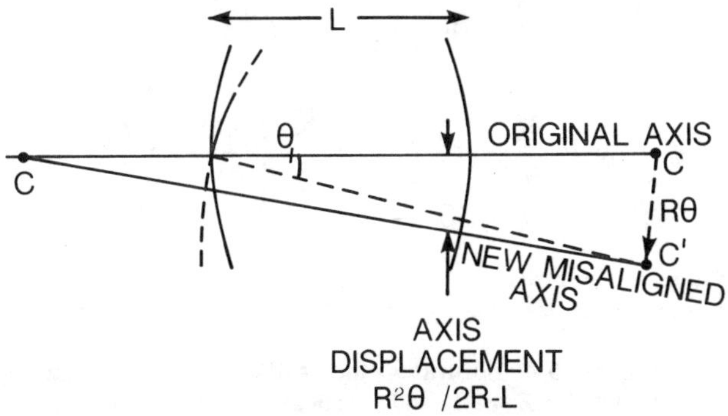


Fig 5.3 Shift in the geometrical cavity axis produced by misalignment of one cavity mirror for a long unfolded resonator.

illustrates the effect of misaligning one mirror of a linear cavity by an angle $\theta$. The axis is shifted by an amount $R^2\theta/(2R-L)$ at the opposite mirror. The ratio of this shift to the TEM00 mode spot size at the mirror is a measure of the alignment tolerance: typically a decentering of the mode in the limiting aperture by >10% of the spot radius constitutes a serious misalignment. For the typical long radius cavity described by equations 5.1 and 5.2, the ratio of shift to spot size is

$$\frac{x}{w_m} = \frac{R^2\theta}{(2R - L)w_m} = 2.69\ \theta\sqrt{L/\lambda} \qquad (5.3)$$

Clearly very long cavities are more strongly affected by support structure alignment drift than short cavities, due to the inherent properties of Gaussian beams within stable resonators. This is compounded by the increasing angular instability of long resonator support structures, with the alignment error due to temperature gradients, creep or gravitational deflection being proportional to length. The sensitivity of the resonator to support structure instability thus scales approximately as $L^{3/2}$.

The introduction of folding reduces the typical deflections of the resonator structure due to the shorter length, and also because the laser designer can often lay out the gain sections in a way which introduces thermal symmetry, reducing transverse temperature gradients. Also resonator designs with curved folding optical elements (equivalent to the unfolded lens guide) can be chosen to maximise tolerence to structural misalignment. Drift associated with the resonator support structure now usually produces an angular misalignment which is common to all the mirrors associated with each end of the folded structure. Analysis of the alignment tolerance is now more difficult, but is

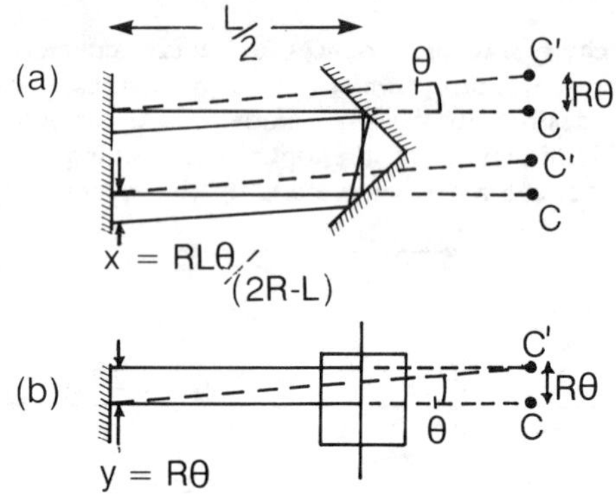


Fig 5.4 Shifts in the cavity axis in a U-fold resonator due to tilt $\theta$ of ***both*** mirrors at one end of the folded cavity Tilt: a) in plane of fold b) normal to fold.

still possible. Figure 5.4 illustrates the shift in axis caused by a common mirror tilt $\theta$, in the horizontal and vertical directions respectively for a two arm, U-fold resonator. For the typical long radius cavity of equations 5.1 and 5.2, the ratios of axis shift to spot size for the two misalignment directions illustrated in Figure 5.4 are:

$$\frac{x}{w_m} = \frac{RL\theta}{(2R-L)w_m} = 0.935\ \theta\sqrt{L/\lambda} \qquad \text{(Horizontal)} \qquad (5.4)$$

$$\frac{x}{w_m} = \frac{R\theta}{w_m} = 4.45\ \theta\sqrt{L/\lambda} \qquad \text{(Vertical)} \qquad (5.5)$$

In comparing these relations with the previous ones for the linear, unfolded cavity (Equation 5.3 and Figure 5.3) it should be remembered that the values of $\theta$ encountered with the folded cavity are likely to be reduced by a factor of two due to the shorter resonator support structure. Taking this into account, the alignment sensitivity for the horizontal direction (tilt axis normal to fold plane) is much improved by a factor of about 6, whilst in the vertical direction, it is about the same as the linear cavity ( Equation 5.3 ).

The examples discussed here serve to demonstrate the basic limitations of plane-mirror folded resonators which should be born in mind during the design of the cavity support structures. The detailed mechanical engineering requires extreme care over choice of materials, control of the thermal environment, and good design practice for strain-free, kinematic mounting of the optical components. A general conclusion is that folding can improve tolerances to angular misalignments, particularly those which are common to all mirrors located at one end of the support structure.

## 5.4 MODE–MIRROR INTERACTIONS

At high powers, mirror coatings can absorb sufficient power from the intracavity beam to produce temperature rises sufficient to distort the optical surface by thermal expansion. This is discussed in detail in Chapter 13, and here we note that the general effect can be minimised using substrates with a large ratio of thermal conductivity to thermal expansion. Clearly, folding a long cavity introduces additional distortion of the cavity mode in proportion to the number of folds, through thermal distortion of the mirrors.

The types of distortion encountered depend on the heat flow patterns within the mirror substrate. For example an edge–cooled mirror with the mode accurately centred will distort to produce a weak convex mirror, due to the axial symmetry, and in general will act to increase the mode diameter. However, an imperfectly seated mirror (with asymmetric heat conduction paths, or an off–set beam will cause effectively a tilt due to non–symmetric thermal bulging, and act as a source of cavity misalignment, in addition to defocusing.

Thermal distortion of fold mirrors is strongly power dependent and is difficult to compensate for by cavity design due to its unpredictability, and presents a fundamental limit to the increase of output power by length scaling.

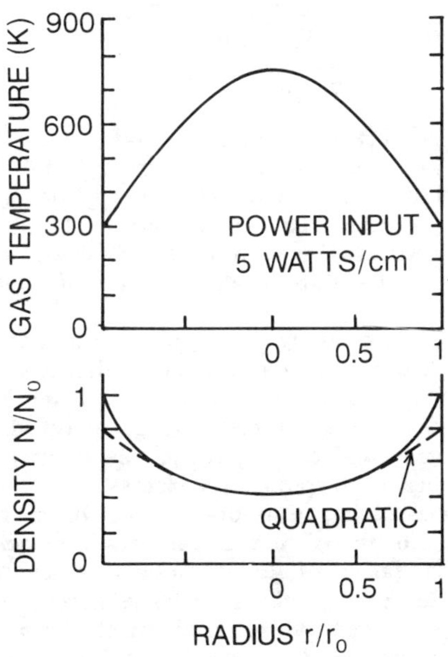


Fig 5.5 Temperature profile in a typical DC excited $CO_2$ laser tube, and the consequent gas density variation.

## 5.5 MODE–MEDIUM INTERACTIONS

Although the refractive index of a gas laser medium is very close to unity, disturbance of the resonator mode may result from *variations* in gas density. For static gas lasers, there is a radial variation of density associated with the temperature distribution across the tube necessary for thermal conduction of heat to the walls. Figure 5.5 illustrates the temperature profile across a longitudinal DC excited $CO_2$ laser tube (Laderman and Byron, 1971) and the resulting normalised variation of gas density, assuming equal pressures throughout the discharge and the ideal gas law. The laser discharge acts as an aberrated diverging lens, and in Figure 5.5, the broken line indicates the quadratic term in a power series describing the gas density. The quadratic part represents an ideal diverging lens, and departures from this ideal are spherical aberrations. Using the data in Figure 5.5 as "typical", the strength of the gas lens can be estimated, in terms of the focal length of one metre of discharge:

$$f \approx - \frac{0.6\ kT_w\ \epsilon_o\ d_o^2}{\bar{\alpha}\ P} \tag{5.6}$$

Here $\bar{\alpha}$ is the mean polarisability of a molecule in the gas mixture (MKS units), $T_w$ is the wall temperature, $P$ is the gas fill pressure (Pa) and $d_o$ the tube diameter. A slightly more general result is obtained if the scaling law for a gas discharge laser, $Pd_o = constant$, is included. Then:

$$f \approx - \frac{0.6\ kT_w\ \epsilon_o\ d_o^3}{\bar{\alpha}\ (Pd_o)} \tag{5.7}$$

The strength of gas thermal lensing is thus strongly dependent on tube diameter. Typical values calculated from equation 5.7 with a $5CO_2$:$1N_2$:80He mixture are $f$ = − 110 m per meter of discharge for a 20 mm tube diameter, reducing to −14 m per m for 10 mm diameter. These should be regarded as estimates, since no account has been made of dissociation effects, or changes in the temperature profiles through the removal of significant power as laser output.

Provided it is not too strong, the gas thermal lens can be accommodated in resonator design, and acts to increase the variation of spot size between the centre of the resonator and the end mirrors. Figure 5.6 presents the results of numerical calculation on a 10 metre long cavity with varying strengths of gas lensing distributed uniformly through the cavity. The lensing was simulated by 20 equally spaced weak lenses, and a gaussian beam was propagated within the cavity, transforming the beam radius of curvature at each lens. Figure 5.6 shows that a gas lens stronger than −100m per m starts to restrict the range of cavity designs which maintain a reasonably uniform beam diameter through the laser. The general effect of gas lensing is to reduce the range of end mirror radii which form stable cavities and to increase both the magnitude and variation of the average spot size. The cavity design is more difficult than may appear in Figure 5.6 since the appropriate value of gas focal length depends strongly on tube diameter, and the tube diameter can only be selected once spot sizes are known. Clearly the calculations must now be iterative, and a compromise design found.

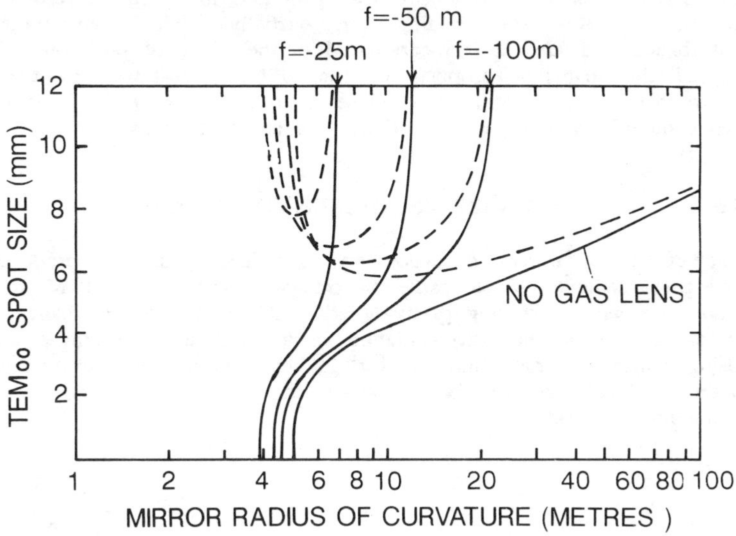


Fig 5.6 Cavity calculations for a 10 metre long mirror spacing with varying degrees of negative thermal lensing. Allowed combinations of spot size at the mirrors and cavity beam waist are shown versus the end mirror radius of curvature. ——— At mirror. – – – – – – – At beam waist.

The thermal lens in a static gas laser is a stationary, reproducible effect, and as a result can be included, and corrected to first order, in the cavity design. In flowing gas lasers, the picture is not so clear, in that density variations are now related to both flow pressure drops and temperature variations. In fast axial flow $CO_2$ lasers, the major temperature gradient associated with discharge heating is along the optical axis, and a transverse density profile resulting in a gas lens is a second order effect. Longitudinal DC excited lasers normally use intensely turbulent flow to disperse the formation of hot discharge filaments and have quite uniform time–averaged temperature distributions. Mode disturbances are of a random, time dependent nature associated with discharge and flow fluctuations, and result in increased, non–correctable beam divergence. Transverse RF excitation does not produce longitudinal discharge filamentation, and dispenses with turbulence generators in the flow. For these conditions, a static optical distortion can result from non–uniformity of pumping associated with the electrode shape. Generally, fast axial flow laser cavities do not need correction for gas lensing, partly because of weaker lensing, and partly because shorter active lengths are needed, relative to slow–flow diffusion cooled lasers.

For transverse flow lasers there is a considerable temperature increase across the resonator in the flow direction, with much smaller variations in the electrode direction. The disturbance of the mode is now due to a distributed prism, with the axis deflected downstream. Often the active region in the flow direction is not well-localised, with a distinct "blow-down" of the discharge, and matching of the positions of the gain and the required resonator mode is uncertain. Resonator folding can partially correct the transverse flow disturbance if a U-fold is incorporated in the plane of the flow. This makes use of the inversion property of the U-fold, illustrated by the arrows in Figure 5.1. An alternative technique is to make use of a U-fold to include two counterflowing discharge sections in a compact cavity.

## 5.6 CURVED MIRROR FOLDING OF CAVITIES

The effects of mirror distortion and gas lensing are to produce unwanted diverging elements which cause progressive loss of control of the resonator mode size with increasing cavity length. Although the disturbances are locally quite weak, it is their accumulation in a long cavity which is the problem. Using concave curved mirrors for folding allows the introduction of focal power to locally correct for the unwanted divergent elements, and prevent an accumulative effect.

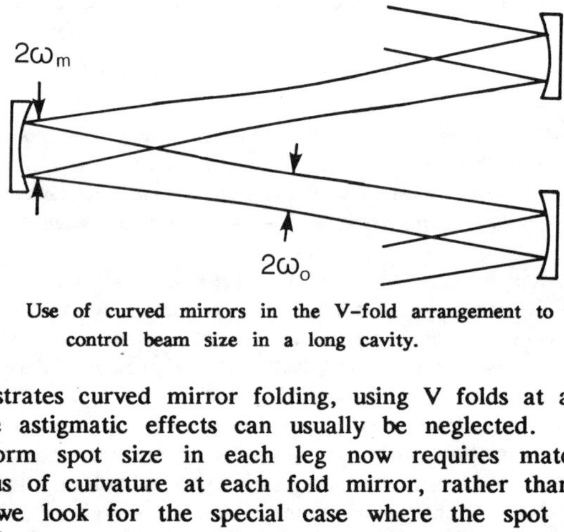


Fig 5.7 Use of curved mirrors in the V-fold arrangement to control beam size in a long cavity.

Figure 5.7 illustrates curved mirror folding, using V folds at a small angle of incidence, where astigmatic effects can usually be neglected. Design of a cavity with a uniform spot size in each leg now requires matching of the gaussian beam radius of curvature at each fold mirror, rather than at the end mirrors only. If we look for the special case where the spot size at each fold mirror is 10% larger than at the beam waist, i.e. $w_m = 1.1w_0$, the required fold and end mirror radius is now:

$$R = 2.88\ L/N \tag{5.8}$$

where $N$ is the number of arms in the folded cavity, and $L$ is the unfolded length. The corresponding beam waist spot size is:

$$w_0 = 0.588\ \sqrt{\lambda L/N} \tag{5.9}$$

Comparing these with equations 5.1 and 5.2 for the plane mirror folded cavity, it is seen that radius of curvatures in the cavity are reduced by $N$ and the spot sizes by $\sqrt{N}$. Gas lensing for example now has a much reduced effect on beam sizes, since it is a weak effect for one arm, competing with the strong focussing effect of the fold mirrors. Curved-mirror folding thus allows much longer folded cavities to be contemplated, although it should be noted that the smaller spot size relative to the equivalent plane-mirror folded cavity means higher power densities and more problems from thermal mirror distortion. Analysis of the misalignment effects with a curved-mirror fold shows that beam centering errors are cyclical, and do not accumulate through the cavity. The set of mirrors behave as a discrete component light guide.

## 5.7 OFF-AXIS CURVED MIRROR FOLDING

The off-axis propagation of rays between a pair of curved mirrors provides opportunities for the construction of very compact folded resonators for both planar and annular gain regions. Figure 5.8 represents the off-axis behaviour

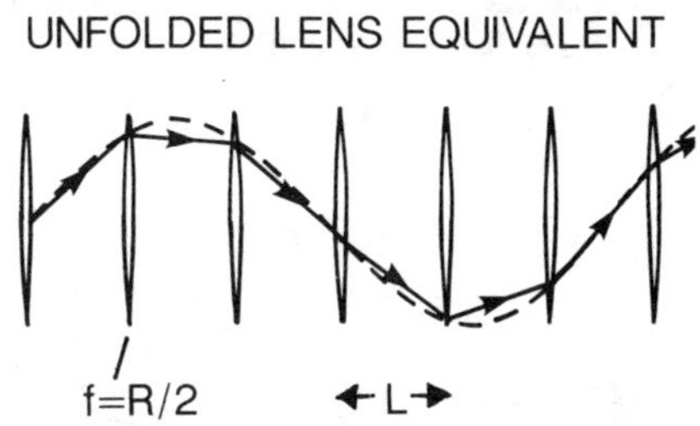


Fig 5.8 Ray paths through the unfolded lens equivalent of a of a pair of concave mirrors.

of a sequence of lenses of focal length $f = R/2$, which can be regarded as the "unfolded" equivalent of a mirror pair spaced by a distance $\ell$. Provided that the cavity stability condition is met, an off-axis ray periodically crosses the axis, and ray positions on the lenses follow a sinusoidal function, as illustrated by the broken line in Figure 5.8. This behaviour is purely geometrical, and it can be shown that the period of the sinusoidal guiding function is $2\pi\ell/\cos^{-1}(1 - \ell/R)$ for a symmetric cavity $(R_1 = R_2 = R)$. The ray illustrated in Figure 5.8 crosses the axis and hence propagates in a plane.

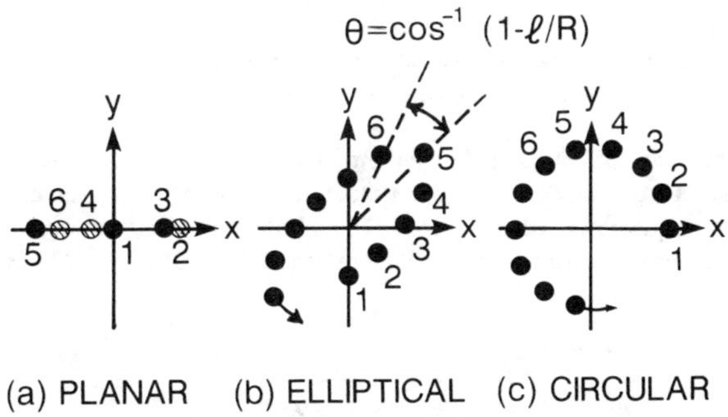


Fig 5.9 End view of rays travelling between two concave mirrors, showing the points of impact of the ray with the mirror: a) is for a ray path which crosses the mirror axis, b) is the general case and c) is a purely azimuthal ray path.

If the cavity is viewed end–on, the ray positions are as shown in Figure 5.9a. A more general ray launched into the lens sequence does not cross the axis at any point, and the end–view is as in Figure 5.9b. The locus of ray impacts is now an ellipse, and ray tracing shows that the ray position rotates around this ellipse by a fixed angle per pass given by:

$$\theta = \cos^{-1} \; [1 - \ell/R] \qquad (5.10)$$

Adjustment of the launch condition for the test ray produces a circular locus as in Figure 5.9c. The off–axis ray behaviour was first demonstrated by Herriott *et al* (1964), using a $TEM_{00}$ laser beam as the test ray. The patterns of particular interest as folded resonators are Figures 5.9a and 5.9c, for planar and annular gain media respectively. An annular folding system between spherical mirrors has been used as a long path absorption cell in spectroscopic applications and as a consequence is often referred to as a "Herriott cell". Figure 5.10 illustrates the construction of a stable long path resonator using the Herriott cell fold, and this has been recently used by Xin and Hall (1987) to extract high quality $TEM_{00}$ output from an annular RF discharge $CO_2$ laser.

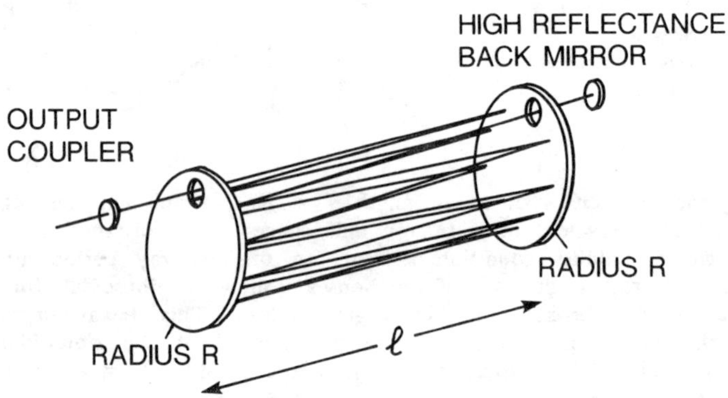


**Fig 5.10 Use of the Herriott cell to produce a long path folded resonator, suitable for annular gain regions.**

Design of the off–axis fold system can proceed for the ideal symmetric case with just a knowledge of geometric ray properties given above and the gaussian beam equations of Chapter 1. The radius of curvature of the fold mirrors determines the number of passes included in the fold system, through the relation:

$$N = \frac{2\pi - \varphi}{\theta} = \frac{2\pi - \varphi}{\cos^{-1} \; (1 - \ell/R)} \qquad (5.11)$$

The angle $\varphi$ represents the angular offset of the entrance and exit holes in Figure 5.10. The radius of curvature and the separation of the fold mirrors determine the spot size of a $TEM_{00}$ beam, for the special case when beam waists occur at the centre of each arm of the fold arrangement, and beams in all arms are the same size. This *mode-matched* case occurs when the two cavity end mirrors are chosen to match the radius of curvature of the beam phase fronts defined by the fold mirror curvature and spacing. Application of equations 1.24 and 1.25 of Chapter 1 give the spot size on a fold mirror as:

$$w_m = \left[\frac{\lambda R}{\pi}\right]^{\frac{1}{2}} \left[\frac{\ell}{2R-\ell}\right]^{\frac{1}{4}} \tag{5.12}$$

The *close-packed* case occurs when the separation of the spots on one of the mirrors is about $\pi w_m$. At this point, the maximum number of passes is achieved whilst still allowing the $TEM_{00}$ beam to be extracted without truncation loss through the holes in the fold mirrors shown in Figure 5.10. Close-packing occurs for $\varphi = \theta$ in Equation 5.11 and for $N$ an odd number, and it produces the maximum overlap of the folded beam with a continuous

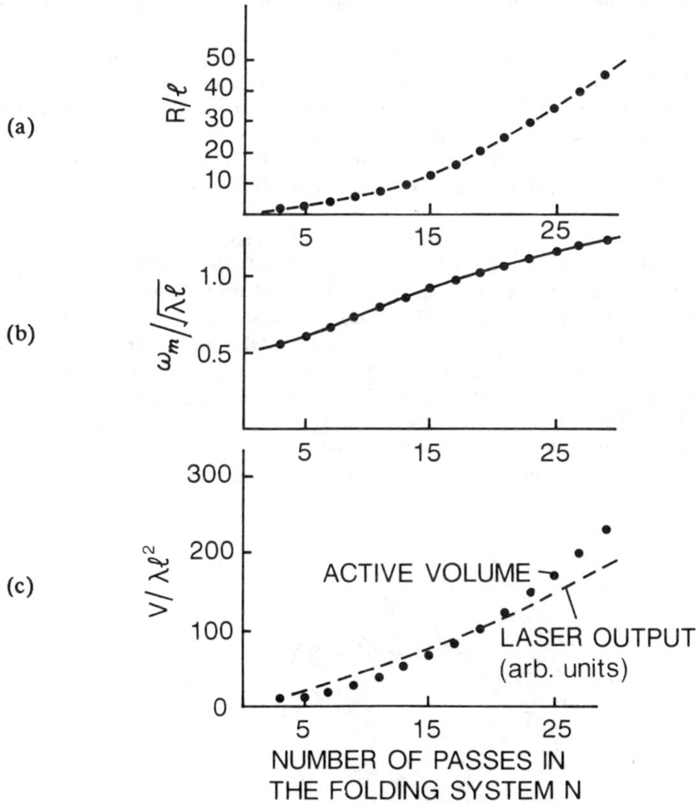


Fig 5.11 Scaling of the Herriott cell cavity with number of passes in the system, presented in dimensionless form. a) Fold mirror radius divided by fold length. b) Spot size at the fold mirror. c) Mode volume where the broken line indicates how the laser output power will scale when discharge scaling laws are included.

annular gain medium, potentially giving the best uniformity of saturation. However, when the finite absorption and scatter in even the best high reflection coatings is taken into account, close-packing may not produce the maximum output power due to the accumulation of mirror losses. The $CO_2$ laser reported by Xin and Hall (1987) uses $N$=19 and about 50% packing factor on the fold mirrors.

To investigate the potential for scaling the active volume of an annular laser using the Herriott cell fold, it is easiest to concentrate on the close-packed case as a limiting example. This is done in a dimensionless form in Figure 5.11 as a function of the number of passes included in the folding system. Equation 5.11 gives the fold mirror radius of curvature, shown as the ratio $R/\ell$ in Figure 5.11a. Equation 5.12 gives the spot size on the fold mirror, plotted as the ratio $w/\sqrt{\lambda\ell}$ in Figure 5.11b. The close-packing condition then specifies the radius of the spot circle on the mirrors and allows calculation of the annular volume required to enclose the folded beam. Figure 5.11c shows this volume in units of $\lambda\ell^2$ assuming an annular gap equal to $\pi w_m$.

It can be seen that the active volume available to a $TEM_{00}$ mode scales super-linearly with $N$ for fixed $\ell$, and also as $\ell^2$ for fixed $N$. Because of the finite losses at the fold mirrors it is preferable to scale the volume using larger $\ell$ rather than larger $N$. Clearly the annular multi-fold concept is an excellent approach to the packaging of long path resonators using the minimum of optical components, given the availability of gain media of annular format.

For $CO_2$ lasers, annular discharges can be produced by transverse radio-frequency discharges between concentric cylindrical electrodes. This *large electrode area* technique allows scaling of the discharge area, whilst the discharge gap remains small, giving good conduction cooling to the electrodes. For this application of the annular resonator, some additional interpretation of the curves in Figure 5.11 is necessary. The power output per unit area of electrode for static gas $CO_2$ discharges is inversely proportional to the electrode separation (Hall and Baker, 1989) and because of the variation of the spot size $w_m$ with fold number, the electrode separation must be a variable. The power output does not directly follow active volume scaling as a

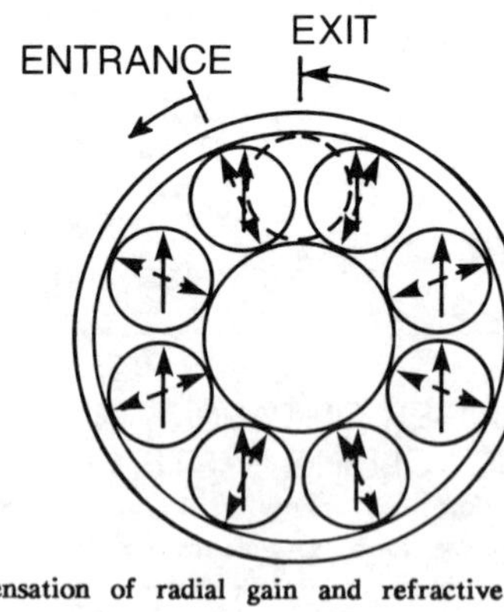


Fig 5.12 Compensation of radial gain and refractive index variations in an annular gain region by the Herriot cell fold.

result. Dividing the active volume by the spot size gives a variable proportional to output power, shown as the broken line in Figure 5.11c, and this scales linearly with fold number *N*. Thus for diffusion cooled $CO_2$ lasers, the output power per unit length of discharge can be scaled in proportion to the number of folds.

The Herriott cell fold has useful properties which correct for some of the distortions produced by gain and refractive index variations in the annular gain medium. In Figure 5.12, the beam spots on one of the fold mirrors are marked with an arrow, indicating the way that a small asymetry on the beam propagates through the resonator. The arrow effectively rotates with respect to the radial gain and index variations, and after a complete pass the beam has sampled the non–uniformities at all angles, averaging out their effect. To first order, the resonator can tolerate the toroidal lens effect generated by the gas lensing in annular geometry. The beam rotation relative to the electrodes is also beneficial in dispersing electrode reflections which degrade beam quality, and also produces a circularly symmetric limiting aperture, through a sequence of truncations of the beam by the electrodes at different angles. As a result the annular laser is able to produce a high quality Gaussian beam output.

## 5.8 SUMMARY

This chapter has sought to bring out some of the limitations on the length scaling of stable resonators, using cavity folding. Folding can be regarded as the traditional route to increasing the output power of particularly the industrial gas lasers, which eventually reaches the physical limitations set by the refractive index variations associated with the gain medium and mirror thermal distortions. The use of focal power at the folding mirrors is a useful way of extending the length of the resonator when there are stationary, axially symmetric distortions, giving the resonator the behaviour of a guided wave structure. Perhaps in the future, long path folded resonators may utilise the techniques of phase conjugate mirrors to further refine the correction of distortions.

Present developments concentrate on the construction of very compact folded cavities, using off–axis spherical mirrors. These give the guided wave advantages of spherical mirror folding, and allow very large fold numbers to be attained with just four cavity mirrors. In the case of an annular gain media, the fold system gives very useful compensation effects for gain medium non–uniformities.

## 5.9 REFERENCES

Hall D R and Baker H J 1989 "RF excitation of diffusion cooled and fast axial flow lasers" in proceedings of Seventh Annual Symposium on Gas Flow and Chemical Lasers, Vienna, 1988 SPIE volume 1031

Herriott D, Kogelnik H and Kompfner R 1964 Appl. Optics 3 523–526

Laderman A J and Byron S R 1971 J. Appl. Phys. 42 3138–44

Xin J G and Hall D R 1987 Appl. Phys. Lett. 51 469–71

# 6 Resonators with Variable Reflectivity Mirrors

*V MAGNI, S DE SILVESTRI and A CYBO-OTTONE*

## 6.1 INTRODUCTION

The generation of diffraction limited light beams with high energy (or power) is one of the central problems in the field of laser science and engineering. For low power lasers, containing active materials of small cross-section, stable resonators operating on the $TEM_{00}$ mode offer a simple and convenient solution. However, unless impracticably long resonators are used, the transverse dimension of $TEM_{00}$ modes are limited to a few millimeters for the visible and the near infrared, and therefore these resonators cannot be effectively used to extract energy from active materials of large cross-section. Unstable resonators, by definition, spread the radiation away from the resonator axis, and are commonly used to obtain modes of large volume in the gain medium. The output beam, extracted around the edge of one of the circular mirrors, presents a doughnut shape with pronounced ripples due to the diffraction wavelets generated at the edge and a far-field pattern with measurable side-lobes. These shortcomings can be avoided by using a transmitting output coupler with a graded reflectivity profile. In particular, unstable resonators with radially variable reflectivity mirrors offer three remarkable advantages:

1) The ripple in the near-field is greatly reduced, as a consequence of the lowered intensity of the edge diffraction waves.
2) The near-field is more uniform, without the hole typical for traditional unstable resonators, and consequently the side-lobes in the far-field are substantially attenuated.
3) The modal loss degeneracy, typical of unstable resonators, is eliminated giving greater transverse mode discrimination.

Mode analysis of unstable resonators with variable reflectivity mirrors of various shape was developed by Kumagai *et al* (1966), Anan'ev and Sherstobitov (1971), Sherstobitov and Vinokurov (1972), and by McAllister *et al* (1974) through the numerical or asymptopic solution of the resonator equations. Resonators containing Gaussian mirrors were treated by Vakhimov (1965) and by Zucker (1970) by solving the mode equation, whereas Arnaud (1974) used the concept of complex point eikonal. Casperson and Lunnam (1975) and Yariv and Yeh (1975) introduced a matrix formalism for describing Gaussian mirrors which allows direct application of the ABCD law for beam propagation. Ganiel and Hardy (1976) and Walsh and Knight (1986) studied the high order modes in Gaussian resonators. McCarthy and Lavigne (1983,

1984) considered the misalignment sensitivity and the output beam characteristics. De Silvestri *et al* (1988b) demonstrated that by employing supergaussian profiles a simple closed-form solution for the modes can be obtained and effective design equations can be derived, indicating that the supergaussian curve has advantages over the Gaussian profile for treating the reflectivity shape of different practical devices.

Although the advantages of unstable resonators employing mirrors with Gaussian reflectivity profiles have been recognized for many years, only a few practical devices with quasi-gaussian shape have been successfully developed. Eggleston *et al* (1981) proposed the radial birefringent element, Armandillo and Giuliani (1985) developed an apoditic filter based on frustrated internal reflection, while De Silvestri *et al* (1987b) introduced the radially variable Fabry-Perot interferometer. The most promising technique, demonstrated first by Lavigne *et al* (1985), is based on the deposition of thin dielectric films of shaped thickness on a transparent substrate. By using Gaussian reflectivity mirrors, the goal of a smooth and diffraction-limited output beam may be achieved. However, since by definition, the modes in Gaussian resonators have a Gaussian profile, the full exploitation of the volume of the gain medium is usually not achieved: a Gaussian curve has long tails and, to prevent diffraction effects due to the aperture of the gain medium, the spot size must be considerably smaller than the aperture itself.

For a greater mode overlap efficiency with the active medium De Silvestri *et al* (1988a and 1988b) proposed a novel *supergaussian* reflectivity profile. The cavity modes resulting from the application of supergaussian mirrors to unstable resonators also present supergaussian profiles, which are bell-shaped and considerably flatter in the centre than a Gaussian curve. This circumstance substantially increases the mode volume in the active material, yet preserving good diffraction and propagation properties of the output beam. Supergaussian mirrors have been fabricated by De Silvestri *et al* (1988a and 1988b) and by Emiliani *et al* (1989) by the deposition of thin dielectric layers on a glass substrate and have been applied with excellent results to Nd:YAG lasers.

## 6.2 UNSTABLE RESONATORS WITH GAUSSIAN MIRRORS

The simplest approach to an analytical treatment of resonators with variable reflectivity mirrors is to consider Gaussian tapering of the reflectivity. Aperture-free resonators employing Gaussian mirrors were first analyzed by Vakhimov (1965) who solved the resonator integral equation for the electromagnetic field through asymptotic expansions. Zucker (1970) found that the modes of unstable resonators with Gaussian mirrors can be represented by Hermite-Gaussian or Laguerre-Gaussian functions in a way similar to stable resonators with uniform reflectivity mirrors. Using the ray matrix (ABCD matrix) formalism Casperson and Lunnam (1975) and Yariv and Yeh (1975) derived the ABCD matrix for Gaussian mirrors and applied the matrix technique to obtain in a straightforward way the fundamental resonator mode and its losses. This procedure is simpler than the manipulation of the diffraction integrals, and allows the treatment of general resonators including also lens-like media. The analysis of higher order modes by matrix formalism, developed by Walsh and Knight (1986), requires the definition of generalized

Hermite–Gaussian beams and it will be not considered here.

It is well known that a Gaussian beam is completely described by the complex parameter $q$ given by the following expression:

$$\frac{1}{q} = \frac{1}{R} - \frac{j\lambda}{\pi w^2} \tag{6.1}$$

where the real part gives the radius of curvature $R$ of the wave front and the imaginary part gives the spot size $w$; $\lambda$ is the wavelength. Thus, if one knows how $q$ transforms, one also knows how to transform Gaussian beams. Given the ray matrix ABCD for an arbitrary system of lenses, mirrors and lens–like media the output beam parameter $q_0$ can be related to the input beam parameter $q_i$ by the following simple relationship:

$$\frac{1}{q_0} = \frac{Cq_i + D}{Aq_i + B} \tag{6.2}$$

The modes of resonators containing only simple optical elements of the type mentioned above may be found from the self consistency requirement that $q$ repeats itself after one round trip through the resonator. If $A$, $B$, $C$ and $D$ represent the round trip matrix elements of the resonator, the steady–state beam parameter $q_s$ at the chosen reference plane must satisfy Equation (6.2) with $q_s = q_0 = q_i$ :

$$\frac{1}{q_s} = \frac{Cq_s + D}{Aq_s + B} \tag{6.3}$$

To apply ray matrix formalism to resonators with Gaussian mirrors we have to find the ray matrix that corresponds to the reflection upon a mirror whose intensity reflectivity profile has the following shape:

$$R = R_0 \exp[-2(r/w_m)^2] \tag{6.4}$$

where $R_0$ is the intensity peak reflectivity, $r$ is the radial coordinate and $w_m$ is the mirror spot size. The ray matrix for a flat Gaussian reflector is given by:

$$\begin{bmatrix} A & B \\ C & D \end{bmatrix} = \begin{bmatrix} 1 & 0 \\ -j\lambda/\pi w_m^2 & 1 \end{bmatrix} \tag{6.5}$$

Application of Equation (6.2) to the above matrix provides the relationship between the spot size $w_r$ of the reflected Gaussian beam to the spot size $w_i$ of the incident beam:

$$\frac{1}{w_r^2} = \frac{1}{w_m^2} + \frac{1}{w_i^2} \tag{6.6}$$

For a curved Gaussian reflector, the ray matrix is obtained by multiplying the ray matrix given by Equation (6.5) by that of a uniform reflectivity mirror of the same radius of curvature.

Figure 6.1(a) shows the schematic of a resonator made of a uniform high reflectivity mirror and a Gaussian reflector at a distance $L$. If the reference plane is chosen near the Gaussian reflector as indicated in Figure 6.1(b), the round trip ABCD matrix of the resonator is given by:

$$\begin{bmatrix} A & B \\ C & D \end{bmatrix} = \begin{bmatrix} A' - j\lambda B'/\pi w_m^2 & B' \\ C' - j\lambda D'/\pi w_m^2 & D' \end{bmatrix} \quad (6.7)$$

where $A'$, $B'$, $C'$, and $D'$ are the elements of the round trip matrix for the same resonator evaluated neglecting the variation of the reflectivity. The

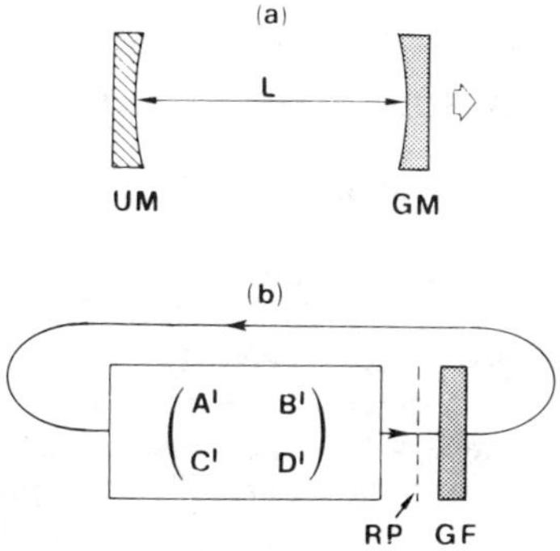


Fig 6.1 (a) Resonator comprising a gaussian mirror, GM, and a uniform mirror, UM, at a distance *L*. (b) Schematic of the resonator described by a ray transfer matrix and a gaussian filter, *GF*, whose transmission is equal to the reflectivity of the gaussian mirror. The reference plane, *RP*, is close to the gaussian filter.

solution of Equation (6.3) for $q_S$ shows the existence of a well defined mode spot size also for $[(A+D)/2]^2 > 1$, a feature which does not occur in the absence of the Gaussian reflector. Thus, the infinite spot size modes which are characteristic of geometrically unstable resonators do not appear when a Gaussian reflectivity element is inserted in the resonator. The fundamental mode remains Gaussian.

For Gaussian resonators the magnification $M^*$ is defined as:

$$M^* = w_i/w_r \quad (6.8)$$

where $w_i$ is the spot size of the Gaussian beam incident upon the Gaussian reflector and $w_r$ is the spot size of the reflected beam. For stable resonators, the magnification is entirely generated by diffraction, whereas for unstable resonators, it is primarily due to geometrical magnification $M$ defined according to the ray analysis developed by Siegman and Arrathoon (1967). It can be shown that geometrical magnification $M$ is a valid approximation for $M^*$ under the condition that:

$$w_m/w_c >> 2/(M-1)^{\frac{1}{2}} \quad (6.9)$$

where $w_c = (\lambda L/\pi)^{\frac{1}{2}}$ is the spot size on the mirrors of a confocal stable resonator of length $L$. Since the whole purpose of using unstable resonators of some magnification is to obtain a mode substantially larger than that of a confocal stable resonator, we conclude that the geometrical optics approximation for a Gaussian resonator may be considered satisfactory, at least in practical cases.

Once Equation (6.9) is satisfied the laborious matrix analysis to calculate the modes in a Gaussian resonator through the solution of Equation (6.3) is no longer necessary. After calculating the geometrical magnification $M$ one can use much simpler expressions. Moreover, the analysis and the design of Gaussian resonators reduce to a particular case of the theory of supergaussian resonators which is presented in the next section.

## 6.3 UNSTABLE RESONATORS WITH SUPERGAUSSIAN MIRRORS

The supergaussian reflectivity profiles for the field intensity are defined by:

$$R(r) = R_0 \exp[-2(r/w_m)^n] \tag{6.10}$$

where $R_0$ is the intensity peak reflectivity, $r$ the radial coordinate, $w_m$ the mirror spot size, and $n$ the order of the profile.

For unstable resonators with supergaussian mirrors the geometrical optics provides for a satisfactory approximation of mode profiles and losses. If $\rho(r) = [R(r)]^{\frac{1}{2}}$ is the amplitude reflectivity and $u(r)$ is the amplitude of the wave incident on the mirror in a resonator of magnification $M$, then the self-consistency equation for the modes is (Siegman and Arrathoon 1967 and Anan'ev and Sherstobitov 1971):

$$\sigma u(r) = (1/M)\ \rho(r/M)\ u(r/M) \tag{6.11}$$

where $\sigma$ is the eigenvalue. Anan'ev and Sherstobitov (1971) expressed the fundamental mode solution for a generic reflectivity shape as the product of an infinite number of terms representing the function $\rho(r)$ with scaled arguments:

$$u_0(r) = A \prod_{k=1}^{\infty} \rho(r/M^k)/\rho(0) \tag{6.12}$$

where $A$ is an arbitrary amplitude factor. The corresponding eigenvalue is $\sigma_0 = \rho(0)/M$, which gives a round trip loss:

$$\Gamma = 1 - \sigma^2 = 1 - R_0/M^2 \tag{6.13}$$

The eigensolutions for high order modes are $u_m(r) = u_0(r)r^m$ with eigenvalues $\sigma = \sigma_0/M^m$. Generally, for an arbitrary shape of $\rho(r)$, Equation (6.12) must be evaluated numerically: however, for a supergaussian profile the infinite product in Equation (6.12) converges to the expression:

$$u_0(r) = A \exp[-(r/w_i)^n] \tag{6.14}$$

where the spot size $w_i$ of the incident beam onto the supergaussian mirror is given by:

$$w_i = w_m(M^n - 1)^{1/n} \tag{6.15}$$

The beam reflected by the mirror is also supergaussian with spot size

$$w_r = w_i / M \tag{6.16}$$

The output beam profile, obtained by multiplying the intensity of the beam incident on the mirror by the transmission $[I_{out} = u_0^2(1 - R)]$, is smooth, but no longer supergaussian and may present a central depression if $R_0 M^n > 1$. The maximally flat output beam is generated when:

$$R_0 M^n = 1 \tag{6.17}$$

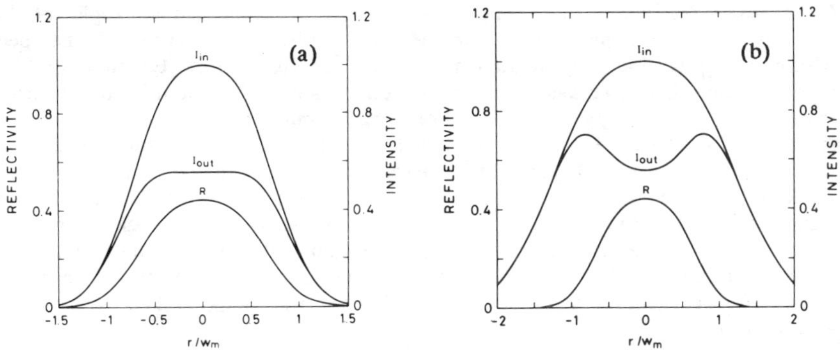


Fig 6.2 Calculated intensity profiles of the incident wave, $I_{in}$, on a supergaussian mirror of order $n$ = 2.8 and of the transmitted beam, $I_{out}$, as a function of the radial coordinate normalized to the mirror spot size $w_m$ for two resonator magnifications: (a) $M$ = 1.36, and (b) $M$ = 2. The mirror intensity reflectivity profile, $R$, is also shown.

As an example, Figure 6.2 shows the intensity profiles $(I_{in} = u_0^2)$ of the mode on the supergaussian mirror $(n = 2.8,\ R_0 = 0.44)$ along with the reflectivity profile as a function of $r/w_m$ for two magnifications. For $M = 2$ the output beam presents a central depression, while for $M = 1.36$, Equation (6.17) is satisfied and the output beam is a flat top.

If the order of the supergaussian curve is not too high (approximately less than 20), the same condition expressed by Equation (6.9) for the validity of the geometrical optics approach to Gaussian resonators, applies to supergaussian resonator as well. Note however, that as $n \to \infty$, one obtains the usual unstable resonator with uniform-reflectivity hard-edge mirror, for which the geometrical optics is a rough approximation.

Equation (6.14) shows that the cavity mode profile is supergaussian of the same order as that of the mirror. Therefore, the supergaussian beam, being in shape somewhere about half way between rectangular and Gaussian, combines the advantages of both profiles. The soft smoothing of the tails avoids the build up of pronounced diffraction rings typical of hard-edge beam profiles, however, while the main portion of the beam diameter is at almost constant intensity resulting in a wide overlap with the gain medium and consequently an efficient energy extraction.

To design a supergaussian resonator, the mode spot size should be optimized for a good balance between efficiently filling the gain medium and low beam perturbations caused by diffraction at the aperture of the active material. Equations (6.10), (6.13), (6.15) and (6.17) allow the design of the supergaussian mirror once the magnification $M$ and the spot size $w_i$ of the beam incident on the mirror are chosen. The choice of the geometrical magnification is somewhat arbitrary and strongly depends on the particular characteristic of the laser performances one is seeking to emphasize. A trade-off has to be achieved between the mode intensity discrimination ratio $1/M^2$, the output losses [see Equation (6.13)] and the angular misalignment sensitivity. This last quantity, defined by Krupke and Sooy (1969) as the ratio of the angle by which the optical axis rotates to the tilting angle of the resonator mirror, is proportional to $M/(M - 1)^2$. The choice of the peak reflectivity $R_0$ is not only related to the output losses [see Equation (6.13)], but also determines the shape of the output beam. The additional constraint between $R_0$ and $M$ given by Equation (6.17) must also be considered if a maximally flat beam is to be obtained. Once the magnification and peak reflectivity have been chosen, the mirror spot size can be calculated from Equation (6.15) if $w_i$ is known. The spot size $w_i$ can be optimized for a good balance between efficient filling of the gain medium by the resonator mode and low beam perturbations caused by diffraction at the resonator apertures. As demonstrated by Parent *et al* (1987) for Gaussian mirrors, this is achieved by choosing the cutting level of the aperture in the range 0.2 – 13% of the peak intensity.

## 6.4 PRACTICAL VARIABLE REFLECTIVITY MIRRORS AND COUPLERS

The main difficulty with tapered reflectivity output couplers is to fabricate practical devices which can withstand high power (if needed), and are not specific to a given laser type. Up to now only a few devices with radially variable reflectivity have been experimentally demonstrated.

The first device with radially variable reflectivity profile, the radial birefringent element (RBE) was proposed by Giuliani *et al* (1980) and by Eggleston *et al* (1981). This apparatus essentially consists of a birefringent crystal, ground to form a lens, and a polarizer inserted inside the resonator near one of the mirrors. When a linearly polarized wave impinges on the birefringent lens, its polarization is rotated by an angle which depends on the thickness of the material and hence on the distance from the lens axis. After traversing the lens the beam goes through the polarizer which transforms the variations of polarization into variations of intensity of the reflected (and transmitted) light. Generally, the polarizer is used as output coupler. Many different reflectivity profiles, bell-shaped or with a central depression, can be obtained with this apparatus. It should be noted that in most cases, the reflectivity curves can be treated more effectively as supergaussian instead of Gaussian, as done originally by the inventors. This device has been applied by Giuliani *et al* (1980) to a pulsed Nd:YAG laser and by Harter and Walling (1986) to a pulsed ruby and to a pulsed alexandrite laser. In each case, the superiority of unstable resonators with variable reflectivity mirror with respect to standard unstable resonators in generating diffraction limited beams has been demonstrated. Although the elements of an RBE can be easily fabricated, this technique does not appear promising for application to high power lasers, especially in the infrared, where finding suitable birefringent materials and polarizers becomes more difficult.

A second radially variable output coupler was described by Armandillo and Giuliani (1985). In its general form, it consists of two rectangular glass prisms nearly in contact along the diagonal surfaces which, instead of being planar, are optically polished to a suitable curvature. The light impinges upon the diagonal surfaces from inside the prism at an angle for which, without the second prism, total internal reflection would occur. At the interface between the curved surfaces part of the light is coupled into the second prism through evanescent waves; since the separation between the surfaces depends on the distance from the centre, the transmitted intensity changes with that distance. The reflected light which is used to couple light out of the cavity, behaves in a complementary way. Up to present, this device has only been applied (Armandillo and Giuliani 1985) to an excimer laser. The critical positioning of the two prisms and their astigmatism seem to hinder widespread practical application of this technique.

The radially variable Fabry-Perot interferometer (RAVI) was proposed by De Silvestri *et al* (1987a, and 1897b) as a novel radially variable output coupler. The apparatus, which closely resembles a Fabry-Perot interferometer, consists essentially of two mirrors of different radii of curvature facing each other with a small separation. The two facing surfaces are curved and coated to a suitable reflectivity, whereas the external surfaces are antireflection coated. The reflectivity profile, determined by the interference of the waves reflected by the two coated surfaces, becomes a function of the distance between the surfaces and therefore varies radially. If the spacing between the two mirrors at the centre is set for constructive interference, the overall reflectivity will there have its maximum value. At the radial position where the spacing increases (or decreases) by $\lambda/4$, a destructive interference occurs and the reflectivity drops to zero. The RAVI has been used in a pulsed Nd:YAG laser to produce diffraction limited output beams with energy up 300 mJ at 28 J input. The main difficulty in using the RAVI lies on on the critical positioning of the two mirrors of the interferometer.

Undoubtedly the most effective and promising technique for implementing radially variable reflectivity mirrors is based on the deposition of a shaped thin layer of a high-refractive-index dielectric on an antireflection coated transparent substrate. The reflectivity is determined by the multiple interference inside the shaped dielectric film and therefore changes with the radial coordinate. In principle the dielectric variable reflectivity mirrors can be adapted in peak reflectivity, size, and wavelength to the requirements of any type of laser.

The first experimental demonstration of dielectric mirrors with Gaussian profile was by Lavigne *et al* (1985) for a wavelength of 10 $\mu$m. These mirrors were fabricated by depositing a variable thickness layer of germanium on antireflection coated substrates of germanium or sodium chloride. McCarthy and Lavigne (1985) and Parent *et al* (1987) used this type of Gaussian mirror in a confocal unstable resonator of an atmospheric pressure $CO_2$ laser. This laser produced output pulses of 12 J in a diffracted limited beam without fringes or side lobes in either near or far field. The corresponding multimode energy extraction was 14J per pulse. The usefulness of dielectric Gaussian mirrors for the operation of $CO_2$ lasers in single transverse and longitudinal mode with unstable resonators has been demonstrated by Parent *et al* (1987) and by Willetts and Harris (1988), who also proved the effectiveness of this solution to the generation of high energy pulses with low chirp rate.

More recently dielectric Gaussian mirrors for the wavelength of 1.06 $\mu$m have been fabricated by Zizzo *et al* (1988) and used by De Silvestri *et al* (1988c). In a pulsed Nd:YAG laser diffraction limited beams of smooth near- and far-field profiles and energy of about 250 mJ (at 17 J input energy) have been extracted. Snell *et al* (1988) also obtained significant results with a pulsed Nd:YAG laser by using mirrors with a parabolic reflectivity profile.

The availability of variable reflectivity mirrors whose parameters (size, peak reflectivity, and shape) can be tuned independently is extremely important in the practical realisation of a number of different design specifications. The supergaussian mirrors, recently introduced by De Silvestri *et al* (1988a, 1988b), besides allowing larger mode volume, can also satisfy this requirement. For this purpose, thin film vacuum evaporation techniques have been developed which allow the fabrication of multidielectric mirrors whose reflectivity profile is entirely under control. This method is essentially based on the shadowing effect of a fixed non-contact mask with a circular aperture, placed between the crucible and the substrate. As an example, the reflectivity profiles of a flat dielectric supergaussian mirror (order $n$ = 2.8, spot size 2.2 mm, and $R_0$ = 0.44) fabricated with this technique is presented in Figure 6.3. The mirror consists of a glass substrate coated with an antireflection $\lambda/4$ layer of magnesium fluoride on which a second layer of zinc selenide is deposited. The thickness of the second layer is $\lambda/4$ at the centre and slowly decreases to zero at the edge. Lastly, to protect the mirror, a very thin film of silicon dioxide has been deposited.

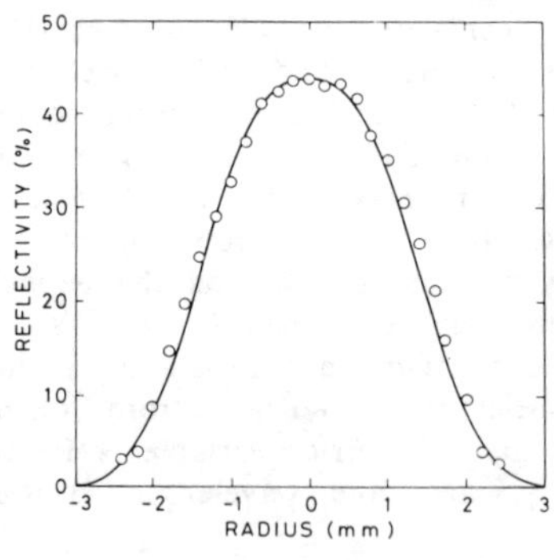


Fig 6.3 Experimental profile of a supergaussian mirror at $\lambda$ = 1.064 $\mu m$ obtained by thin film deposition technique (dots). The solid line is the interpolating curve with $n$ = 2.8, $w_m$ = 2.2 mm, and $R_0$ = 0.44.

As an example of a supergaussian resonator we consider the use of mirror shown in Figure 6.3, which was used in a flashlamp-pumped Nd:YAG laser with a rod 76 mm in length, 6.35 mm in diameter. The near- and far-field intensity distributions of the laser output were recorded by a linear array of 512 photodiodes spaced by 25 $\mu$m. To compensate for the geometrical divergence of the output beam, the far-field was taken at the plane of minimum spot size of a converging lens. The performances of unstable resonators with supergaussian mirrors were compared at two resonator magnifications $M = 1.36$ and $M = 2$; the first was obtained with a resonator 360 mm long and a convex rear mirror of 15 m curvature radius, the second with a resonator of 375 mm and a convex 3 m mirror. The laser rod was placed close to the rear mirror. The choice of a high magnification improves the mechanical stability with respect to the output energy, while the opposite happens for low magnification. For $M = 1.36$, where the peak reflectivity of the mirror is $R_0 = 0.44$, the output losses are $\Gamma = 1 - R_0/M^2 = 0.76$ and the condition [see Equation (6.17)] for maximally flat beam is fulfilled. The calculated beam profiles for this resonator are shown in Figure 6.2. At the rod position the mode is apertured by the rod diameter at 0.15% of the peak intensity which falls close to the lower limit established by the criterion

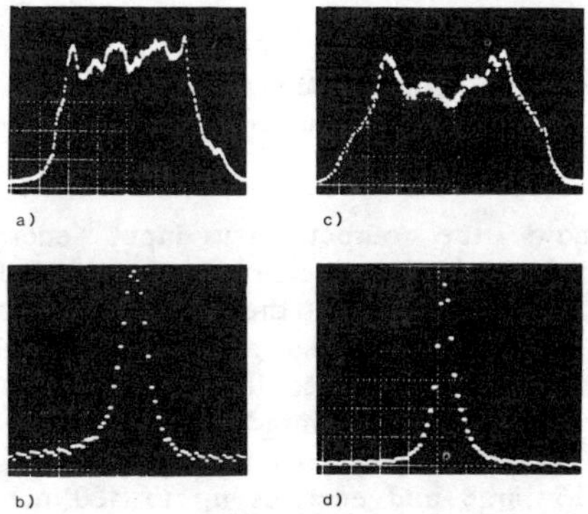


Fig 6.4 Profiles of the beams emitted by a Nd:YAG laser with two different supergaussian resonators employing the mirror of Fig 6.3: (a) and (c) are the near-fields for $M = 1.36$ and $M = 2$ respectively (horizontal scale 1.1 mm/div), (b) and (d) are the corresponding far-fields (horizontal scale 0.36 mrad/div).

reported in the previous section. Figures 6.4(a) and 6.4(b) show the shape of typical near- and far-field patterns of the output beam. The near-field intensity distribution is quite uniform and almost flat with a diameter at half maximum of 5.4 mm, in satisfactory agreement with the theoretical curve of Figure 6.2(a). From the far-field pattern [see Figure 6.4(b)] a divergence angle (half width at half maximum) of 0.2 mrad is measured, which is very close to the diffraction limit. At higher magnification ($M = 2$) the output losses increase up to $\Gamma = 0.89$ and the resonator does not satisfy the condition for maximally flat output beam, as seen also from the theoretical plots shown in Figure 6.2(b). In this case the mode is apertured by the rod diameter at 5.4% of the peak intensity, which falls again within the criterion considered in the previous section. The shapes of typical near- and far-field intensity patterns are shown in Figures 6.4(c) and 4(d). According to the theoretical plot of Figure 6.2(b) the output beam shows a central depression in the near-field and the diameter of the beam at half maximum (6.8 mm) is larger than that obtained at lower magnification. The divergence angle, 0.12 mrad, is slighter smaller as a consequence of a larger output beam and/or better mode discrimination. Also in this case the beam is diffraction limited.

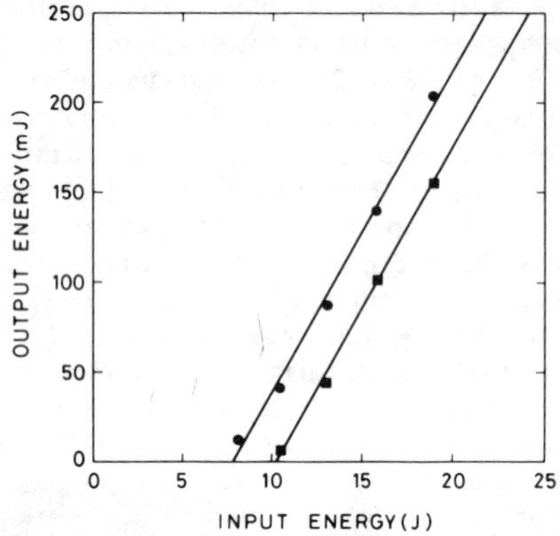


Fig 6.5 Output energy from a pulsed Nd:YAG laser as a function of the electrical input energy to the lamp for the two resonators also considered in Fig 6.4. O for $M = 1.36$, □ for $M = 2$.

Figure 6.5 shows the output-versus-input energy curves for both resonators. It can be seen that increasing the magnification decreases the output energy, but also decreases the divergence and the misalignment sensitivity. The last quantity, measured as the mirror tilt angle which halves the output power, turns out in practice to be practically the same for the two resonator mirrors and equal to 0.25 mrad for $M = 1.36$ and 0.8 mrad for $M = 2$. Recently the authors obtained from a pulsed Nd:YAG laser diffraction limited beams of 0.15 mrad and energies up to 400 mJ at 23 J input power using a supergaussian mirror of order $n = 12$.

## 6.5 REFERENCES


Anan'ev Y A and Sherstobitov V E 1971 Sov. J. Quantum Electron. 1 263

Armandillo E and Giuliani G 1985 Opt. Lett. 10 445

Arnaud J A 1974 Proc. IEEE 62 1561

Casperson L W and Lunnam S D 1975 Appl. Opt. 14 1193

De Silvestri S, Laporta P and Magni V 1987a J. Opt. Soc. Am. A 4 1413

De Silvestri S, Laporta P, Magni V and Svelto O 1987b Opt. Lett 12 84

De Silvestri S, Laporta P, Magni V and Svelto O 1988a IEEE J. Quantum Electron. QE-24 1172

De Silvestri S, Laporta P, Magni V, Svelto O and Majocchi B 1988b Opt. Lett. 13 201

De Silvestri S, Laporta P, Magni V, Svelto O, Arnone C, Calì C, Sciortino S and Zizzo C 1988c Opt. Commun. 67 229

Eggleston J M, Giuliani G and Byer R L 1981 J. Opt. Soc. Am. 71 1264

Emiliani G, Piegari A, De Silvestri S, Laporta P and Magni V 1989 Appl.

Opt. in press
Ganiel U and Hardy A 1976 Appl. Opt. 15 2145
Giuliani G, Park Y K and Byer R L 1980 Opt. Lett. 5 491
Harter D J and Walling J C 1986 Opt. Lett. 11 706
Krupke W F and Sooy W R 1969 IEEE J. Quantum Electron. QE–5 575
Kumagai N, Mori H and Shiozawa T 1966 Electron. Commun. Jap. 49 1
Lavigne P, McCarthy N and Demers J G 1985 Appl. Opt. 24 2581
Lavigne P, Parent A, Pascale D and McCarthy N 1986 IEEE J. Quant. Electron. QE–22 2200
McAllister G L, Steier W H and Lacina W B 1974 IEEE J. Quant. Electron. QE–10 346
McCarthy N and Lavigne P 1983 Appl. Opt. 22 2704
McCarthy N and Lavigne P 1984 Appl. Opt. 23 3845
McCarthy N and Lavigne P 1985 Opt. Lett. 10 553
Parent A, McCarthy N and Lavigne P 1987 IEEE J. Quantum Electron. QE–23 222
Sherstobitov V E and Vinokurov G N 1972 Sov. J. Quantum Electron. 2 224
Siegman A E 1986 Lasers (Oxford: Oxford University Press) 913–921
Siegman A E and Arrathoon R 1967 IEEE J. Quantum Electron. QE–3 156
Snell K J, McCarthy N, Pich M and Lavigne P 1988 Opt. Commun. 65 377
Vakhimov N G 1965 Radio Eng. Electron. Phys. 10 1439
Walsh D M and Knight L V 1986 Appl. Opt. 25 2947
Willetts D V and Harris M R 1988 IEEE J. Quantum Electron. QE–24 849
Yariv A and Yeh P 1975 Opt. Commun. 13 370
Zizzo C, Arnone C, Calì C and Sciortino S 1988 Opt. Lett. 13 342
Zucker H 1970 Bell Syst. Tech. J. 49 2349

# 7 Numerical Resonator Calculations

*P E JACKSON*

## 7.1 INTRODUCTION

The determination of the resonator transverse mode parameters is perhaps the most important calculation necessary in the design of a laser. After all, the resonator mode structure and spatial extent largely determine most of the important characteristics of a laser. For example they control the overlap efficiency with the active medium (and hence the laser output power), the size of the cavity optics, the wavelength and spectral purity of the oscillating field with the highest gain–to–loss ratio, and the output beam intensity profile and propagation characteristics.

This chapter will outline some of the techniques which have been applied to determine the resonator modes of a variety of cavity configurations. It will not include the derivation of the closed–form solutions that describe Gaussian resonator modes as these have already been covered in Chapter 1. Instead, we shall turn our attention to resonator mode calculations which are too intractable analytically and which require numerical and iterative computation. In particular, section 7.2 introduces the scalar diffraction type calculations pioneered by Fox and Li in 1960 and the development of more sophisticated procedures and applications thereafter. Briefly described in section 7.3 are the matrix techniques to solve the optical resonator eigen problem. Finally, outlined in section 7.4 is the open–ended waveguide approach (sometimes called the coupled–mode approach) to optical resonator theory.

## 7.2 SCALAR–DIFFRACTION ITERATION ALGORITHMS

### 7.2.1 Fox and Li Calculations

In 1960, A G Fox and T Li at Bell Telephone Laboratories published the results of their pioneering theoretical investigation to determine the lowest order transverse modes in Fabry–Perot interferometers, which constitute the basic resonators in laser oscillators. Their calculations involved numerical computation on a digital computer which was programmed to simulate the multimode consecutive round trips of a beam inside the laser cavity. The process begins with an assumed transverse electromagnetic field distribution at one mirror. The propagation of this field back and forth between the mirrors is determined using standard diffraction calculations. The simplest approach assumes *strip* mirrors and is therefore 1–dimensional. However, the modes of two–dimensional resonators can be obtained by taking the product of the modes for strip mirrors in orthogonal directions. The algorithm they used may be summarised as follows:

The electromagnetic field $U$ at a point $x$ on one mirror can be expressed in terms of an integral of the linearly polarised field at the other, using the scalar formulation of Huygens' principle. For a pair of infinite strip mirrors of width $2a$ and separated by $L$, the iterative equation reduces to (equation 27 Fox and Li, 1961):

$$U_{q+1}(x_2) = \frac{e^{j[(\pi/4)-kL]}}{\sqrt{\lambda L}} \int_{-a}^{a} U_q(x_1)\, e^{-jk(x_1-x_2)^2/2L}\, dx_1 \qquad (7.1)$$

where $q$ is the previous iteration number which designates the launched field, $k = 2\pi/\lambda$ is the propagation constant of the medium between the mirrors, and $\lambda$ is the wavelength of light in the medium. This equation is valid for mirror dimensions which are large in terms of the wavelength and for Fresnel numbers $N_0 = a^2/\lambda L \ll (L/a)^2$.

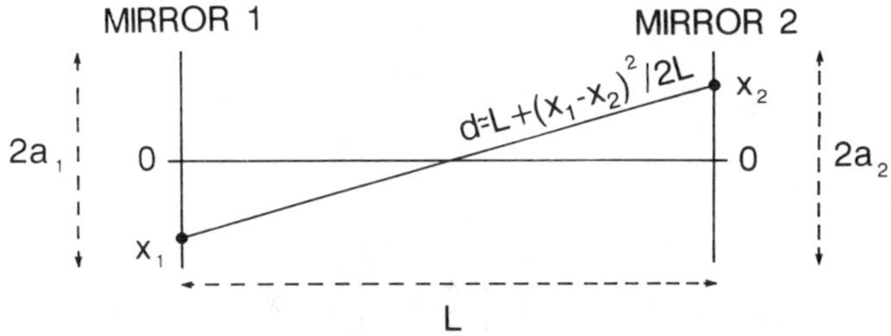


Fig 7.1 Geometry of plane strip mirrors

Equation (7.1) may be evaluated, for example, using Simpson's rule with a few tens or hundreds of increments (points across each mirror). Often the initial field $U_0(x_1)$ is defined as an equiphase uniform field with unity amplitude. However, a uniform plane wave excitation can never give rise to a mode with odd symmetry, and in order to investigate the possibility of modes of this type, it is necessary to start with an initial field over one–half of the mirror equal in magnitude but opposite in sign to the field over the other half. Alternatively, and in order to test the ability of the resonators to select a single mode, a pseudorandom distribution of amplitude and phase is introduced at the start, with the addition of further random noise (typically 1% of the average field amplitude) after each round trip. The noise represents the effects of spontaneous emission from the active medium and ensures the excitation of all the possible modes in the cavity. For the moment we will assume an empty resonator and introduce the effects of adding a saturable gain medium later in section 7.2.3.

The iterative integral calculation is deemed to have converged when the normalised iterative field amplitude and relative phase is reproduced to within some specified tolerence after each round trip. For empty resonator calculations, it is often sufficient to monitor the normalisation factor required to maintain a constant area under the intensity distribution after each transit round the resonator. Similarily, in the presence of gain, convergence is accompanied by a stabilisation of the circulating intracavity power.

Equation (7.1) may be suitably normalised by defining $X_i = x_i/a$ and the Fresnel number, $N_0 = a^2/\lambda L$, giving the general form:

$$U_{q+1}(X_2) = (1+j)\sqrt{\frac{N_0}{2}} \int_{-1}^{1} U_q(X_1)e^{-j\pi N_0 (X_1 - X_2)^2} dX_1 \qquad (7.2)$$

where the term $e^{-jkL}$ outside the integral has been ignored. Equation (7.2) may be modified to describe resonators with curved mirrors by including a term in the exponent within the integral, which accounts for the phase shift corresponding to the change in mirror spacing caused by mirror curvature. Mirrors with tilt can be treated in a similar way. Modes in resonators with mirror pairs of unequal widths, such as those found in unstable resonators, can be calculated by restricting the range of integration where appropriate. For example, the iterative equation for a confocal positive branch unstable strip resonator, such as that described in section 2.5 is given by:

$$U_{q+1}(X_2) = (1+j)\left[\frac{N_0}{2}\right]^{\frac{1}{2}} \int_{-1}^{1} U_q(X_1)K_1 \, dX_1 \qquad (7.3a)$$

$$U_{q+1}(X_1) = (1+j)\left[\frac{N_0}{2}\right]^{\frac{1}{2}} \int_{-\frac{1}{M}}^{\frac{1}{M}} U_q(X_2)K_2 \, dX_2 \qquad (7.3b)$$

where the kernel: $K_{1,2} = \exp(-j\pi N_0 [(X_1 - X_2)^2 - \Delta_{1,2}X_1^2 - \Delta_{2,1}X_2^2])$

and: $\Delta_1 = (1-M^2)/2M \qquad \Delta_2 = (M-1)/2M \qquad M = \frac{a_2}{a_1} = -\frac{R_2}{R_1}$

with mirror widths $2a_1$ and $2a_2$, Radii of curvature $-R_1$ (secondary convex output mirror) and $R_2$ (primary concave rear mirror) respectively.

Hence, to calculate the steady-state resonator modes of the confocal unstable resonator, it is sufficient to chose values for the magnification $M$ and the Fresnel number $N_0$, or $N_0$ and $\Delta_i = L/R_i$ for the general resonator, with no special relation between $L$ and $R_i$. Figure 7.2 shows the relative intensity and phase distributions for the steady-state field obtained using equations (7.3a) and (7.3b) with $M = \surd 2$ and $N_0 = 10.0$. Here a total of one hundred increments was used for the numerical integration, and the iterated field was normalised after each round trip to set the average amplitude to unity and the relative phase at $X_1 = 0$ to zero. Figure 7.3 shows the fluctuation of the output coupling (diffraction loss) as a function of the number of round trips for the three cases of an even-symmetric, odd-symmetric, and pseudorandom initial field distribution. Evidently, the higher-order modal components, which give rise to the periodic beating behaviour in the curves, take many transits to *'fade away'*. The power loss of the resultant mode is 17.9 % per round trip in each case, which is considerably lower than that predicted by geometrical considerations, namely $(1-{}^{1}/M)$ 29.3% .

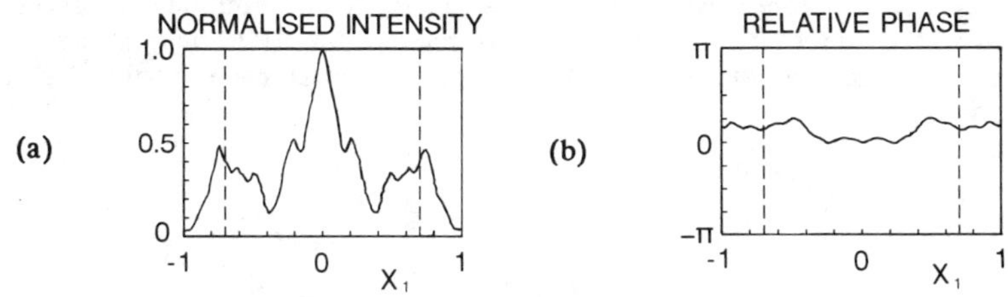


Fig 7.2 Steady-state empty resonator mode for a confocal unstable resonator with $M = \sqrt{2}$ and $N_0 = 10.0$ a) Normalised intensity incident on the output mirror. The extent of the rear mirror is over the entire $X$ range, while that of the output mirror is indicated by the vertical dashed lines b) Relative phase adjusted to zero at $X_1 = 0$.

The results shown in figures 7.2 and 7.3 (and also the figures to come) were obtained at the author's facility with a VAX mainframe computer code in Fortran. The Fortran programming language was chosen because of its complex number handling capabilities which greatly simplify the code needed to manipulate the iterated fields. There is, however, no reason why the calculations could not be performed with the convenience of a personal computer and suitable complex number routines. The computational time

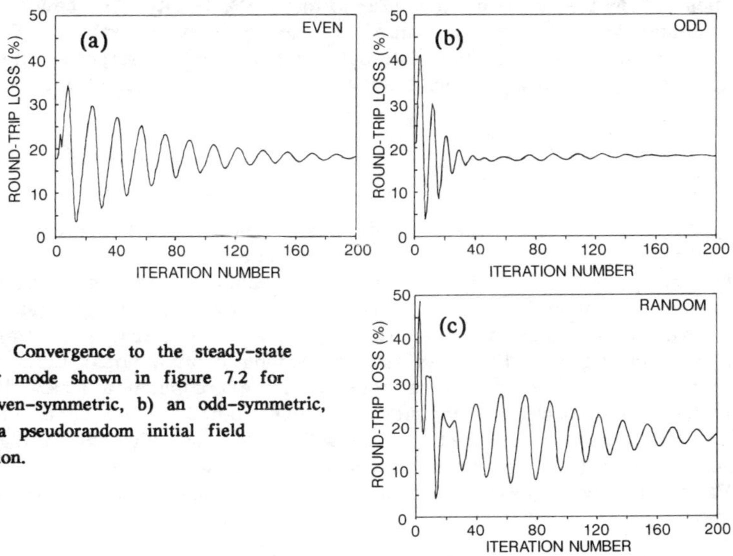


Fig 7.3 Convergence to the steady-state resonator mode shown in figure 7.2 for a) an even-symmetric, b) an odd-symmetric, and c) a pseudorandom initial field distribution.

required to reach a steady-state solution to the resonator mode problem is a product of the calculation time per round-trip and the number of trips to reach a field distribution within some user-defined acceptable convergence tolerance. The calculation time per trip is governed by the number of terms (points across each mirror) in the evaluation of the iteration integral, which in turn is determined by the Fresnel number. High Fresnel number resonators contain $N_0$ Fresnel zones with perhaps as many field amplitude reversals in curvature in the amplitude distibution. To resolve the fine field structure, it is generally necessary to have a number of points across each mirror which is several times $N_0$.

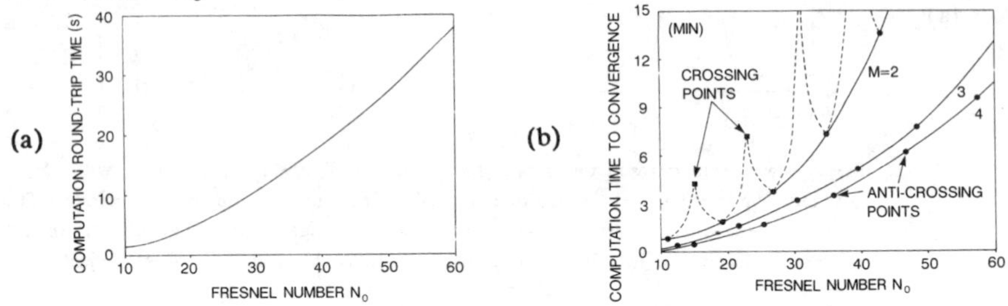


Fig 7.4 (a) Computation round-trip time and (b) convergence time for the confocal unstable resonator.

Figure 7.4 shows the computation round-trip time and convergence time recorded in reaching a steady-state solution for the confocal unstable resonator as a function of Fresnel number for various magnifications. Here a pseudo-random initial field distribution was used with $5N_0$ and $5N_0/M$ points across the primary and secondary mirror respectively. Convergence was confirmed when the area of the pre-normalised intensity distribution on the output mirror was consistently within $10^{-5}$ of its previous value. The solid lines in Figure 7.4b are the interpolation of the convergence times for resonators with equivalent Fresnel numbers that satisfy the condition:

$$N_{eq} = \frac{M-1}{2M^2} N_0 = n + \frac{3}{8} \qquad n \text{ (integer)} > 0 \qquad (7.4)$$

corresponding to the anti-crossing points (see section 2.8 and Figure 2.9) for strip resonators. Evidently, the computation time rises sharply with increasing $N_0$ and with decreasing $M$, and for some combinations of these values, convergence is intolerably slow. The convergence time in general will be greater than shown, and maximum for $N_{eq} = n + 7/8$ (crossing points). The dashed line in Figure 7.4b shows the intervening computational times for the resonator with $M$=2. It should be noted that the computation times shown in Figure 7.4 relate specifically to the mainframe computer and code employed, and the absolute computation times will vary for differing initial field distribution inputs (although the trends shown are general ones.)

Faster and more powerful (if not so elegant) algorithms will be discussed in section 7.2.4.

### 7.2.2 High-Order Modes

The computational technique described in the proceeding section yields a steady–state solution to the resonator round trip iterative equation. The predominating mode is always the lowest–order (by definition), lowest–loss mode for the particular mode symmetry under investigation. This simple and elegant numerical iterative method does not yield readily the higher order modes without additional computational techniques. Information about the higher order modes is essential to determine the resonator transverse mode discrimination, and to avoid configurations which result in mode degeneracy or mode beating. Mode degeneracy occurs when two modes of different order experience the same round trip loss (these points of mode degeneracy, sometimes called *crossing* or *cusping* points, are one of the striking properties of unstable resonators discussed in Chapter 2). Mode beating occurs during the covergence to a steady–state solution, and is responsible for the fluctuations in Figure 7.3 which are caused by the periodic intereference of different modes with different phases on successive round trips. Note that the number of modes excited in the first instance is dependent on the initial field distibution. Mode beating can persist in the presence of gain inside the resonator if incomplete mode competition arises because two or more modes

(a)

(b)

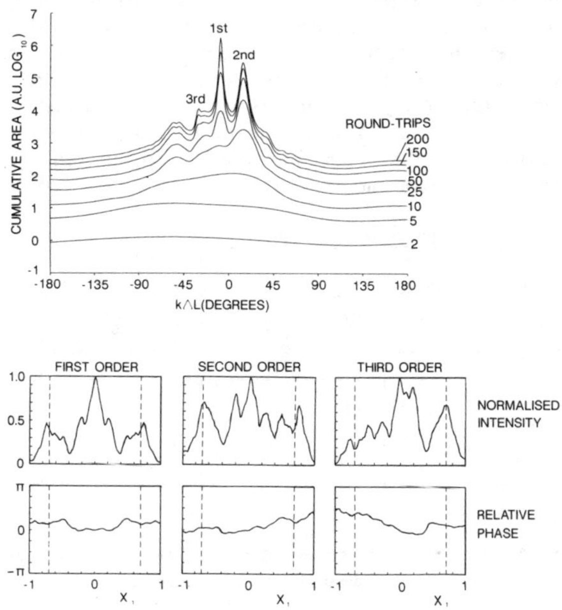


Fig 7.5 Higher–order mode extraction a) Cumulative area of the resultant summed fields recorded after each transit for various transit totals as a function of the mirror spacing of the empty cavity. b) The first three lowest–order modal fields obtained after 200 transits. The second and third order modes are noticeably inpure because of interference from other modes.

experience similar gain–to–loss ratios above the threshold for oscillation. This can occur if they interact with different parts of the gain medium as a consequence of their different spatial distributions, or if they sample a nonuniform transverse gain distribution or a non linear saturable gain.

To obtain solutions for the higher order modes of the Fabry–Perot resonator, Fox and Li modified their earlier algorithm (section 7.2.1) to compute the optical resonator modes by the method of resonance excitation (Fox and Li, 1968). Whereas previously, the main interest was in reaching a steady-state field distribution after many transits, this time the interest is in recording the field after every round trip and in taking the complex sum of these recorded fields. The resultant summed field then represents the output of the externally excited resonator for that particular mirror spacing. For an arbitrary mirror spacing, the iterated fields do not add in phase, so the resultant sum is generally small, but for certain mirror spacings the field components for one or another of the modes will add in phase to produce an appreciable resultant resembling that mode. Their frequency spacings correspond to the resonance of the various modes. Further rounds of calculations in an identical tandem resonator starting with the resultant field from the previous round can be used to *filter* the mode to obtain a closer approximation to a pure mode. Hence, it is possible to *tune* the length of the resonator to extract the different resonant modes. Figure 7.5 shows the transmitted power and distribution functions of the resultant iterated fields obtained by *tuning* the resonator which gave rise to the steady-state field distribution in Figure 7.2. Figure 7.6 shows the purified second-order mode after successive round-trips in three identical tandem resonators. After further rounds of purification, the mode distribution did not change noticeably. An indication of the purity of modes extracted in this way can be obtained by

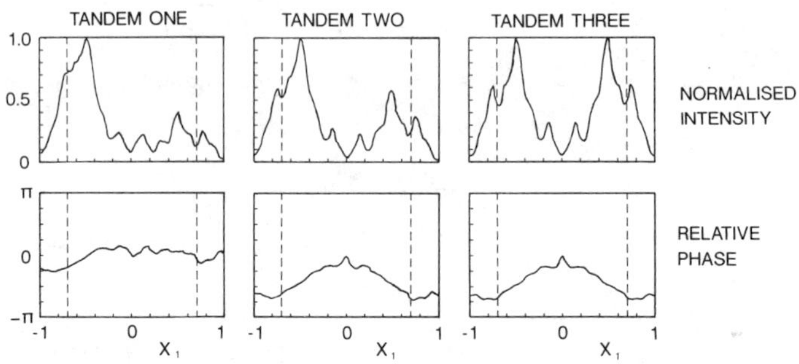


**Fig 7.6 Purification of the second order mode by resonant excitation in identical tandom resonators.**

testing for the orthogonality of their normalised field distributions with the integral evaluation of their product. For purer modes this integral approaches zero.

### 7.2.3 Modes in the Presence of Saturable Gain

In the presence of a saturable gain medium, a mode having larger losses can compensate for its higher diffraction loss by adjusting itself to have a lower flux intensity, thus saturating the medium less and still allowing that mode to reach oscillating threshold. For example, Rensch and Chester (1973) modified the Fox and Li strip resonator calculation to include a saturable gain sheet positioned right at the primary mirror of an unstable resonator. The amplitude wave was propagated to the primary mirror, amplified and saturated in a homogeneous fashion using the Rigrod formula (with local intensity doubled to account for incoming and outgoing waves) and propagated back to the feedback mirror. More formally, the field amplitude is multiplied by $\exp(g\ell)$, where $2g\ell$ is the net round-trip intensity gain, prior to reflecting it from the primary mirror. At each point on the mirror, the gain is determined using $g = g_0/(1 + 2I/I_0)$ where $g_0$ is the small-signal gain of the medium, $I$ is the local field intensity, and $I_0$ is the saturation intensity. Obviously, nonuniform or nonlinear transverse gain and saturation distributions arising from effects such as nonuniform pumping and gas flow, require a more intricate gain equation, although its application is the same. This ***thin gain sheet*** model will only accurately predict the oscillating field and the power output from the resonator, for modes with negligible variations in the shape of their transverse intensity profile during transit through the gain medium (such as those produced in low magnification unstable resonators). For laser resonators which do not satisfy this criterion, the computational effort required to perform the Fox and Li calculations at suitable intervals along the propagation path is generally too great, and more intricate and faster algorithms are required (some of these are described in section 7.2.4).

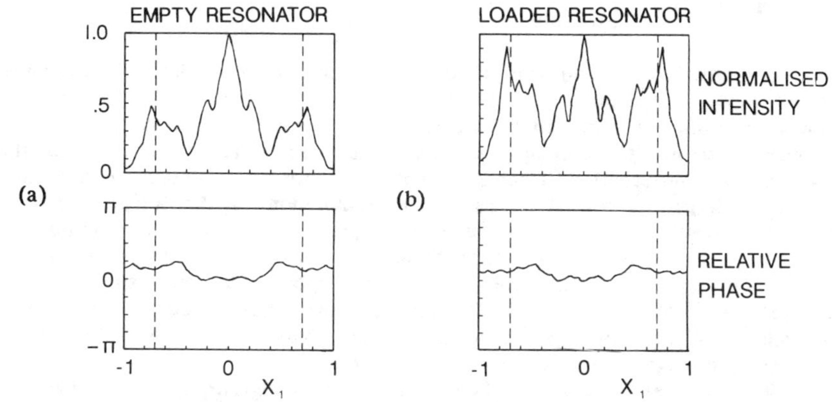


Fig 7.7 Normalised intensity and relative phase for the unstable resonator (see figure 7.2) with and without the thin sheet gain approximation.
a) Empty resonator steady-state mode. b) Active resonator steady-state mode with a thin sheet gain twice the geometric threshold value.

Figure 7.7 shows an example of an unstable resonator mode calculation with the presence of saturable gain. As can be seen the *loaded resonator* distribution is broadely similar to the *empty resonator mode*. The major difference is that the higher peaks are pushed down in amplitude by the local gain saturation that they produce. In general, for unstable resonators, the addition of gain causes the round-trip resonator loss (output coupling) to increase, since the gain tends to spread out the intracavity distributions to fill the resonator more uniformly, distorting the cavity modes until they more closely approximate the geometric case. Although the addition of gain has only subtle effects on the near field pattern, as indicated in Figure 7.7, the far field pattern tends to spread out. Figure 7.8 shows the build-up of the intra-cavity flux and its saturation. A converging mode sampling small-signal gain will continuously adjust its field distribution with the onset of gain saturation, to maintain the lowest round-trip loss.

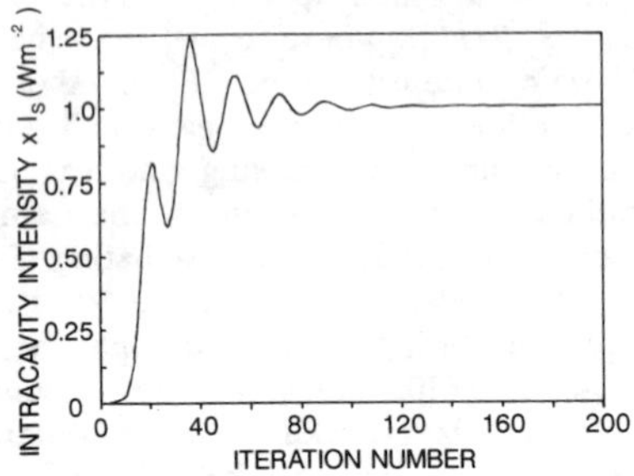


Fig 7.8 Intensity build-up and saturation for the resonator mode shown in figure 7.6b

### 7.2.4 Other Algorithms

The mathematical manipulation needed to propagate an initial field distribution stored in a numerical array through repeated round trips of the resonator cavity is prone to cumulative errors. This reduced precision arises through round-off errors in computer calculation and as a result of approximations used in the propagation scheme. The necessity to maintain high accuracy implies a large number of terms in the summation of the integral equations, with the result that the computer storage requirements and computational processing time limit the application of this approach (especially for two-dimensional resonator calculations as opposed to the strip resonator calculations described here). In general, the approximations used in the derivation of the propagation algorithm are only strictly valid for Fresnel numbers too large to be handled numerically.

Since the early work by Fox and Li, new algorithms have evolved for solving the iterative integral equations which require less computational effort, permitting large Fresnel number resonator calculations in the region of validity where standard methods fail. One such approach is the asymptotic analysis first introduced by P. Horwitz in 1973 (Horwitz 1973). This analysis involves the asymptotic expansion of the integral equation in terms of orders $1/N$ and the determination of the roots of a complex polynomial derived from the resonator eigenequation. The fundamental difference between this approach and the Fox and Li scheme, is that in the latter, the resonator integral equation itself is iterated to convergence (starting with some assumed initial mode shape), whereas in the asymptotic algorithm only the eigenvalue is passed from iteration to iteration. This procedure calculates the lowest-order and higher-order resonator modes simultaneously, converging in a few iterations only. The asymptotic analysis technique was later extended (see for example Smith 1981) to include the effects of a saturable gain distributed along the propagation path (as opposed to the *thin gain sheet* method).

For a comprehensive list of references on the Fox and Li iteration algorithm and its developments, the asymptotic approach, and other powerful techniques (including algorithms to study the effects of mirror misalignment on resonator modes), the reader is referred to the book by Siegman (1986).

## 7.3 MATRIX TECHNIQUES

An alternative to the scalar–diffraction iteration algorithm for exploring resonator mode behaviour is the matrix eigenequation formulation. In this approach, the optical resonator modes are found by solving numerically for the eigenvectors $U_n$ and eigenvalues $\gamma_n$ which satisfy the matrix eigenequation.

$$MU = \gamma U \tag{7.5}$$

where $M$ is a complex matrix derived from optical diffraction theory (for example, the Fresnel–Kirchhoff integral equation discretised using Gaussian quadratures). The complex eigenvectors $U_n$ correspond to the optical resonator mode patterns. The fractional round–trip power loss of a given resonator mode is equal to $1-\gamma_n^2$ (or $[1-\gamma_n]$ for strip resonators) where $\gamma_n$ is the corresponding eigenvalue, and the exact resonant frequency is given by the phase angle of $\gamma_n$. The iteration algorithm most commonly applied to equation 7.2 is the *power* method and may be summarised as follows:

An arbitrary initial vector: $$U = V_0 = \sum^{n} K_n U_n \tag{7.6}$$

(a multimode summation weighted by $K_n$) is repeatedly multiplied by the matrix $M$ to obtain a sequence of iterated vectors $V_1$, $V_2$, $V_3$ etc. After many iterations the higher–order higher–loss mode components $U_n$ *drop out* of the iterated vector leaving the lowest–loss resonator mode $U_1$ (by definition) and the corresponding eigenvalue $\lambda_1$. In practice this may require 10's to 100's of iterations with large order matrix computation, and for large Fresnel numbers (as is the case for unstable resonators) or for closely spaced *dominant* eigenvalues, convergence is intolerably slow.

The desire for rapidly convergent routines that extract all the dominant eigenvalues simultaneously has led to the development of more powerful iteration algorithms. The Krylov method, for example, reduces the initial vector $V_0$ to $N$ dominant eigenvectors. The $N$ dominant eigenvalues are then extracted simultaneously by iterating simultaneously on $N$ different initial vectors like $V_0$. Another similar approach called the Prony method (Siegman and Miller, 1970) takes advantage of the orthogonality properties of optical resonator eigenmodes to convert the discretised Matrix $M$ into a complex symmetric nonhermitian matrix. The eigen problem then takes the form of a set of $N$ coupled linear simultaneous equations that can be solved by standard matrix symmetrization techniques.

## 7.4 THE OPEN-ENDED WAVEGUIDE APPROACH

An alternative to the integral–equation approach to describe the resonator round–trip propagation is the waveguide approach pioneered by Weinstein (1969), wherein the resonator is viewed as a guiding structure in the direction transverse to the resonator axis. The mirror surfaces constitute the waveguide boundaries, and the edges of the mirrors form the open ends of the waveguide. The waveguide can support many modes although the modes relevant to the resonator problem are of very high order and near the *cut–off* frequency. Oscillating modes are reflected at the two open ends and are weakly coupled by leakage at the mirror edges. The self–consistency requirement leads to an eigenequation which may be solved numerically to derive the resonant modal fields. This waveguide approach has been successfully applied to stable and unstable resonators (Chen and Felsen, 1973).

## 7.5 REFERENCES

Chen L–W and Felsen L B 1973 IEEE J. Quantum Electron. QE–9 11 1102–1113

Fox A G and Li T 1961 Bell Syst. Tech. J. 453–488

Fox A G and Li T 1968 IEEE J. Quantum Electron. QE–4 7 460–465

Horwitz P 1973 J. Opt. Soc. Am. 63 12 1528–1543

Rensch D B and Chester A N 1973 Appl. Opt. 12 5 997–1010

Siegman A E 1986 Lasers (California: University Science Books)

Siegman A E and H Y Miller 1970 Appl. Opt. 9 12 2719–2736

Weinstein L A 1969 (Vaynshtein) Open Resonators and Open Waveguides Transl. P Beckmann (Colorado: The Golem Press)

# 8 Frequency Stabilisation of Lasers

*K M ABRAMSKI and D R HALL*

## 8.1 INTRODUCTION

The recognition of lasers as high frequency oscillators, whose spectral purity and relative frequency stability could be extremely precise, is as old as the laser itself. In their classic paper proposing "Infrared and Optical Masers", Schawlow and Townes (1958) suggested the possibility of achieving single frequency ($\nu_0$) output with a spectral bandwidth $(\Delta\nu)$ of very much less than 1 Hz at optical frequencies ($10^{13}$ - $10^{15}$ Hz). In fact, if we consider *only* the limitations on linewidth which derive from the quantum nature of the light generating process, then values of relative frequency stability, $(\Delta\nu/\nu_0)$ of less than $10^{-19}$ might be anticipated. However, because of additional environmental factors, the values which are actually achieved are many orders of magnitude worse than this. One of the aims of this chapter is to consider these additional factors, to see how they affect frequency stability, and how their impact may be reduced by improved *resonator design*.

Before doing so however, it is instructive to briefly review some of the many application areas for frequency stable lasers. Stabilisation techniques have been applied to many laser types, with the gas lasers, particularly He–Ne and $CO_2$ attracting the most attention. The application area which has generated perhaps the largest volume of research and associated literature is that related to the development of frequency, wavelength and length standards. This work, much of which has been done at the various national standards laboratories, (e.g. Hall *et al*, 1987) has given rise to many of the significant developments in the field. In spectroscopy, advances in frequency stabilisation techniques, coupled with increased tunability of laser sources at a number of wavelengths, has made possible dramatic advances in high resolution spectroscopy (Persson and Svanberg, 1987)). In the field of laser radar, advances in the relative frequency stability in both cw (Freed, 1988)) and pulsed $CO_2$ lasers have made possible the development of a number of impressive high resolution laser Doppler radar systems (Vaughan 1987), while the use of stabilised, external mirror laser cavities for diode lasers have facilitated the application of heterodyne techniques to fibre–optic communication systems. In the future, frequency–stabilised all–solid–state lasers are set to provide stable optical sources for metrology systems and other engineering applications of coherent radiation, such as in fibre–optic sensor systems.

The recent achievement of frequency stabilisation at the level of a few tens of millihertz (Salomon, Hils and Hall 1988) opens the way to the possibility of a number of fundamental scientific experiments in gravity wave detection (Drever *et al*, 1981), squeezed states, relativity and astrophysics (Faller *et al* 1985)

## 8.2 LASER FREQUENCY FLUCTUATIONS

### 8.2.1 Elementary Considerations

Virtually all the main applications of frequency stable optical sources require, as a pre-requisite, that the spectral output of the laser consist of a *single* narrow band of frequencies, which is stabilised, either in absolute or relative terms with some level of precision. Thus, the laser designer is faced with the dual objective of ensuring single frequency output *and* of stabilising this frequency. In order to see what is involved in this task, it is nécessary to examine which factors control the output spectrum of lasers. It is clear that at one extreme the fundamental limitation on the maximum spectral bandwidth of the laser output (assuming no harmonic conversion) is the linewidth of the laser transition itself, i.e. the spectral output must lie within the gain–bandwidth of the laser amplifier, illustrated in Figure 8.1 (a). Within

Fig 8.1

Single longitudinal mode operation of a laser.

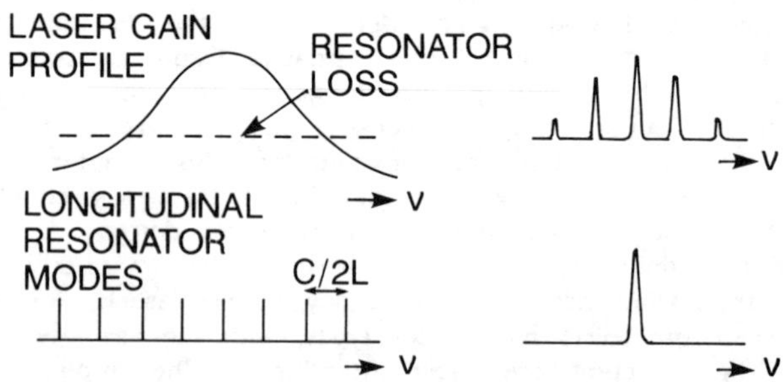


this restriction, the frequency content of the laser emission depends on the spectral and loss characteristics of the laser resonator design, in other words on its longitudinal and transverse mode structure. If we now assume that the cavity design ensures single mode ($TEM_{00}$) operation, then the frequency components in the output of the laser are further restricted to one or more of the longitudinal cavity modes, illustrated in Figure 8.1 (b). For the case of inhomogeneously (e.g Doppler) broadened transitions and for homogeneously broadened lasers where significant spatial inhomogeneity exist, there may be a number of longitudinal mode frequencies above threshold, as depicted in Figure 8.1 (c). The *average* values of these frequencies, $\nu_q$ , ignoring frequency pulling effects (caused by an interaction with the gain medium), are given by the equation:

$$\nu_q = q \frac{c}{2\, n_0 L_0} \tag{8.1}$$

where $q$ is the number of half waves along the resonator of length $L_0$, $c$ is the velocity of light, while $n_0$ represents some average of the refractive indices of the various media comprising the laser resonator. Since, as indicated earlier, multiple frequencies (even stabilised) are unsuitable for most applications, the first task is to design the laser so as to ensure single longitudinal (axial) mode operation. This topic is rather extensive in itself and well covered in text books and in the literature (e.g. Smith 1973), so for the purposes of the present discussion, it is assumed that steps have been taken to ensure that the laser stabilisation techniques which will be discussed here, are applied to lasers which oscillate on a single axial mode, as illustrated in Figure 8.1 (d). With this assumption, we can now turn our attention to an examination of the factors which may cause fluctuations in the frequency of our single axial mode laser.

When considering the instabilities in the frequency of an oscillator, it is usual to distinguish two sources of such fluctuations, those which are *intrinsic* and those which are the consequence of external or *environmental* factors.

### 8.2.2 Intrinsic Fluctuations

At the most fundamental level, the extent to which the oscillation bandwidth of a single mode laser may be narrowed is restricted to a theoretical lower limit resulting from the quantum nature of radiation which introduces a broadening of the spectral width of the radiation emitted. For a two-level system, the fundamental limit on laser linewidth $\Delta\upsilon_i$ is given by the well known Schawlow-Townes formula (Schawlow and Townes, 1958):

$$\Delta\upsilon_i = \frac{2\pi\ (\Delta\upsilon_c)^2 h\upsilon}{P} = \frac{h\nu}{2\pi\tau_c^2 P} \qquad (8.2)$$

where $h$ is Planck's constant, $P$ is the laser output power, $\Delta\upsilon_c$ is the cavity spectral linewidth, and $\tau_c$ the decay time of the empty laser cavity. The intrinsic linewidth is dominated by frequency fluctuations caused by the random walk of the oscillation phase under the influence of spontaneous emission, (or quantum noise). Estimates of the intrinsic linewidth for four important laser types, calculated from equation (8.2), are given in Table 8.1. As suggested above, the ratio $\Delta\nu/\nu_0$ which is called the fractional spectral width, implies

| | $CO_2$ | HeNe | Nd:YAG | CaAlAs |
|---|---|---|---|---|
| $\lambda$ | 10.6$\mu$m | 0.63$\mu$m | 1.06$\mu$m | 0.83$\mu$m |
| $\upsilon_0$ | 28x10$^{12}$Hz | 4.7x10$^{14}$Hz | 2.8x10$^{14}$Hz | 3.6x10$^{14}$Hz |
| $\Delta\upsilon_c$ | 5MHz | 0.5MHz | 5MHz | 50MHz |
| $P$ | 1W | 1mW | 100mW | 10mW |
| $\Delta\upsilon_i$ | 3x10$^{-6}$Hz | 5x10$^{-4}$Hz | 3x10$^{-4}$Hz | 0.4Hz |

Table 8.1 Intrinsic linewidths of laser lines calculated from the Schawlow-Townes formula for different lasers.

that the intrinsic limit on the relative frequency stability is less than $10^{-19}$.

### 8.2.3 Environmental Fluctuations

In practice, because of instabilities caused by *environmental* fluctuations, the relative frequency stability of real lasers, even when measured over short integration times, is many orders of magnitude larger than the value imposed by intrinsic quantum considerations. Nevertheless, using techniques of passive and active frequency stabilisation, to be discussed below, there has been extraordinary progress over the past 25 years, particularly during the past few years, when linewidths as narrow as 50 millihertz have been demonstrated (Salomon, Hils and Hall, 1988; Freed, 1988).

From equation (8.1), it may be easily shown that the temporal variation in the cavity mode frequency, $\Delta\nu(t)$, which occurs as a consequence of geometrical length fluctuations $\Delta L(t)$ and/or fluctuations, $\Delta n(t)$, in the refractive index of the media, may be expressed as

$$\Delta\nu(t) = -\frac{\nu_0}{n_0 L_0}\left[n_0 \Delta L(t) + \Delta n(t) L_0\right] \tag{8.3}$$

Thus, fluctuations in cavity length and refractive index caused by acoustical and mechanical vibrations, power supply instabilities, variations of temperature and pressure etc are called environmental fluctuations. These factors dominate the spectrum of the laser emission, because the random deviations caused by environmental fluctuations are much larger than the intrinsic spectral width. They may range from a few kHz for some gas lasers, to hundreds of megahertz for semiconductor lasers. For example, for $CO_2$ lasers excited by dc discharges, frequency shifts in the order of 0.5–1MHz/mA and 5MHz/Torr are expected as a result of changes in refractive index due to current or pressure fluctuations. For semiconductor diode lasers, the frequency modulation caused by injection current modulation is typically 300MHz/mA. As regards the impact of cavity length changes, equation (8.3) indicates that a linear displacement of 1 Angstrom ($10^{-10}$ m), or about 1 hydrogen atom diameter, in the position of a mirror in a 1m long $CO_2$ laser cavity will cause a frequency shift of about 12kHz, while for a 0.1mm long semiconductor laser this level of displacement would give rise to a frequency shift of 120MHz. Thus, in summary, the main problem in laser frequency stabilisation is to minimise the effects of environmental fluctuations using passive stabilisation techniques, and to compensate for these variations using active stabilisation techniques.

## 8.3 CHARACTERISATION OF FREQUENCY STABILITY

### 8.3.1 Measurement Techniques

The most convenient way to analyse the frequency stability of a laser output is to downshift its frequency to the radiofrequency region, where *electronic* frequency measurement is possible. A measurement of stability implies a comparison with another signal, which, in a sense, acts as a frequency standard. Both of these requirements, downshifting and comparison, may be accomplished by the use of ***heterodyne techniques***. This involves the mixing of signals from two laser sources on the surface of a fast photodetector, enabling the difference (or beat) frequency to be monitored. Figure 8.2 shows the basic experimental arrangement of a heterodyne measurement system with the

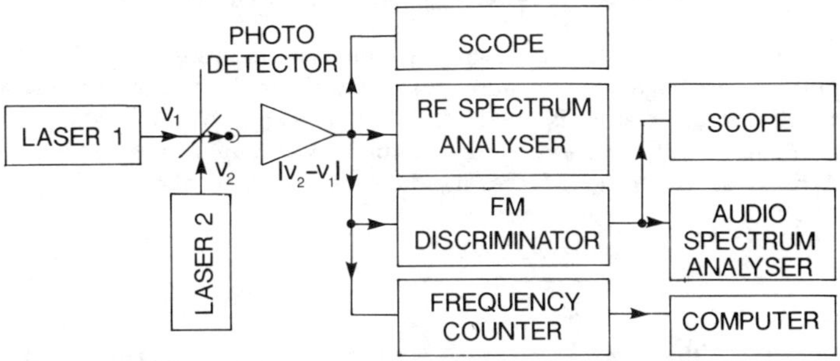


Fig 8.2 Typical heterodyne laser system with diagnostic equipment for measuring the spectral properties of the beat signal.

type of diagnostic equipment used for monitoring the beat signal. Two views of this process are possible. For certain levels of frequency precision, it may be acceptable to consider one of the lasers as a frequency standard, whose stability relative to the test laser is very high. In this case, the fluctuations in the beat signal can all be attributed to instabilities in the test laser. At higher levels of precision however, no such standards exist, and heterodyne measurements are usually performed between two lasers which are independent but, in principle, identical with each other. Assuming the fluctuations of the two lasers are statistically independent, the mean–square fluctuation in the beat signal is twice as large as the mean–square fluctuation in each individual laser. Thus, the requisite information on the frequency stability of a single laser may be obtained. The beat signal may be observed directly on an oscilloscope or RF spectrum analyser, or may be analysed on an audio spectrum analyser following an FM discriminator. The frequency variation over a range of averaging times, $\tau$, can be measured with a counter, allowing an assessment of the frequency stability to be determined, as indicated in the next section. An alternative experimental technique, employing the so–called ***homodyne*** principle, is very useful in some cases. Here, the output of the laser under test is split using some form of beamsplitter. One signal is routed directly to the detector, where it is mixed with the second beam, which meanwhile, has traversed a much longer path, (for example in a long length of fibre). This technique allows the mixing of the laser signal with its output at some time, $\tau$ later ($\tau$ depends on the path difference). The fluctuations in the resulting beat signal provide an indication of the laser frequency stability.

### 8.3.2 Definition of Frequency Stability

The analytical representation for the amplitude of a single frequency laser signal may be written:

$$v(t) = [V_0 + \epsilon(t)] \exp \{ j[\omega_0 t + \varphi(t)] \} \tag{8.4}$$

where $\omega_0$ is the average angular frequency and $\epsilon(t)$, $\varphi(t)$ are random variables representing amplitude and phase fluctuations, respectively. The instantaneous angular frequency, $\omega(t)$ is given by:

$$\omega(t) = \frac{d}{dt}\left[\,\omega_0 t + \varphi(t)\,\right] = \omega_0 + \Omega(t) \tag{8.5}$$

where $\Omega(t) = \dot{\varphi}(t)$ is the random variable representing the deviation of the frequency of the signal. Because this parameter is a random variable, some measure of its dispersion is needed. For this purpose the standard deviation $\sigma[\Omega(t)]$ of the $[\Omega(t)]$ process is determined, whereupon the frequency stability, $S$, is defined as the ratio of the standard deviation to the average frequency.

$$S = \frac{\sigma[\Omega(t)]}{\omega_0} \tag{8.6}$$

Assuming that the $[\Omega(t)]$ process is stationary, its power spectral density, $G_\Omega(\omega)$ can be introduced so that the variance of the frequency fluctuation can be expressed (Cutler and Searle 1966):

$$\sigma^2\,[\Omega(t)] = \frac{1}{\pi}\int_0^\infty G_\Omega(\omega)\,d\omega \tag{8.7}$$

Substituting equation (8.7) into (8.6), we obtain a definition of the frequency stability in the *frequency* domain, which is rather impractical and has only theoretical interest. A much simpler determination of $[\Omega(t)]$ can be performed in the *time* domain, by measuring the number of cycles of the analysed signal in some time period, $\tau$, using a conventional frequency counter (Barnes *et al* 1971). The average frequency $\bar{\nu}_k$ over the measurement period, $\tau$ may be written as:

$$\overline{\nu_k} = \frac{1}{2\pi\tau}\int_{t_k}^{t_k+\tau} \dot{\varphi}\,(t)\,dt = \frac{\varphi\,(t_k + \tau) - \varphi\,(t_k)}{2\pi\tau} \tag{8.8}$$

The statistical variance for $N$ samples of the time period, $\tau$, with periodicity $T$ may be expressed as:

$$\langle \sigma^2(N,\tau,T) \rangle = \left\langle \frac{1}{N-1} \sum_{n=1}^{N} \left[ \bar{\nu}_k - \frac{1}{N} \sum_{k=1}^{N} \bar{\nu}_k \right]^2 \right\rangle \tag{8.9}$$

where $\langle\ \rangle$ denotes averaging of many measurements.

A particularly useful and frequently quoted measure of the variance may be obtained from equation (8.9) by substituting $N = 2$ and $T = \tau$. The resulting expression is a simple and convenient formula, which is widely known as the "Allan variance" (Allan, 1966). It may be written as:

$$\langle \sigma^2\,(2,\,T = \tau) \rangle = \left\langle \frac{(\bar{\upsilon}_{k+1} - \bar{\upsilon}_k)^2}{2} \right\rangle \tag{8.10}$$

A measure of the laser frequency stability, $S(\tau)$ may conveniently be represented by a diagram in which the square root of the Allan variance is plotted against the averaging times $\tau$, which are typically in the range 1ms – 1000s. In other words, $S(\tau)$ may be written as:

$$S(\tau) = \frac{< \sigma^2(2, T = \tau) >^{\frac{1}{2}}}{\nu_0} \qquad (8.11)$$

Thus, any specification of the stability of an oscillator should always be accompanied by an indication of the value of the associated averaging time, $\tau$. The general shape of a typical frequency stability (Allan Variance) curve is illustrated in Figure 8.3. This may be understood by reference to the power spectral density, $G_\Omega(\omega)$ of the heterodyne signal. For small averaging times the spectrum is dominated by white noise ($G_\Omega(\omega) \sim$ *const*), while for lower frequencies it is characterised by 1/f or "flicker" noise. For longer averaging times, the frequency stability is subject to long term drift. In practice, in specifying frequency stability, it is common to distinguish between *short term* ($\tau$ < 1 sec) and *long term* ($\tau$ > 1 sec) stability.

If the laser is switched on and off a number of times, then unless the laser is *actively* frequency locked to some stable reference, the centre frequency at which the laser oscillates is not repeatable from experiment to experiment. If $\Delta\nu_R$ represents the spread in the oscillation frequency during a series of such experiments, and $\nu_0$ its average value, then we may define $R$, the *reproducibility or re-settability* of the frequency such that $R = \Delta\nu_R/\nu_0$.

Fig 8.3 Typical plot of laser stability as a function of averaging time $\tau$.

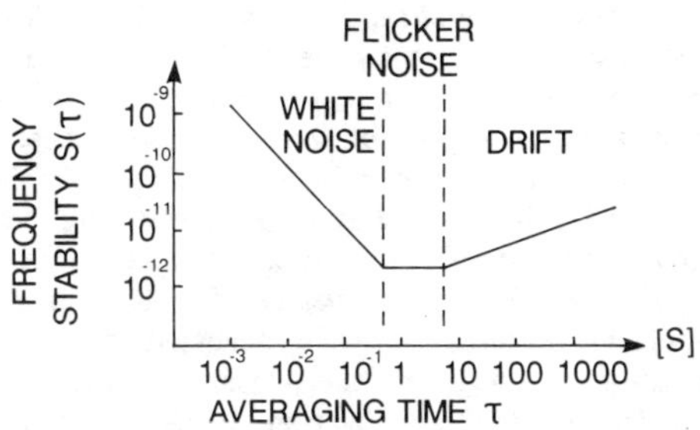


## 8.4 PRINCIPLES OF LASER FREQUENCY STABILISATION

### 8.4.1 Passive Stabilisation

As indicated earlier, the major limitation on the frequency stability of single mode lasers is imposed by fluctuations in environmental conditions. As a rule, the first step towards offsetting these effects is to design lasers which minimise their impact on resonator length and medium refractive index. This is sometimes referred to as "passive" or "open-loop" stabilisation. The usual approach adopted is the construction of rigid, mechanically stable structures, acoustically and magnetically shielded, employing low thermal expansion materials (such as super invar or one of the proprietory forms of ultra low expansion quartz, whose expansion may be as low as 2 x $10^{-8}$ per Kelvin) to define the resonator mirror spacing. In gas lasers, close control of the gas pressure (preferably by sealed–off operation) is necessary as is the minimising of fluctuations in the input electrical power to the discharge. Where active cooling is involved, precise control of the temperature and flow rate of the coolant is needed to minimise variations in the *optical* length of the resonator.

Although these passive techniques are capable of very precise values of fractional spectral width (for example, values of 1 part in $10^{13}$ have been reported by Freed 1988), such stability can only be maintained for small fractions of a second without some level of external control. Moreover, they do not provide an *absolute* frequency standard and are not re–settable to a particular frequency. A high degree of passive stability is usually an important pre–requisite to the realisation of actively stabilised systems. To achieve stabilisation of the frequency in an absolute sense, an active stabilisation system based on a feedback loop is required to lock the laser frequency to some absolute frequency standard such as a resonant atomic or molecular spectral line. The basic principles of such systems are discussed in the next section, and several examples are outlined in the final section.

### 8.4.2 Active Frequency Stabilisation

During the past three decades many different approaches have been investigated for active stabilisation of the laser oscillation frequency, some of which are described below. These techniques, all of which address the basic problem of stabilising the frequency within the laser gain bandwidth, have certain characteristics in common. They include *three* basic elements:–

1. *A Stable Reference Frequency* which lies within the gain bandwidth of the laser. This may be an "absolute" reference point, for example that provided by an atomic or molecular spectral feature such as an absorption resonance. Representative examples of such reference frequencies are the centre of the laser gain line, Lamb dip in the gain line (e.g. Soohoo *et al* 1985), inverted Lamb dips resulting from saturated absorption (Barger and Hall, 1969), the spectral lines of ion traps (Janik *et al* 1985) or laser cooled atomic beams (Persson and Svanberg, 1987)). All these spectral features may be accessed by using additional cells placed either inside or outside the laser resonator, depending on the optimum conditions for stabilisation. An alternative type of stable reference, although not one which is absolute or re–settable is provided by a spectral resonance with a high finesse external reference cavity, etalon or interferometer, maintained under strictly controlled environmental conditions.
2. *An Electronic Servo Loop* which senses the instantaneous frequency difference between the laser and the reference point, and generates a signal which has the right sign and is proportional to this difference. The precision achieved depends on the frequency width of the reference line, which translates into the slope of the frequency discriminant.
3. *A Frequency Transducer* which takes the signal from the servo loop as an input and controls the instantaneous frequency of the laser. Piezoelectric transducers are the most common devices used for laser cavity tuning. One of the laser mirrors is mounted on a piezoelectric device which elongates with applied voltage so providing a mechanical tuning of the

laser cavity length. Their two basic parameters, the elongation with voltage (typically a few $\mu$m per 100V) and the first mechanical resonance, limit the stabilisation range and speed of the associated control loop. Large tuning (although very slow) may be obtained by controlling the temperature in the resonator (for example for He–Ne lasers with internal mirrors). Somewhat faster frequency tuning may be achieved by modulation of the laser discharge current ( Hall *et al* 1987). Electro–optic modulators provide a much higher frequency tuning element and are very effective well into the megahertz range. In some systems where more than one control loop is required, it is quite common to use both a slow PZT driver and a faster electro–optic modulator (Leeb *et al* 1982, Drever *et al* 1983).

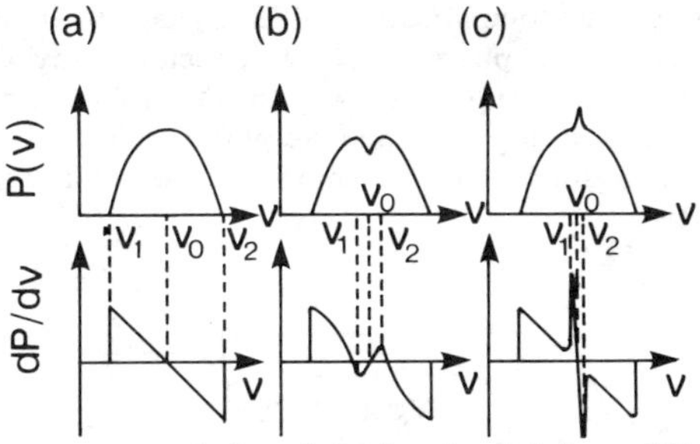


Fig 8.4 Laser output power and first derivative (needed for stabilisation) versus frequency for three important lineshapes. Discriminant region (capture range) is between $\nu_I$ and $\nu_{II}$). a) Gaussian or Lorentzian shape of laser line b) Gaussian line with "Lamb dip" in its centre c) Laser line with absorption peak power when absorption cell (external or internal) is used.

As an example, curves showing the shape of the laser power versus frequency profile for three important cases, the Lorentzian/Gaussian laser gain profile, the Lamb dip and the inverted Lamb dip are shown in Figure 8.4. Also shown in each case are the discriminant shapes (first derivatives) which are used as error signals for the feedback control system. These discriminant shapes may be obtained using the technique of synchronous detection. The precision with which the frequency may be stabilised depends on the slope $D(\nu) = d^2f(\nu)/d\nu^2$ in the zero crossing/discriminant region. Clearly, steeper slopes, corresponding to narrower spectral features give rise to better stabilisation because of the larger signal–noise ratios produced. Thus in summary, stable, very narrow and high contrast spectral features, which are resonant with the laser gain line are the best candidates for reference points in active stabilisation.

Having outlined the fundamental principles of active frequency stabilisation, three specific types of systems are briefly described in the following sections.

### 8.4.2.1 Frequency Stabilisation System Based on Electronic Synchronous Detection

A block diagram of the classic system for the frequency stabilisation of gas lasers (He–Ne or $CO_2$) in which the laser frequency is locked to the centre of the power versus frequency profile is shown in Figure 8.5.

An audio frequency modulation voltage (sometimes refered to as a *dither* signal) at some frequency between a a few hundred hertz and a few kilohertz is used to drive a piezoelectric (PZT) transducer, to which one of the cavity mirrors is firmly mounted. This causes a corresponding modulation in the cavity length, and because of the frequency dependent gain, in the laser power output. The magnitude of the modulation and the phase relative to the modulation drive signal depend on the relative position in frequency space of the cavity mode frequency and the centre of the gain line of the particular laser transition. This modulation in laser power, and its phase relative to the PZT drive signal, is monitored with the photodetector, which provides the necessary signal for phase comparison. This signal is amplified, filtered and synchronously detected in the phase sensitive detector, providing an error signal for control purposes. The loop is closed to the PZT cavity control element via a low pass filter and a high voltage amplifier. The function of the filter is to protect the loop against self oscillation on the first mechanical resonance of the PZT transducer.

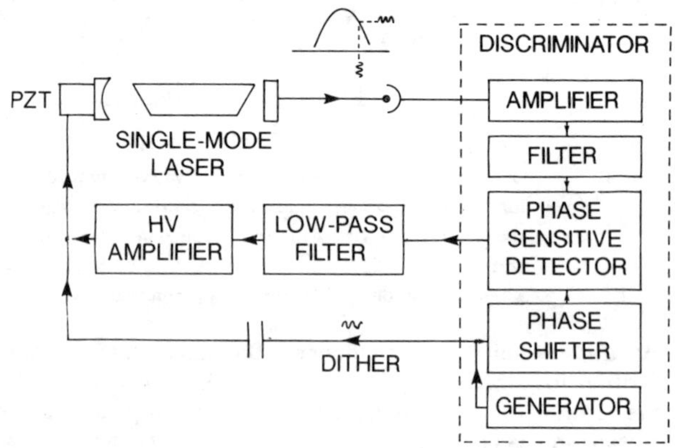


**Fig 8.5**

**Block diagram of a typical electronic control loop for the frequency stabilisation of a single-mode gas laser.**

### 8.4.2.2 Offset–Frequency Stabiliation

An inevitable consequence of systems such as that shown in Figure 8.5 is that the final stabilised output from the the laser still "carries" the frequency modulation caused by the dither voltage. For many applications, this FM of the output signal is undesirable. Fortunately, a number of ways exist to avoid the problem. One approach is to dither the reference frequency about the laser frequency rather than the other way round as in Figure 8.5. This may be achieved by using Stark effect modulation of a resonant absorber in an extra–cavity cell to derive the error signal (Hall *et al*, 1978). An alternative technique, involving the offset stabilisation, approach is shown in Figure 8.6.

The master oscillator in this system has its own independent frequency control loop with an output at a frequency $\nu_1$. This is heterodyned with the output at a frequency $\nu_2$ from the laser to be stabilised and the output from the photodetector is amplified and fed into one port of a double balanced mixer. The frequency offset $f_0$, equal to $\nu_2 - \nu_1$, can then be selected by the reference frequency generator. Offset frequency methods of this type are based on phase locking techniques (Hall *et al* 1987), or on frequency–to–voltage converters (Padgett *et al* 1988). The stabilised laser frequency produced does not suffer from the problem of associated AM and FM, and may be tuned continuously within the range of the laser gain profile. This technique is widely used in laser spectroscopy and as a technique for laser frequency synthesis.

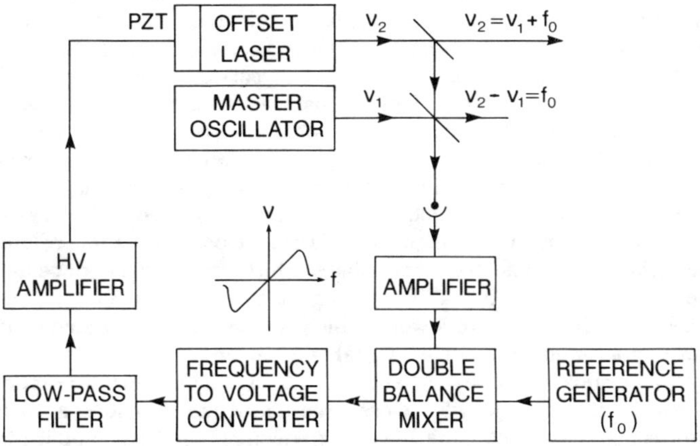


Fig 8.6 Typical block diagram of offset-frequency system with double-balance mixer and frequency to voltage convertor forming tunable frequency discriminator.

### 8.4.2.3 Laser Frequency Stablisation Using Fast Optical Discriminators

Frequency stabilisation systems which are based on electronic discrimination, such as those described above, are not fast enough to compensate for rapid frequency fluctuations (faster than perhaps a kilohertz) of the laser spectrum. However, remarkable advances in the techniques of laser frequency stabilisation have been made in the last few years with the introduction of systems based on a new very fast *optical* frequency discriminator (Drever *et al*, 1983). This new idea of "optical" stabilisation, illustrated in Figure 8.7, is called the

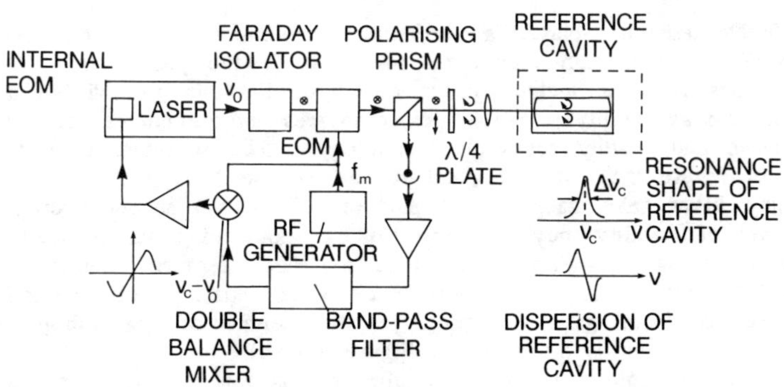


Fig 8.7 Pound–Drever method for laser frequency stabilisation.

"Pound–Drever method". The reference frequency for stabilisation is the central frequency $\nu_c$ of the resonant line of an extra–cavity, stable Fabry–Perot interferometer (the reference cavity). The beam from the laser to be stabilised, of carrier frequency $\nu_0$, is externally phase modulated by an electro–optic modulator at a frequency $f_m$ of perhaps 15 MHz, which is chosen to be larger than the few hundred kilohertz linewidth, $\Delta\nu_c$ of the passive reference cavity. The carrier and two sidebands are transmitted by the polarising beamsplitter and are incident on the coupling mirror of the reference cavity. At this point, the sidebands, which are outside the passband of the reference cavity, (plus some of the carrier signal) are reflected back towards the Thompson prism. Double passage through the quarter wave plate and the polariser directs this signal, which effectively represents the local oscillator, towards the optical detector. The other signal propagating along this path is some of the carrier frequency signal, which has "leaked out" of the reference cavity, and which is now phase shifted with respect to the incident signal. In the frequency range close to the reference cavity resonance, the magnitude of this phase shift is strongly dependent on the difference frequency $\nu_c - \nu_0$. The two signals are heterodyned on the detector, processed in the IF stage, following which phase sensitive demodulation against the electro–optic modulator drive signal $f_m$ occurs. This provides the desired frequency discriminant curve. Feedback to a second electro–optic modulator which is within the laser cavity, closes the loop and provides a high frequency correction signal to lock the laser frequency. The application of the Pound–Drever technique, which has now been demonstrated for all the important laser types, is having a very significant impact on the whole field of laser frequency stabilisation.

## 8.5 REPRESENTATIVE FREQUENCY STABILISATION SCHEMES

In this section, we present selected information on frequency stabilisation schemes as they apply to four types of lasers, He–Ne and $CO_2$ (gas lasers), semiconductor lasers and solid–state lasers.

### 8.5.1 He–Ne Lasers

The He–Ne laser has played a particularly prominent role in the development of frequency stabilisation techniques for most of the past twenty five years. The reasons for this include its spectrum of lines in the visible and near infrared, the availability of stable power sources and discharges, as well as the inexpensive and readily accessible technology. The simplest active technique, shown in Figure 8.5, when applied to the He–Ne laser with a few MHz of frequency dither gives stability of 1 part in $10^8$ for a 1 second averaging time. This level of stability may be improved by about an order of magnitude if, instead of using the gain line centre as the reference frequency, this is provided by the Lamb dip (about 15–20 MHz wide). This spectral feature which results from gain saturation at the centre of the inhomogeneously broadened gain line, is illustrated in Figure 8.4 (b).

The best absolute frequency stability results for He–Ne lasers have been achieved with saturable absorbers. Using the visible, 0.63$\mu$m laser transition with an intra-cavity absorption cell, Wallard (1972) used the hyperfine absorption components in Iodine 127 to produce very narrow inverted Lamb dips (see Figure 8.4) and stabilised frequencies to 1 part in $10^{11}$ for 1 second integration times. Perhaps the best candidate for a laser frequency standard is the He–Ne laser 3.39$\mu$m line stabilised with the help of an intracavity methane cell (Barger and Hall, 1969), as illustrated in Figure 8.8. The absorption

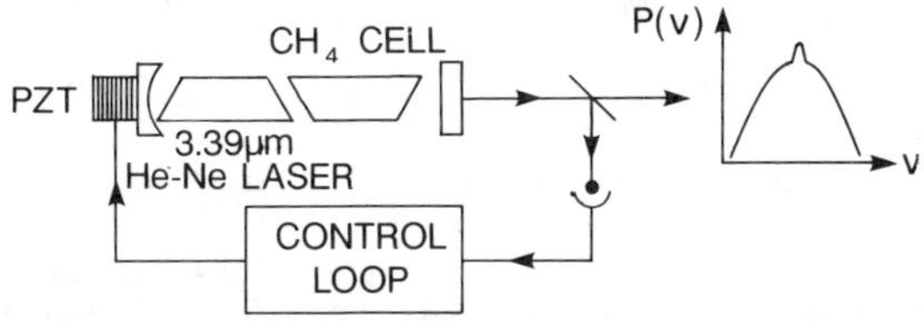


Fig 8.8 He–Ne 3.39$\mu$m laser frequency stabilised by saturation absorption in methane cell

line of the $CH_4$ molecule coincides almost exactly with the $Ne^{22}$ isotope, and a high contrast inverted Lamb dip peak may be achieved, with a linewidth as narrow as 10 kHz. Frequency stability may reach 1 part in $10^{13}$ for 1 second averaging times (Letokhov and Chebotayev, 1977)

Recently as described earlier, the Pound–Drever method has been used (Salomon *et al*, 1988) to narrow the linewidth of cw He–Ne laser to a point approach the Schawlow–Townes intrinsic limit defined by equation (8.3).

### 8.5.2 $CO_2$ Lasers

There are a number of applications of $CO_2$ lasers for laser radar and high resolution absorption spectroscopy which require a high degree of frequency stability. The stabilisation techniques appropriate for $CO_2$ lasers are very similar to these described for He–Ne lasers. A simple dc laser tube designed with a mechanically stable resonator incorporating a grating for vibrational line selection and employing the classic gain line centre synchronous-detection stabilisation technique (Figure 8.5) can achieve a stability of 1 part in $10^9$ for 1 second averaging times (Freed, 1969). Much better stability may be obtained ($10^{-12}$/1sec) when the extracavity saturable absorption method is applied, for example using $SF_6$ (Van Lerberghe, 1978), and recently Freed (1988) has reported heterodyne signals with a linewidth of less than $10^{-5}$ Hz.

### 8.5.3 Semiconductor Lasers

The requirements for noise reduction and frequency stabilisation of semiconductor lasers are stimulated mainly by the requirements of coherent optical communication systems (see also Chapter 14). Current investigations are generally concentrated on efficient line narrowing of diode lasers, which may be achieved by coupling the diode to a high Q passive resonator. For example, the use of an external grating reflector spaced about 10–20cm from the laser gain region, as illustrated in Figure 8.9, leads to a reduction in laser linewidth to a few tens of kHz. Although the coupling of the diode laser gain medium to external optics such as mirrors, diffraction gratings, Fabry–Perot interferometers (Li and Telle, 1989), produces significant

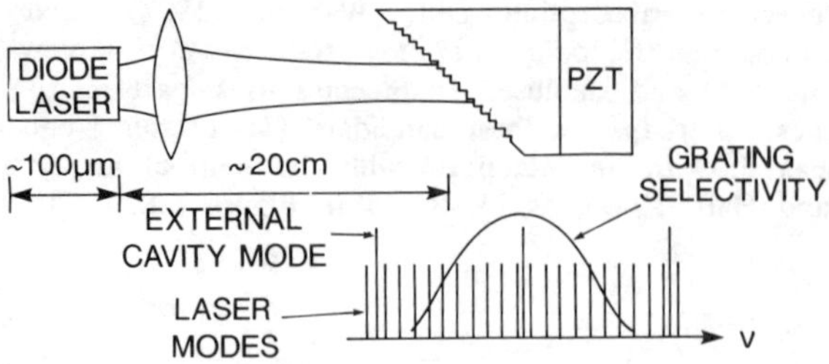


Fig 8.9 Diode laser with external grating cavity with schematic of axial mode structure.

improvements in frequency control, it is the utilisation of additional fast control loops (electronic or optical), taking advantage of the high frequency response capability of diode lasers, coupled with the use of ultra low expansion cavities or saturable absorbers as frequency references, which is perhaps the most important feature in improving frequency stability. Frequency stabilities of the order of $10^{-10}$ for 1 second have been demonstrated for semiconductor lasers using these techniques (Suzuki and Yamaguchi, 1988).

## 8.5.4 Solid State Lasers

The recent developments of all solid-state lasers (see Chapter 15) in which media such as neodymium YAG have been optically pumped by the new generation of high power diode lasers has made these devices very attractive as tunable, frequency stable and efficient optical sources (Tso Yee Fan and Byer 1988). The selective and very efficient pumping of solid-state lasers goes a long way towards eliminating the major source of frequency noise, namely thermal instabilities. The linewidths of a free-running diode end-pumped solid-state laser may be less than 50kHz and it is comparable with linewidths of gas lasers.

Many techniques of frequency stabilisation may be applied to this laser system. Using the Pound-Drever method of stabilisation and an out-of-plane ring resonator (see Section 4.4.2), Byer and his coworkers (Nilsson et al, 1989) have obtained linewidths less than 30Hz for Nd:YAG diode-pumped laser. Further improvements can be expected for these lasers as the technology matures.

## 8.6 CONCLUSIONS

The determination of oscillator frequency represents one of mankind's most accurate measurement techniques, and its measurement and control in the optical domain has been central to many important advances. Thus, for the past twenty five years, the development of increasingly more effective techniques for the stabilisation of lasers has been an extremely important and fruitful technology, allowing the exploitation of laser sources in many diverse areas of science and engineering. Recent improvements involving optical discriminators and the realisation of ultra narrow frequency reference lines, from trapped-ion resonances and atomic beam cooling, indicate that the trend is set to continue, so that in the future, lasers with stability at the milli–hertz level, approaching the limits set by intrinsic photon generation noise, will be available. These optical sources of such extraordinary stability and spectral purity will constitute key components both for new experimental tests of fundamental physical principles, and for applications in communications and engineering measurement systems.

## 8.7 REFERENCES


Allan D V 1966 Proc. IEEE 54 2 221–30

Barger R L, Hall J L 1969 Phys. Rev. Lett. 22 4–8

Barnes J A *et al* 1971 IEEE Trans. Inst. and Meas. IM–20 2 105–20

Cutler L S and Searle C L 1966 Proc. IEEE 54 2 136–54

Drever R W P *et al* 1981 Laser Spectroscopy V (Berlin, Heidelberg, New York: Springer) 33–40

Drever R W P, Hall J L, Kowalski F V, Hough J, Ford G M, Munley A J and Ward H 1983 Appl. Phys. B 31 97–105

Faller J E, Bender P L, Hall J L, Hils D and Vincent M A 1985 Kilometric optical arrays in space (Paris: European Space Agency) 157

Freed Ch 1988 SPIE vol 894 Gas Laser Technology 86–95

Hall J L, Sheng M L and Kramer G 1987 IEEE J. Quantum Electron. QE–23 4 427–37

Hall D R, Jenkins R M and Gorton E K 1978 J. Phys. D: Appl. Phys. 11 859–69

Janic G, Nagourney W and Dehmett H 1985 J. Opt. Soc. Am. B 2 1251–7

Kuo Chien–Yu and van der Ziel J P 1986 Appl. Phys. Lett. 48 14 885–7

Leeb W R, Philipp H K, Scholtz A L and Bonek E 1982 Appl. Phys. Lett. 41 592–4

Letokhov V S and Chebotayev V P 1977 Nonlinear laser spectroscopy (Berlin, Heidelberg, New York: Springer Verlag)

Li H and Telle H R 1989 IEEE J. Quantum Electron QE–25 3 257–64

Li H and Telle H R 1989 IEEE J. Quantum Electron. QE–25 257–264

Nilsson A C, Day T, Farines A D, Gustafson E K and Byer R L 1989 Paper WDD27 Quantum Electronics and Laser Science Conference, Baltimore

Padgett M J, Bett N and Butcher R J 1988 J. Phys. E: Sci. Inst. 21 554–57

Persson W and Svanberg S 1987 Laser spectroscopy VIII (Berlin: Springer Verlag)

Schawlow A L and Townes C H 1958 Phys. Rev. 112 1940–49

Shimuzu F, 1969 Appl. Phys. Lett. 14 378

Smith P W 1972 Proc. IEEE 60 422–40

Salomon Ch, Hils D and Hall J L 1988 J. Opt. Soc. Am. B 5 1576–87

Soohoo K L, Freed C, Thomas J E and Hans H A 1985 IEEE J. Quantum Electron. QE–21 1159–71

Suzuki M and Yamaguchi S IEEE J. Quantum Electron. QE–24 2392–2399

Tso Yee Fan and Byer R L 1988 IEEE J Quantum Electron. QE–24 895–909

Van Lerberghe A, Avrillier S and Borde C J 1978 IEEE J. Quantum Electron. QE–19 471–486

Vaughan J M, Keeler R J, Serafin R J, Schwiesow R L, Lenschow D H and Woodfield A A 1987 J. Atmospheric and Oceanic Technology 4 113–127

Wallard A J 1972 J. Phys. E: Scientific Instrum. 5 926–930

# 9 Propagation of Multimode Laser Beams - The $M^2$ Factor

*M W SASNETT*

## 9.1 INTRODUCTION

Gaussian beam theory was introduced in Chapter 1 and developed to give equations governing the variation with propagation distance of the beam radius (Equation 1.27) and the wavefront radius of curvature (Equation 1.28) for the fundamental, $TEM_{00}$ mode. However, many lasers produce beams that comprise modes of higher order than the $TEM_{00}$ mode. In fact most lasers, data sheet claims notwithstanding, do not produce beams that are pure $TEM_{00}$, and consequently equations 1.27 and 1.28 do not adequately describe how these beams propagate. The properties of the higher order laser modes are discussed in the literature (See references at the end of this chapter). The *objective* of this chapter is to extend earlier work by developing propagation equations for higher order modes and mixtures of modes, which are adequate in terms of accuracy and yet retain the simple utility of the fundamental Gaussian mode beam equations.

An early paper by Marshall introduced a simple way of quantifying beam quality by the use of a single number which neither depends on, nor offers any detailed knowledge of specific high-order modes that might make up a real laser beam (Marshall, 1971). This method has much in common with the familiar practice of specifying "number of times diffraction limit" (Siegman, 1986). It also invites inquiry into the possiblity of having a set of relationships similar to those governing Gaussian beam propagation, but generalized to make use of a *beam quality number* which takes account of the less than perfect propagation characteristics.

## 9.2 HIGH-ORDER BEAMS

The field amplitude distribution for high-order beams, $U_{mn}(x,y,z)$, can be written as the product of the fundamental, Gaussian beam field amplitude$^{\oplus}$, $u(x,y,z)$ introduced in Chapter 1 (Equation 1.18) and the functions $F_{mn}$ that provide radial and azimuthal modulation of that beam (Kogelnik and Li, 1966 and Siegman, 1986):

$$U_{mn}(r,z) = F_{mn}(r/w)\ u(r,z) \tag{9.1}$$

(This equation introduces the convention of using upper case letters to represent the attributes of high-order modes that for Gaussian beams are expressed in lower case). The functions $F_{mn}(r/w)$ are either Laguerre polynomials for circularly symmetric geometry or Hermite polynomials for rectangular geometry, and they depend on the same normalized beam radius, $(r/w)$, introduced in writing Equation 1.23. Figure 9.1 shows the six lowest

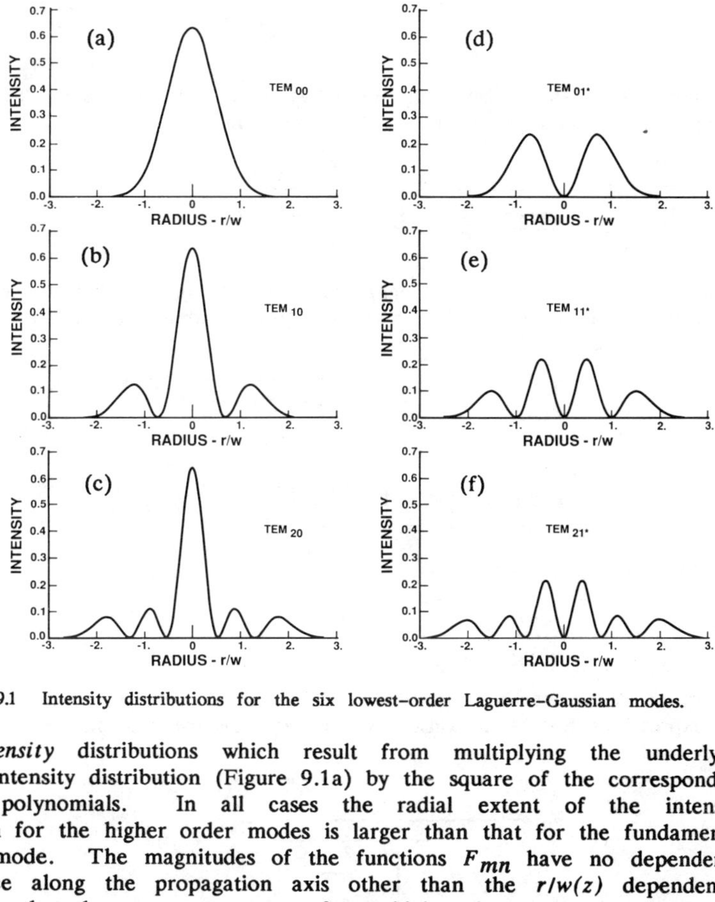


Fig 9.1 Intensity distributions for the six lowest-order Laguerre–Gaussian modes.

order *intensity* distributions which result from multiplying the underlying Gaussian intensity distribution (Figure 9.1a) by the square of the corresponding Laguerre polynomials. In all cases the radial extent of the intensity distribution for the higher order modes is larger than that for the fundamental Gaussian mode. The magnitudes of the functions $F_{mn}$ have no dependence on distance along the propagation axis other than the $r/w(z)$ dependence. This means that the transverse extent of any high-order mode, expressed here as $W(z)$, is everywhere larger than the underlying Gaussian beam by a constant factor, $M$:

$$W(z) = M\, w(z) \tag{9.2}$$

Making use of (9.2) to substitute for $w(z)$ in Equation 1.22, gives an expression for the fundamental mode complex radius of curvature in terms of the measureable properties of the overlying high order mode; $M^2$ and $W(z)$:

$$\frac{1}{q(z)} = \frac{1}{R} - \frac{jM^2\lambda}{\pi W^2(z)} \tag{9.3}$$

As before, $$q(z) = q_0 + z \tag{9.4}$$

and $$q_0 = \frac{j\pi W_0{}^2}{M^2\lambda} \tag{9.5}$$

Inverting (9.4) and substituting in (9.3) gives an expression analogous to Equation (1.26) in terms of the higher–order mode radius:

$$\frac{1}{\dfrac{j\pi W_0{}^2}{M^2\lambda} + z} = \frac{1}{R(z)} - \frac{jM^2\lambda}{\pi W^2(z)} \tag{9.6}$$

Equating the real and imaginary parts of (9.6), as before, now yields expressions for beam radius and wavefront radius of curvature for the high–order mode:

$$W(z) = W_0\sqrt{1 + \left[\frac{M^2\lambda z}{\pi W_0{}^2}\right]^2} \tag{9.7}$$

$$R(z) = z\left[1 + \left[\frac{\pi W_0{}^2}{M^2\lambda z}\right]^2\right] \tag{9.8}$$

The identity:

$$\frac{\pi W^2}{M^2\lambda R} = \frac{M^2\lambda z}{\pi W_0{}^2} \tag{9.9}$$

is easy to derive starting with (9.4). Then by substitution in (9.5) and (9.6), the expressions for beam waist diameter and distance to the beam waist may be derived:

$$W_0 = \frac{W(z)}{\sqrt{1 + \left[\dfrac{\pi W^2(z)}{M^2\lambda R}\right]}} \tag{9.10}$$

$$Z_0 = \frac{R(z)}{1 + \left[\dfrac{M^2\lambda R(z)}{\pi W^2}\right]^2} \tag{9.11}$$

Equations (9.7), (9.8), (9.10) and (9.11) are the propagation equations which allow mathematical treatment of the propagation of high–order modes and mixtures of modes.

## 9.3 A NUMERICAL EVALUATION OF BEAM QUALITY

It can readily be seen that Equations (9.7) through (9.11) have the same form as their fundamental Gaussian beam counterparts except for the quantity $M^2$ which multiplies the wavelength. For $M^2 = 1$, they reduce to the fundamental Gaussian beam equations. The quantity $M^2$ is then a numerical expression of (inverse) beam quality with 1 being a perfect Gaussian beam, and higher values indicating "poorer" quality. As will be shown, it is this number which expresses how many times larger the diameter of a focused beam is, compared with the focus-diameter for a pure fundamental Gaussian beam. Equivalently, $M^2$ is the factor by which the far-field Gaussian beam divergence angle for a given beam waist diameter is multiplied to give the actual far-field divergence angle. As such, it is entirely equivalent to the "number of times diffraction limit" quantity (Siegman, 1986). This beam quality number does not by itself give any information about the details of high-order mode content in the beam, but correspondingly neither is any detailed knowledge about mode content required to determine its value. This opens the possibility that relatively simple measurements can yield a number which has meaningful utility in propagation calculations *for beams of any mixture of modes.* Also, having a generalized complex radius of curvature means that matrix methods can be used to calculate the propagation of beams with any mix of modes, through optical systems that can be characterized by ray transfer (*ABCD*) matrices. The calculations can be represented on circle diagrams, (ie. Collins charts) (Kogelnik and Li, 1966), or treated with existing Gaussian beam computer programs after substituting $M^2\lambda$ for $\lambda$.

## 9.4 EXPERIMENTAL MEASUREMENTS

A vital issue which arises when considering actual measurements, is the question about what constitutes the *"diameter"* of an intensity distribution profile made up of an unknown mixture of modes, and having decidedly sloping sides. It has been suggested that an unassailable determination of the diameter of a cotton ball using vernier callipers is a task of comparable difficulty! The intensity distributions shown in Figure 9.1 are those of the lowest orders of the pure, unmixed, Laguerre–Gaussian modes. Beams from real lasers rarely consist of single, pure modes. The pure modes which make up a mixed-mode inside a laser cavity each oscillate at an optical frequency which is different from the others by a substantial fraction of the cavity axial mode separation, typically tens of megahertz or more. Ordinary means for measuring intensity distributions of the scanning-pinhole type are not capable of sorting out the individual contributions of each pure mode. What is registered is the incoherent sum of the individual intensities.

It is interesting to note that the set of Laguerre–Gaussian or Hermite–Gaussian functions which describe the field amplitude and phase for each pure mode, form an orthogonal set. Therefore, if one could measure the distribution of amplitude and phase across a mixed-mode laser beam, it would be possible to perform an integral transform that would yield the amplitude of each pure mode making up that distribution. However, the set of functions which describe the *intensities* of these modes, are not an orthogonal set. Consequently, it is not possible to recover exact information about the composition of a mixed-mode beam from its intensity distribution. One need look no further than this fact to recognise that claims in laser data sheets about mode content which are based on a measurement of intensity distribution at one location in front of the laser, are to be used with caution. Figure 9.2 illustrates this point. Part (a) shows the intensity distribution for a $TEM_{00}$ mode, and part (b) shows a mixture of the first three modes shown in Figure 9.1. Only half of the total power is contributed by the $TEM_{00}$ mode. The $TEM_{01*}$ and $TEM_{10}$ modes contribute 30% and 20% respectively, yet the intensity distribution in B would be difficult to distinguish from the one above it in the absense of relative intensity and radius scales.

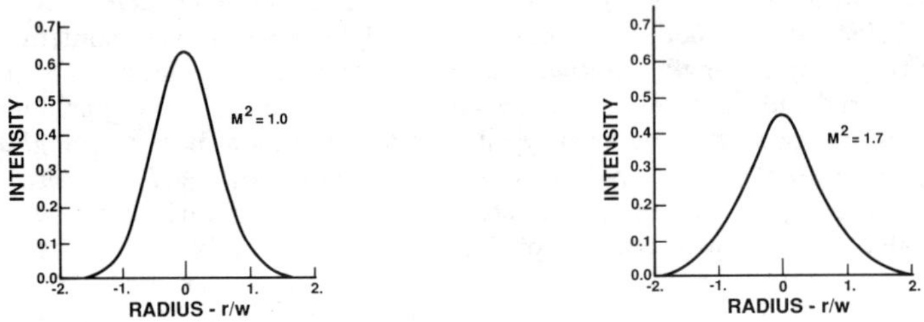


(a) (b)

Fig 9.2 a) Intensity distributions for a $TEM_{00}$ mode b) Intensity distribution for a mixture of modes.

Several methods for computing characteristic spot radius and corresponding $M^2$ values for laser modes have been studied. These include;

- finding the radius at which the intensity falls to $1/e^2$ of some peak intensity,
- finding the radius of a circular aperture that passes all but $1/e^2$ of the total power in the beam,
- computing radius from the second moment of the intensity distribution.

Results of such calculations for the Laguerre–Gaussian modes shown in Figure 9.1 are summarized in Table 9.1. The use of the technique which involves the measurement of $1/e^2$ of the highest peak intensity inadequately characterizes the higher order modes when the outer peaks begin to fall below this level. The $TEM_{20}$ mode is an example. Measurement of $1/e^2$ of the outermost peak might seem to eliminate this problem. However, it is easy to create mixtures of two or more modes wherein the outer peaks are eliminated, thereby causing anomalous changes in the mode spot radius arrived at by this definition. Methods which involve integration of the intensity distribution over the beam area solve these problems. This is achieved by defining beam radius as the radius of an aperture that transmits some standard fraction of the total power. Another such method involves in computation of the second moment $\overline{r^2}$ of the intensity distribution $I(r)$, defined by equation 9.12:

| Mode Designation | | Radius based on second moment of $I(r)$ | Radius passing 86.5% of total power | Radius at $I=I/e^{-2}$ of outermost peak | Radius at $I=I/e^{-2}$ of highest peak |
|---|---|---|---|---|---|
| $TEM_{00}$ | radius: | 1 $w$ | 1 $w$ | 1 $w$ | 1 $w$ |
| | $M^2$: | 1 | 1 | 1 | 1 |
| $TEM_{01*}$ | radius: | $\surd 2$ $w$ | 1.32 $w$ | 1.50 $w$ | 1.50 $w$ |
| | $M^2$: | 2 | 1.75 | 2.25 | 2.25 |
| $TEM_{10}$ | radius: | $\surd 3$ $w$ | 1.64 $w$ | 1.91 $w$ | 1.50 $w$ |
| | $M^2$: | 3 | 2.70 | 3.65 | 2.26 |
| $TEM_{11*}$ | radius: | 2 $w$ | 1.88 $w$ | 2.17 $w$ | 2.01 $w$ |
| | $M^2$: | 4 | 3.54 | 4.71 | 4.03 |
| $TEM_{20}$ | radius: | $\surd 5$ $w$ | 2.12 $w$ | 2.43 $w$ | 1.00 $w$ |
| | $M^2$: | 5 | 4.48 | 5.88 | 1.00 |
| $TEM_{21*}$ | radius: | $\surd 6$ $w$ | 2.31 $w$ | 2.63 $w$ | 2.40 $w$ |
| | $M^2$: | 6 | 5.35 | 6.92 | 5.76 |

Table 9.1 Mode radius and $M^2$ values for the fundamental and higher-order Gaussian modes. $w$ = radius at $I = I_{peak}/e^2$ for the underlying fundamental ($TEM_{00}$) mode. Radius based on second moment (Equation 9.14).

$$\overline{r^2} = \int_0^\infty r^2 I(r)\, dA \qquad (9.12)$$

where I(r) is normalized $\int_0^\infty I(r)dA = 1$

and the area, $A$, over which the integration is performed is centred on the beam axis. For modes with circular symmetry this becomes:

$$\overline{r^2} = 2\pi \int_{r=0}^\infty r^3 I(r)dr \qquad (9.13)$$

The beam radius, $W$, based on the second moment of the intensity distribution is defined here as:

$$W = \sqrt{2\overline{r^2}} \tag{9.14}$$

This definition provides a result for the case of a pure $TEM_{00}$ mode, which is consistent with the standard definition of this radius as the radial distance at which the intensity falls to $1/e^2$ of the peak intensity.

Values for mode radius and $M^2$ which result from this definition for the six modes of Figure 9.1 are shown in Table 9.1. Circular symmetry has been assumed. The integer values of $M^2$ are an interesting result. At the very least, these are the easy values to remember! There are two further recommendations for basing the radius definition on the second moment. First, for the pure modes, these values fall midway in the range of values resulting from the other methods. Secondly, it is easy to show that the diameter computed in this way for mixed-mode intensity distributions is just equal to the power-weighted sum of $M^2$ values of the component pure modes. This is not true of the other methods mentioned above.

While the second moment method of diameter determination has much to recommend it, the beam diameter experimental data shown in this section was actually obtained by the method of measuring diameters at the $1/e^2$ of peak intensity level. This choice was expedient at the time and has proven justifiable as the two methods give nearly equal results for mixed-mode intensity distributions that have near-Gaussian shape.

A simple test of the generalized propagation equations has been to measure beam diameters at several distances in front of various lasers and to discover that by choosing suitable values for $M^2$, Equation (9.7) provides an excellent fit to the actual data in each case. Figure 9.3 shows an example

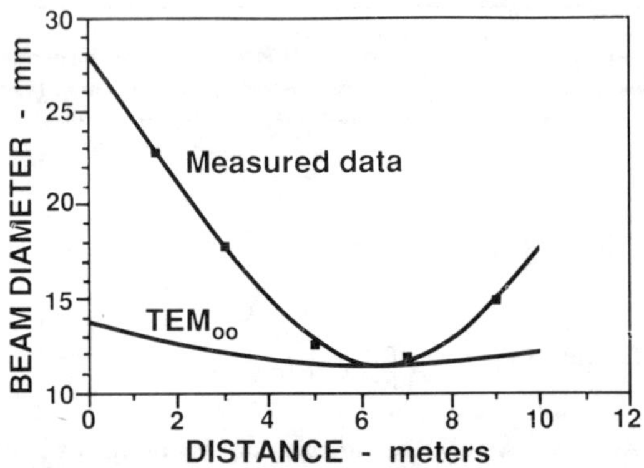


Fig 9.3 Beam diameter vs distance for a 750 watt $CO_2$ laser with $TEM_{00}$ beam diameter vs distance shown for comparison.

for the case of a commercial 750 watt industrial $CO_2$ laser. $M^2 = 3.4$ provides the fit to the measured data, as shown. Also shown in Figure 9.3 is a plot of diameter versus distance for a $TEM_{00}$ beam having the same waist diameter and location. It is clear that specifying the mode quality for this laser as $TEM_{00}$ would be inaccurate and misleading.

A further test of this concept was conducted using a second commercially available $CO_2$ laser operating at 1250 watts output power. Measurements of the beam properties made with no optical elements in the beam after the output coupler yielded the beam waist location, diameter and $M^2$. Next, a reflective telescope consisting of two spherical, concave mirrors, labeled $R_1$ and

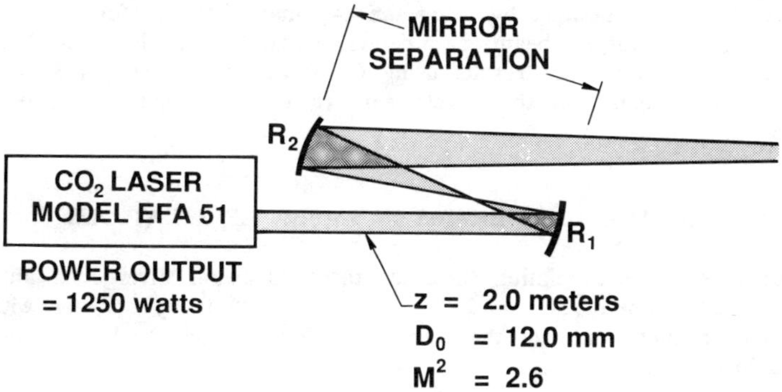


Fig 9.4 Diagram showing laser and reflective telescope.

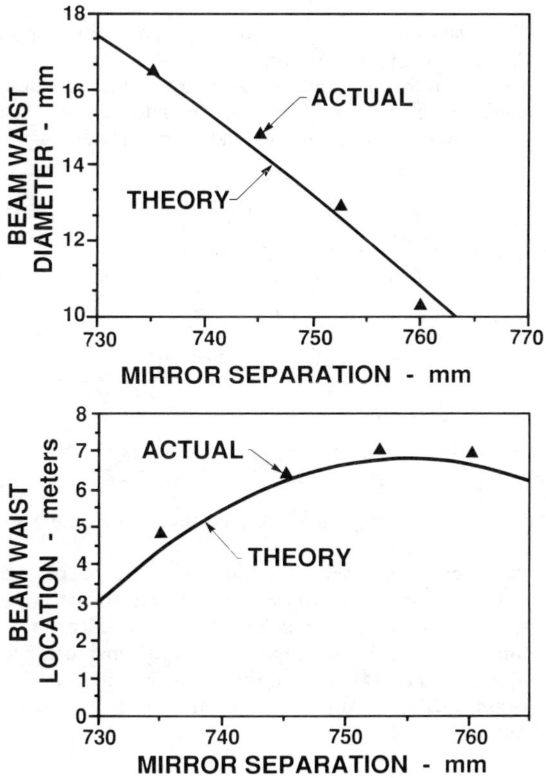


a) Beam waist diameter vs telescope mirror separation.

b) Beam waist location vs telescope mirror separation.

Fig 9.5

$R_2$ in Figure 9.4, with radii of curvature 600mm and 900mm respectively was constructed and mounted in front of the laser. The included angle between the input beam centreline and the beam centreline between $R_1$ and $R_2$ was $6^0$. Measurements of beam waist location and diameter were again taken for several different settings of the distance between the telescope mirrors. These data are plotted in Figure 9.5. Also shown in these plots are theoretically predicted curves obtained by using the ray matrix model for the telescope and predicting the output beam parameters according to the *ABCD* law. Any attempt to predict these results using Gaussian beam theory ($M^2 = 1$) fails as soon as it is seen that the input beam to the telescope cannot be adequately modelled.

## 9.5 HIGH-ORDER BEAMS IN LASER APPLICATIONS

A useful expression relating the full, far-field beam divergence angle, $\theta$, and beam waist diameter, $D_0 = 2W_0$, may be written by starting with Equation 9.7 and noting that in the far-field, $M^2\lambda z/\pi W_0^2 \gg 1$ so that to close approximation,

$$D_0\,\theta = \frac{4M^2\lambda}{\pi} \qquad (9.15)$$

This shows the price to be paid in larger focus-spot diameter and/or diffraction angle for beams of poor quality. For example, a lens of given focal length, $f$, and a beam of given diameter, $D$, at this lens produce a cone angle $\theta = D/f$. The diameter at the focus, $D_0$, then varies directly in proportion to $M^2$, and as stated previously, $M^2$ is just the familiar

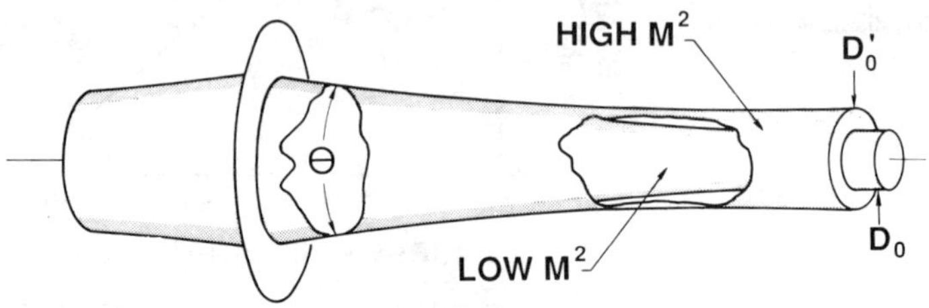


$$D_0 = \frac{4M^2\lambda}{\pi\theta}$$

Fig 9.6 Effect of beam quality ($M^2$) on focus-spot diameter.

"number-of-times-diffraction-limit" figure of merit. Figure 9.6 illustrates this point. For laser applications where it is important to focus the beam to a small diameter to achieve high intensity, a low value of $M^2$ is essential. Since the peak focus-spot intensity increases with the inverse square of spot diameter ($1/M^4$) and only linearly with total power, it is likely that better beam quality from a given laser will be worth some sacrifice in total power output.

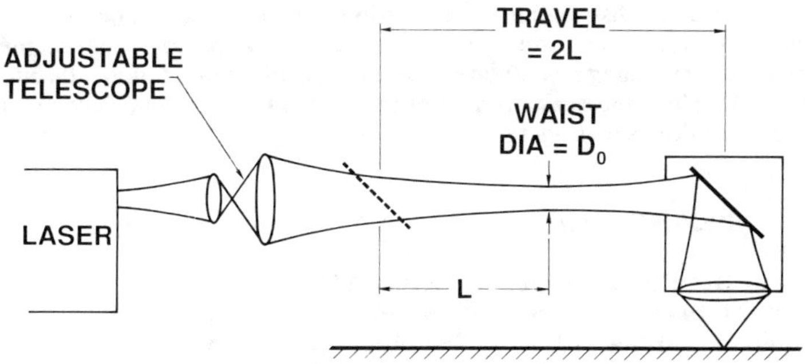


$$D_0 = 2\sqrt{\frac{M^2 \lambda L}{\pi}}$$

Fig 9.7 Gantry-type beam delivery system.

Often, laser applications require delivery of the beam over long and variable length optical paths. An example, shown in Figure 9.7, is a gantry beam delivery system where the laser remains stationary and the final beam-bender/lens assembly may move over a range of a few meters to 20 meters or more. The most economical design for such an optical system uses optical components of the smallest possible diameter consistent with allowing the entire beam to pass without significant aperture loss. This in turn requires that a beam waist of optimum diameter be positioned in the middle of the variable portion of the optical path, at a distance *L* from each end of the travel. This optimum waist diameter is determined by setting the first derivative of $W(z)$ with respect to $W_0$ in Equation (9.7) equal to zero, letting $z = L$ and solving for the value of beam waist diameter, $W_0$, that mimimizes the beam diameter at the ends of the variable part of the path.

$$D_{0\ opt} = 2\ W_{0\ opt} = 2\sqrt{M^2 \lambda L/\pi} \tag{9.16}$$

Here, L is half the range over which the moving lens travels. Substitution of (9.13) for $W_0$ in (9.7) shows the minimum beam diameter at the extremes of the travel to be $\sqrt{2}$ times larger than the optimum beam waist diameter. Again the importance of good beam quality is seen by the decrease in diameters directly with the factor *M*.

## 9.6 SUMMARY AND CONCLUSION

A set of beam propagation equations are derived that permit beam diameter calculations to be made for beams made up of high-order modes or mixtures of modes. These equations are analogous to the Gaussian beam equations, but include a parameter, $M^2$, which serves as a measure of beam quality and permits calculations to be performed for less than perfect beams for which detailed intensity distribution information is unknown and/or unimportant. Experimental results confirm the continued utility of ray matrix methods for mathematical modeling of non–Gaussian beam propagation through optical systems. Finally, the design of optical systems for lasers having mixed and high–order modes is considered.

## 9.7 REFERENCES

Bridges W B 1975 Appl. Opt. 14 2345–2347
Carter W H 1980 Appl. Opt. 19 1027–1029
Freiberg R J and Halsted A S 1969 Appl. Opt. 8 355–362
Kogelnik H and Li T 1966 Appl. Opt. 5 1550–1567
Luxon J T, Parker D E and Karkheck J 1984 Appl. Opt. 23 2088–2090
Marshall L April 1971 Laser Focus, 26–28
Siegman A 1986 Lasers (Mill Valley California: University Science Books)

⊕ The terms $TEM_{00}$ fundamental mode, and Gaussian beam are used interchangeably here for brevity.

# 10 Laser Beam Analysis by Geometrical Optics

*W A E GOETHALS*

## 10.1 INTRODUCTION

The paper by Kogelnik and Li (1966) describing a theory of laser optics, which is based on paraxial ray tracing (*ABCD* matrices) and the wave analysis of laser beams and resonators, is widely regarded as the standard reference on this subject. An alternative treatment, based purely on geometrical optics, is presented in this chapter. This approach relies on a formalism, derived from classical mechanics, which is called the *configuration phase space* (Goldstein, 1980). It may be used to describe the dynamics of an ensemble of particles, such as a particle beam. For example, it has proven to be an invaluable practical method in the design of charged particle optics. Because of the deep underlying similarity between geometrical optics and classical mechanics, this same formalism may be applied to the analysis of laser beams, as outlined in this chapter.

First of all, it is shown that the paraxial ray analysis in phase space leads to the same results as those given by Kogelnik and Li (1966). Then, a method is presented for the description of laser beams in phase space in combination with exact ray tracing algorithms. With this method, it is possible to trace single mode and multimode Gaussian beams, as well as beams from unstable resonators. The calculated parameters include beam waist diameter, divergence, Rayleigh range, pointing angle and beam quality, i.e. the parameters of interest for laser optics and system designers. The effects of beam quality, lens aberrations and misalignments of optical components can also be easily determined.

## 10.2 PRESENTATION OF AN OPTICAL SYSTEM IN THE CONFIGURATION PHASE SPACE

Let us assume that the z–axis of a Cartesian co–ordinate system represents the optic axis of a bundle of rays. A single ray can then be described by its deviation, $x(z)$ and $y(z)$, and its inclination angles, $x'(z)$ and $y'(z)$, in each direction transverse to the optic axis. These can be considered as the co–ordinates of the position vectors $(x,x')$ and $(y,y')$, or of one 4–dimensional position vector $(x,y,x',y')$ in the so–called *phase space*. Because the path in the x–direction can be treated in a similar way to the path in the y–direction, and because the two paths are not coupled, only the 2–dimensional $x,x'$ phase space need be considered here, in combination with exact meridional ray traces in the plane containing the x–axis and z–axis.

### 10.2.1 Paraxial Single Ray Tracing

When a ray passes through an optical system its position vector $(x,x')$ is transformed and the resulting vector describes a certain trajectory in phase space (see Figure 10.1). When the paraxial approximation is assumed ($\tan \alpha = \alpha$), these transformations are linear and can be determined using the well known *ABCD* transfer matrices (Kogelnik and Li, 1966), for the cases of free-space propagation (drift space) and transformation/focussing by a lens:

Drift space $L$: $$\begin{vmatrix} 1 & L \\ 0 & 1 \end{vmatrix} \qquad (10.1)$$

Focussing action $f$: $$\begin{vmatrix} 1 & 0 \\ -1/f & 1 \end{vmatrix} \qquad (10.2)$$

The matrix method is a very powerful tool for the initial design of optical systems, since the action of a complete system can be represented by a single matrix operating on an ensemble of input rays.

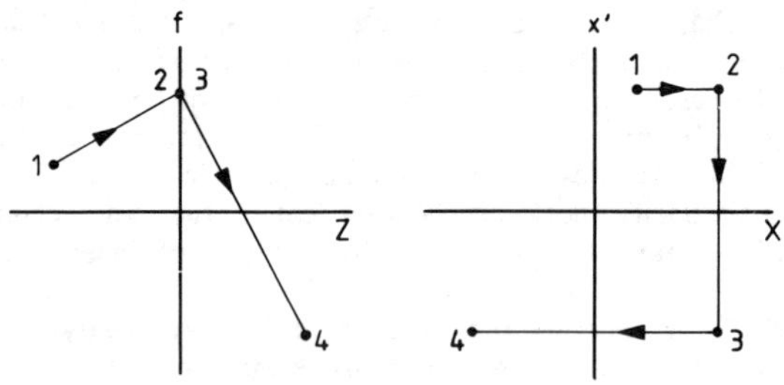


Fig 10.1 Ray trajectory in real space and its equivalent in phase space.

### 10.2.2 Ray Ensembles

Progressing from single ray trajectories, we can now investigate the propagation of ray ensembles. A simple example is a beam of parallel rays being focussed to a spot by a single lens (see Figure 10.2).

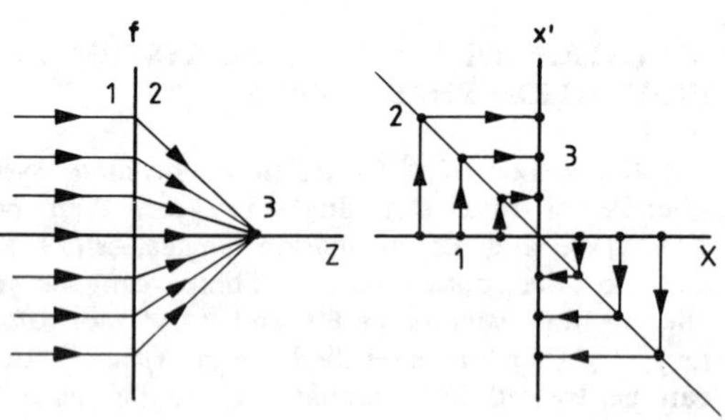


Fig 10.2 Parallel beam focussed by an ideal thin lens in real space and in phase space.

In matrix notation this is represented by the product of the matrices above (10.1 and 10.2) and the ray position vector, as follows:

$$\begin{vmatrix} 0 & f \\ -1/f & 1-L/f \end{vmatrix} \cdot \begin{vmatrix} x \\ 0 \end{vmatrix} = \begin{vmatrix} 0 \\ -x/f \end{vmatrix} \qquad (10.3)$$

Equation (10.3) shows that the parallel beam is being focussed to an infinitely small spot at the focal point of the ideal thin lens.

With exact ray tracing, the effects of spherical aberrations introduced by the shape and the orientation of the lens can be shown (see Figure 10.3),

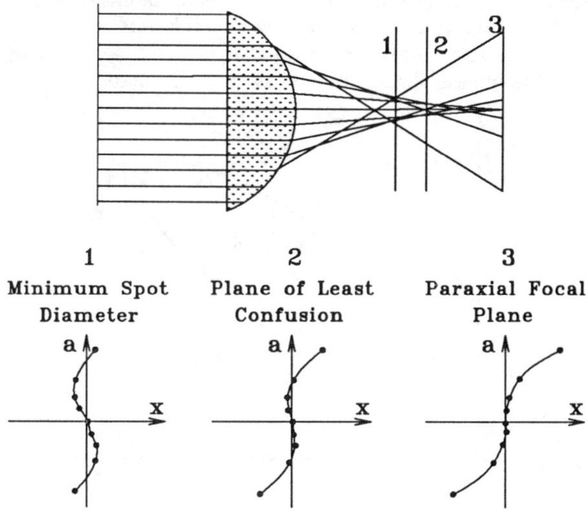


Fig 10.3 Parallel rays focussed by a lens in real space and in phase space.

indicating that the phase space respresentation gives a good insight into the actual optical performance of the lens.

Thus, when the performance of a real optical system is to be determined in terms of aberrations, alignment sensitivity etc., exact ray tracing, based on Snell's law and the precise shape and properties of the optical elements, must be applied. For this purpose, several computer programs for the analysis and optimization of optical systems are available. These programs however do not always offer the possibility of phase space representation.

A parallel light beam only exists in theory and an exact ray tracing analysis of an optical system is as good or as bad as the description of the ray ensemble of the actual light source. It is therefore important to know the phase space distribution of real light sources, such as laser beams. Knowledge of this distribution, and its motion in phase space, enables many parameter values to be derived, such as, beam diameter, divergence, offset from the optic axis and pointing angle (see Figure 10.4).

### 10.2.3 Liouville's Theorem

Perhaps the most interesting property of the motion of a beam in phase space is described by Liouville's theorem. According to this theorem, an ensemble or beam of rays fills a certain volume in phase space, which may change its shape at further positions $z$, but not the magnitude of its volume. Any attempt to reduce the size of this volume by optical elements is in vain (athough it may often be desirable!). In two dimensions this implies that the area of the beam in the $x,x'$ plane remains constant. This theorem agrees with the fact that the determinant of any *ABCD* matrix equals unity: $AD - BC = 1$, as stated by Kogelnik and Li (1966). Note that this theorem has general validity: it is not restricted to paraxial systems. The magnitude of the phase space volume is also known as the ***beam quality***, and the energy of the beam divided by this volume as the ***beam brightness***.

The analaysis of laser beam propagation involves an investigation of the transformation of the phase space volume as a function of $z$.

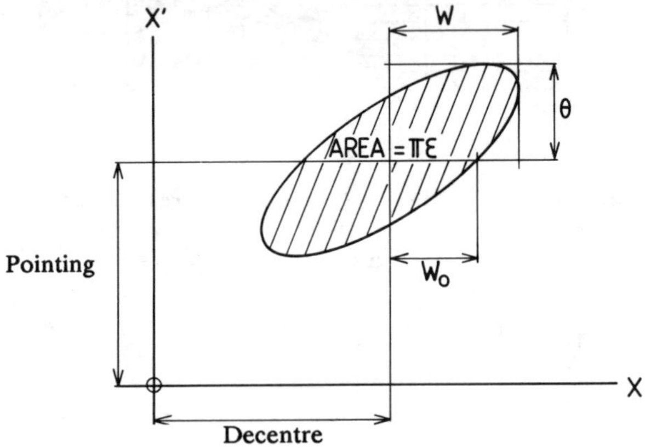


Fig 10.4 Parameters of a (TEM) beam in phase space.

## 10.3 PARAXIAL OPTICS OF GAUSSIAN AND HIGHER ORDER LASER BEAMS

From the geometrical analysis of stable resonators, it appears that the $x,x'$ phase space distribution of Gaussian and higher order mode laser beams covers an area with an elliptical envelope (Schildbach, 1988), as shown in Figure 10.4. The dimensions of the ellipse ($w$, $w_0$ and $\theta$) are determined by the *ABCD* matrix of the resonator (Kogelnik and Li, 1966) and by the mode diameter allowed by the intracavity aperture size.

The theory of ellipses in phase space is well known in the area of charged particle optics and is used in the design of beam guiding systems for particle accelerators (Wollnik, 1987).

It may be shown that the description of a single mode Gaussian laser beam is given by a classical Gaussian ray ensemble having an elliptical envelope in the $x,x'$ plane which is determined by the $e^{-2}$ values of the brightness, and with an area equal to the wavelength $\lambda$ (Verster, 1983). This thesis is applied here, and expanded to include higher order, multimode and unstable resonator beams. Again, there are two ways of studying the transformation of the beam, using matrix methods and exact ray tracing.

### 10.3.1 Paraxial Beam Tracing

A major advantage of the matrix method is that the transformations can be performed at once on the complete ray ensemble. A technique method for doing this was proposed by Kogelnik & Li (1966), using the complex beam parameter $q$ and the $ABCD$–law for transformation, (see also 1.38) defined as:

$$\frac{1}{q} = \frac{1}{R} - \frac{j\lambda}{\pi w^2} \tag{10.4}$$

$$q_2 = \frac{Aq_1 + B}{Cq_1 + D} \tag{10.5}$$

where $R$ is the radius of curvature of the wavefront. However, the physical meaning of $q$ is a difficult concept to grasp for most designers. Verster (1983) has shown the relationship between phase space ellipses and the $q$ parameter, using the Lissajous description of an ellipse:

For all values of $\Phi$: $$\begin{cases} x = w.\sin(\Phi + \chi) & (-\pi/2 < \chi < \pi/2) \\ x' = \theta.\cos\Phi \end{cases} \tag{10.6}$$

Assuming that $\epsilon$ is the area of the ellipse, then $w$ and $\theta$ are given by:

$$w = \sqrt{\epsilon a} \qquad \theta = \sqrt{\frac{\epsilon}{a.\cos\chi}} \tag{10.7}$$

Moreover, it can be shown that: $$q = \frac{a}{\tan\chi + j} \tag{10.8}$$

allowing the transformation of phase space ellipses using the $ABCD$–law.

### 10.3.2 Twiss Parameters

A more direct method of transformation uses the so-called Twiss parameters $A_T$, $B_T$ and $C_T$ which characterize a phase space ellipse of area $\epsilon$ in the $x,x'$ plane (Wollnik, 1987):

$$x^2C_T + 2xx'A_T + x'^2B_T = \pi\epsilon \tag{10.9}$$

When the total $ABCD$ matrix of the optical system is known the transformation of the ellipse is given by:

$$\begin{vmatrix} B_T \\ A_T \\ C_T \end{vmatrix}_2 = \begin{vmatrix} A^2 & -2AB & B^2 \\ -AC & AD+BC & -BD \\ C^2 & -2CD & D^2 \end{vmatrix} \cdot \begin{vmatrix} B_T \\ A_T \\ C_T \end{vmatrix}_1 \tag{10.10}$$

The maximum values of $x$ (= beam radius) and $x'$ (= beam divergence) are given by:

$$x_{max} = \pm\sqrt{B_T \epsilon} = w \tag{10.11}$$

$$x'_{max} = \pm\sqrt{C_T \epsilon} = \theta \tag{10.12}$$

Calculations are usually started with an upright ellipse (= beam waist) having the values:

$$\pi\epsilon = \pi w_{01}\theta_1 \tag{10.13}$$

$$A_{T1} = 0, \quad B_{T1} = 1/C_{T1} = w_{01}/\theta_1 \tag{10.14}$$

where $w_0$ and $\theta$ are the beam waist radius and the beam divergence.

Equation (10.10) then simplifies to:

$$\epsilon B_{T2} = A^2 w_{01}^2 + B^2\theta_1^2 \tag{10.15a}$$

$$-\epsilon A_{T2} = AC w_{01}^2 + BD\theta_1^2 \tag{10.15b}$$

$$\epsilon C_{T2} = C^2 w_{01}^2 + D^2\theta_1^2 \tag{10.15c}$$

### 10.3.3 Examples

By way of example, we will calculate the envelope, $w$ of a laser beam having its waist at $z$=0. Using the matrix ($ABCD$) = (1, $z$, 0, 1) and equations (10.11) and (10.15), we find:

$$\epsilon B_T(z) = w_0^2 + z^2\theta^2$$

Hence:

$$w(z) = \pm\sqrt{w_0^2 + z^2\theta^2} \tag{10.16}$$

which is the well known hyperbola-shaped envelope.

In the same way, we can derive the formulae for the focussing of a laser beam with a lens of focal length $f$. For example, the position of the new waist after the lens ($z_2$) as a function of the position of the initial waist ($w_0$ at $z_1$) is calculated using the $ABCD$ matrix:

$$\begin{vmatrix} A & B \\ C & D \end{vmatrix} = \begin{vmatrix} 1-\frac{z_2}{f} & z_1+z_2-\frac{z_1 z_2}{f} \\ -\frac{1}{f} & 1-\frac{z_1}{f} \end{vmatrix} \tag{10.17}$$

The condition for waist-to-waist transformation (upright ellipses) applied to equation (10.15b) gives:

$$A_{T1} = A_{T2} = 0$$

so that:

$$AC.w_{01}^2 + BD.\theta_1^2 = 0$$

whereupon:

$$z_2 = f + \frac{f^2(z_1-f)}{(z_1-f)^2 + (w_{01}/\theta_1)^2} \tag{10.18}$$

Once again, this is a result which looks familiar. In the same way, the equation for the new waist size may be derived, giving:

$$w_2^2 = \frac{f_{01}^2 \; f^2}{(z_1 - f)^2 + (w_{01}/\theta_1)^2} \tag{10.19}$$

Since no aberrations are taken into account this waist size is called *'diffraction limited'*.

These equations (10.16, 10.18, 10.19) can be found in most textbooks on laser optics (e.g O'Shea, 1985), however the derivations given here differ in as much as they are based on simple geometrical principles instead of complex wavefront theory (Kogelnik and Li, 1966). Also they are not restricted to $TEM_{00}$ beams but valid for any TEM mode beam.

## 10.4 BEAM QUALITY

The area of the ellipse, scaled with the laser wavelength $\lambda$, is known as the beam quality factor and equals $\pi w_0 \theta / \lambda$.

For a $TEM_{00}$ beam it is known that:

$$\frac{\pi w_0 \theta}{\lambda} = 1 \qquad (TEM_{00}) \tag{10.20}$$

For multimode lasers, the beam quality may be described by the $C^2{}_{nm}$ factor (Nonhof, 1988 and Goethals, 1988), also called the $M^2$ factor (Chapter 9). It is derived from the relationship between higher order mode beams (Koechner, 1976/1988):

Multiplying: $$w_{nm} = C_{nm} \cdot w_{00} \tag{10.21}$$

with: $$\theta_{nm} = C_{nm} \cdot \theta_{00} \tag{10.22}$$

we have: $$w_{nm}\theta_{nm} = C^2{}_{nm} \cdot w_{00}\,\theta_{00} \tag{10.23}$$

With (10.20) this gives us: $$C^2{}_{nm} = w_{nm}\theta_{nm}\pi/\lambda \tag{10.24}$$

Which is equivalent to equation (9.3) in Chapter 9:

$$\frac{1}{q} = \frac{1}{R} + \frac{jM^2\lambda}{\pi w^2} \tag{10.25}$$

where the quantity $M^2\lambda$ is called the *'augmented wavelength'*.

The ratio $w_0/\theta$ is known as the Rayleigh range of the beam, and is independent of beam mode, as is the wavefront radius of curvature, $R$. The fact that $\pi w_0 \theta / \lambda$ (=$C^2{}_{nm}$) remains constant (Liouvilles theorem) implies thjat we may state a *"law of conservation of beam quality"*. However, Liouville's theorem is only valid if the optical system is conservative, i.e. if no rays are added or removed. When apertures remove rays, the phase space area decreases and so the beam quality may be improved.

## 10.5 EXACT BEAM TRACING

In order to determine the effects of aberrations on (for example) the laser spot size, some laser optics designers apply exact ray tracing using a parallel beam of rays, based on the idea that the divergence of the laser beam is very small anyway. The inclusion of aberrations results in a finite spotsize $S_A$, instead of the ideal infinitely small spot of equation (10.3), which is then assumed to be the laser spotsize. This is of course wrong! The correct approach is to also calculate the diffraction limited spotsize $S_D$, using the laser beam parameters, as in equation (10.10). The total spotsize $S_T$ is then estimated by:

$$S_T{}^2 = S_A{}^2 + S_D{}^2 \tag{10.26}$$

which sometimes gives reasonable results, provided that the beam waist and the focal plane coincide.

For a complete and accurate analysis however, it is necessary to apply the exact beam tracing method. This requires separate tracing of the rays that determine the phase space ellipse. The effects of the optics on the beam parameters can then be calculated easily from the position vector coordinates of the final ray ensemble. These types of calculations however, are not included in most of the existing optical design programs. Nor do these programs allow the possibily of automatically setting up a phase space ray ensemble from (measurable) data, such as beam waist size and the divergence of an actual laser beam. For this reason, a computer program (LaserTrace) has been written, which uses the phase space description in combination with exact ray trace algorithms as found in O'Shea (1985). From the calculated phase space distributions of a beam passing through an optical system, the program derives beam radius, offset, divergence, pointing, quality and trend (converging, diverging or waist) as a function of $z$. It also calculates the amount of aberration in terms of an augmented beam quality factor.

### 10.5.1 Laser Beams from Stable Resonators

Examples of exact beam tracing analyses are shown in Figures 10.5 to 10.7.

### 10.5.2 Laser Beams from Unstable Resonators

The simplest type of unstable resonator is formed by two plane mirrors and a lasing medium limited by apertures, as is common in commercial (pulsed)

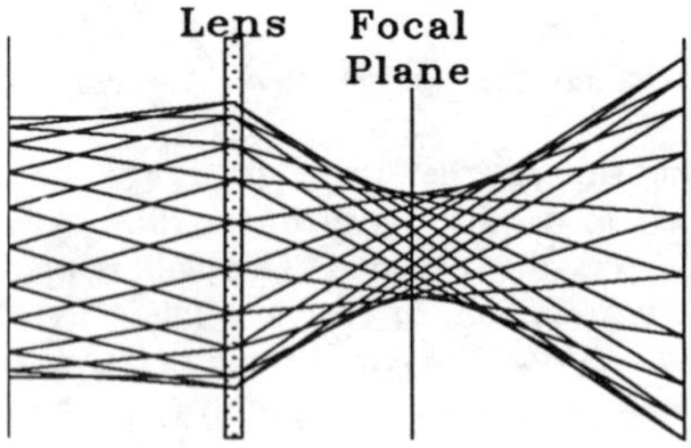


Fig 10.5 Exact ray tracing of a laser beam focussed to a waist by a single lens.

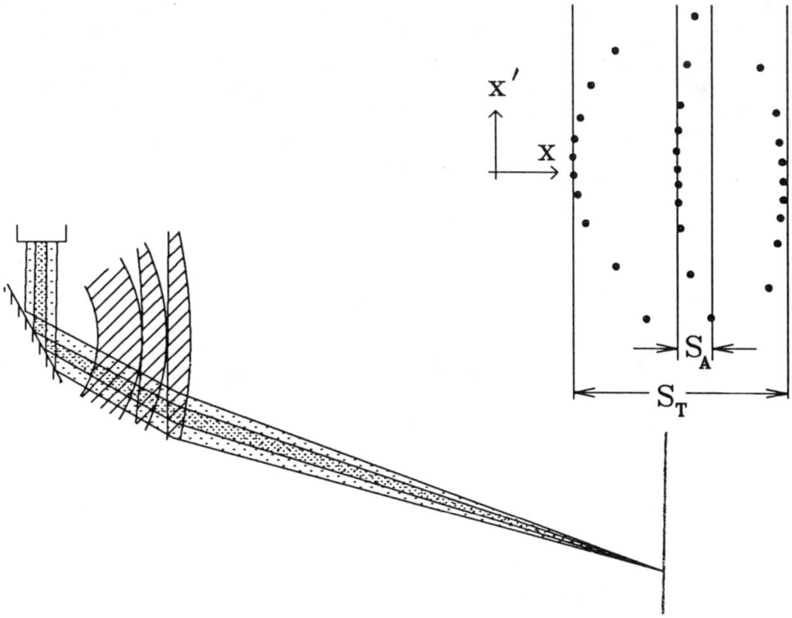


Fig 10.6 Exact ray trace of a flat field scanning lens for a multimode Nd:YAG laser Analysis of spot size at the scanning surface.

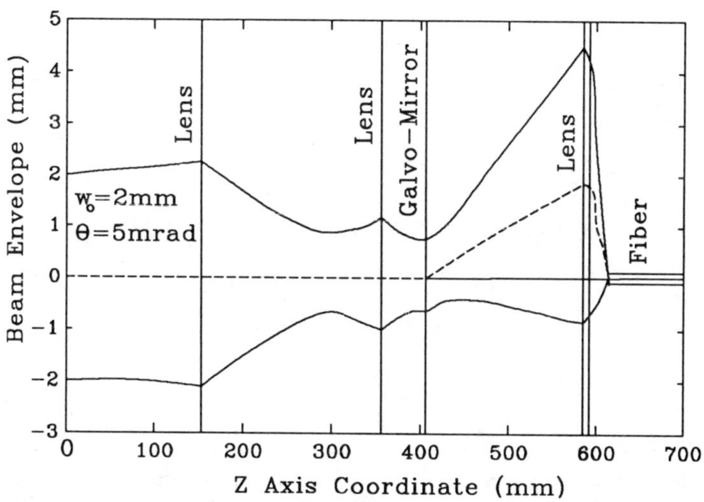


Fig 10.7 Calculated beam envelope in a time-sharing fiber multiplexer used for coupling a Nd:YAG beam into different optical fibers (Goethals 1988).

excimer lasers. It can be geometrically modelled if we assume a random ray ensemble emerging from the medium which is then spatially limited several times by the apertures. The number of effective apertures depends on the number of passes of the light pulse through the resonator. This implies that the phase space distribution of the beam has the shape of a parallelogram (see Figure 10.8) or, in case where there are several apertures of different diameters, the shape of an (irregular) polygon.

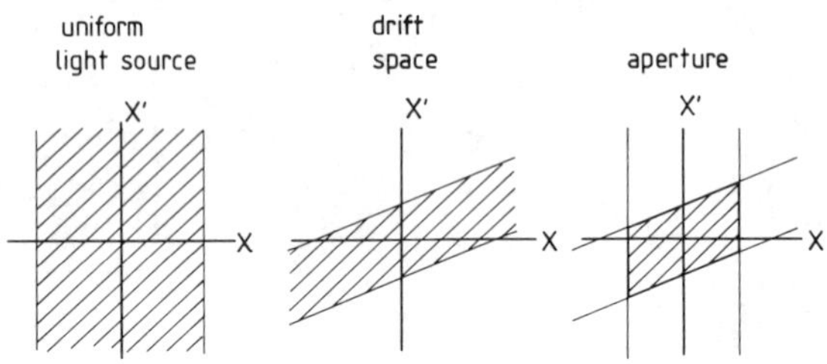


Fig 10.8 Formation of phase space distribution in an unstable resonator with plane mirrors.

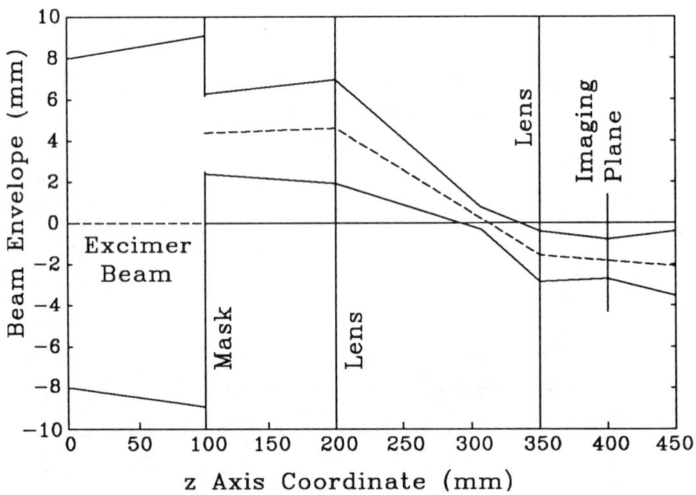


Fig 10.9 Example of imaging optics for a mask-marking application with a beam from an unstable resonator.

Paraxial tracing of these types of ray ensembles is not as simple as in the case of ellipses, and has to be performed by tracing the different "corner points". For this reason it is more difficult to derive general formulae for optical transformations of such beams. Exact phase space tracing however, is a very powerful tool for analysis in such cases and may be applied in the same way as described in section 10.5.1 (see Figure 10.9).

## 10.6 DISCUSSION

It has been shown that the combination of two familiar concepts, exact ray tracing and phase space, can be applied successfully for the analysis of laser beam propagation characteristics.

The reliability of the results from beam tracing depends on the accuracy of the definition of the input ray ensemble. When one has to design optics for a specific laser it is necessary to determine the characteristics of the beam. In order to determine the phase space properties, one needs to know the intensity distribution $I(x,y,x',y')$ at one position, $z$. However, commercially available beam profilers only measure $I(x,y)$, since this measurement is easier to visualize and to understand. Nevertheless, it may have become clear now that this is only half the information necessary for a complete beam characterization. Thus, what the laser world needs now, is a measurement method which gives the intensity distribution of a laser as function of four phase space co-ordinates, from which all the beam parameters, and subsequent beam propagation characteristics can then be derived.

Another remaining challenge is to incorporate optimizing algorithms into ray tracing software based on the phase space analysis presented here. A further problem is then to find the final intensity distribution $I(x,y)$ from the phase space contours which are calculated by tracing a beam through an optical system. This interesting area of theoretical research would also benefit from the integration of spatial coherence aspects with the phase space concept.

## 10.7 REFERENCES

Kogelnik H and Li T Oct 1966 Proc. IEEE 54 1312–29

Herbert Goldstein 1980 Classical Mechanics, Addison Wesley

Verster N F 1983 Lasers in fysische experimenten, Lecture notes, Eindhoven University of Technology, Department of Applied Physics

Hermann Wollnik 1987 Optics of Charged Particles, Academic Press

O'Shea D C 1985 Elements of Modern Optical Design, John Wiley & Sons

Nonhof C J 1988 Material Processing with Nd Lasers, Electrochemical Publications

Walter Koechner 1976/1988 Solid State Laser Engineering, Springer Verlag

Klaus Schildbach 1988 Excimer Lasers and Geometric Optics, Proceedings of the International Congress on Optical Science and Engineering, Hamburg

Walther A E Goethals 1988 Optical Fiber Multiplexer for Industrial Nd:YAG Lasers, Proceedings of the International Congress on Optical Science and Engineering, Hamburg

# PART TWO LASER RESONATOR TECHNOLOGY

# 11 YAG Laser Resonator Technology

*G BURROWS*

## 11.1 INTRODUCTION

The design of optimised resonators for use with solid-state lasers has been a topic of extensive research since the operation of the first ruby laser in 1960. Many of the generic developments which have occurred in resonator physics during these intervening years, as described in earlier chapters, have been also applied to solid-state lasers. However, the issue of greatest significance in addressing the subject of effective resonator design for solid-state lasers is the distortion produced in the intracavity optical beam as it transits the gain medium; this distortion is derived from thermal effects. Much of the emphasis in this chapter concerns the temperature distribution in the laser material caused by the optical pump power, the corresponding optical effects and the consequences in terms of limitations in output beam quality and laser power. To facilitate the use of specific practical examples, discussion will concentrate on the use of the laser material Nd:YAG in rod geometry; however, similar arguments can be applied to other materials such as Nd:glass and Ruby.

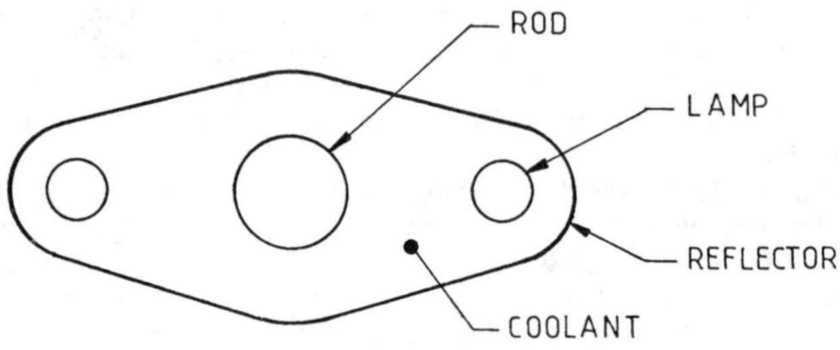


Fig 11.1 Cross sectional view of a simple pumping chamber suitable for use with a solid-state laser.

Figure 11.1 shows a simple arrangement used to excite a solid-state laser rod by flashlamp optical pumping. As a consequence of the optical pumping scheme, heat is deposited in the host crystal material. This occurs as a consequence of three particular physical mechanisms;

a) Radiationless transfer as active ions decay from the pump bands to the upper laser level.
b) Loss of upper laser level population by energy transfer to the lattice (i.e. the quantum efficiency for fluorescence is less than unity).

c) Direct excitation of the lattice by the optical pumping radiation; this occurs especially by UV and IR wavelengths contained in broad band flashlamp emissions.

Heat is removed from the system through the circumferential cooling of the rods by the cooling fluid which is pumped through the chamber. The corresponding temperature distribution which is generated in the rod by the removal of this heat can be calculated by solving Poisson's equation. Temperature gradients influence optical behaviour through *two* main effects, *firstly* as a consequence of the variation of refractive index with temperature, which gives rise to a lensing effect; and *secondly*, due to the formation of internal stresses which locally modify the refractive index and produce birefringence effects.

## 11.2 THERMAL EFFECTS AND THE ROD GEOMETRY

Assuming that the absorption length for the wavelength of the pump radiation is reasonably long, then the power dissipation in the laser rod will be fairly uniform; as shown in Figure 11.1, removal of this heat may be accomplished by flowing a liquid coolant over the rod barrel, thus holding that surface at a low temperature.

Assuming continuous operation and uniform volumetric heat generation ($Q$) in the rod (units of $\mathrm{Wcm^{-3}}$), then the steady–state temperature distribution which develops is parabolic. It may be expressed by the equation (Koechner 1976):

$$T(r) = T(r_0) + \frac{Q}{4K}(r_0^2 - r^2) \tag{11.1}$$

where $r_0$ is the radius of the rod and $K$ the thermal conductivity.

For the case of pulsed operation, Koechner (1973) has described the transient thermal profile of a pulse pumped rod using the general solution for radial heat flow given by Carslaw and Jaeger (1948); the radial dependence of temperature $T(r,t)$ is given by the sum of a series of Bessel functions. The complexity of this general solution is reduced for the two limiting cases of single 'shot' and repetitively pulsed operations.

In general, heat dissipation is slow compared to the duration of the pump pulse and thus, for single pulse operation, the temperature distribution is determined by the non–uniform absorption of pump radiation. For repetitively pulsed systems a quasi–steady state regime is reached in which the temperature profile is given by the sum of the particular profile produced after a single pulse and that remaining after thermal relaxation from the previous pulses.

We define a thermal relaxation time, $\tau$ given by Koechner (1976):

$$\tau = \frac{r_0^2 c\gamma}{4K} \tag{11.2}$$

where $c$ and $\gamma$ are the specific heat capacity and mass density respectively, and where $\tau$ is defined as the time taken for the rod centre temperature to decrease to $1/e$ of its initial value after the heating source is switched off.

At high repetition rates, where the pulse–to–pulse interval is short compared to the thermal relaxation time, the pulsed behaviour approaches that of the steady state (equation 11.1), with heating determined by the average power.

Having discussed the factors which establish a temperature profile, we now consider the consequent effects.

### 11.2.1 Thermal Birefringence

Temperature gradients inside an optical element such as a laser rod give rise to internal stresses. Using the temperature profile of equation (11.1) and the equations of Timoshenko and Goodier (1951) the radial $\sigma_r(r)$, tangential $\sigma_\varphi(r)$, and the axial $\sigma_z(r)$ components of stress for an isotropic rod with free ends are given by:

$$\sigma_r(r) = QS(r^2 - r_0^2)$$

$$\sigma_\varphi(r) = QS(3r^2 - r_0^2) \tag{11.3}$$

$$\sigma_z(z) = 2QS(2r^2 - r_0^2)$$

$$S = \frac{\alpha Y}{16K(1-\nu)}$$

where $\alpha$, $Y$ and $\nu$ are the thermal expansion coefficient, Young's Modulus and Poisson's ratio, respectively and where $S$ is a material parameter. These stresses induce changes in the refractive index by the photoelastic effect; the orientation of the indicatrix which describes these changes, is shown in Figure 11.2 for a point $(r,\varphi)$.

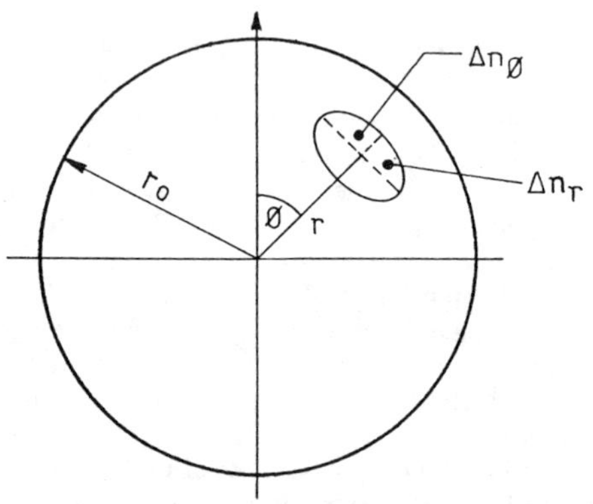


Fig 11.2 Orientation of birefringence effects.

The calculation of the birefringence ($\Delta n_r - \Delta n_\varphi$) for crystals requires an involved tensor analysis. However, for cubic systems such as YAG with optical propagation along the [111] direction (the orientation of commercially grown material), Koechner (1976) has shown that:

$$\Delta n_r - \Delta n_\varphi = n_0^3 \frac{\alpha Q}{K} \frac{1}{48} \left[ \frac{1+\nu}{1-\nu} \right] [P_{11} - P_{12} + 4P_{44}] r^2$$
$$= \delta Q r^2 \qquad (11.4)$$

Where $n_0$ is the initial unperturbed refractive index and $P_{nm}$ are components of the photoelastic tensor (see for example Nye (1964)). The constant $\delta$ for Nd:Yag has a value of $-3.2 \times 10^{-6}$ cm/W.

For a homogeneous material (glass) with stress optic coefficient, $\psi$ we have:

$$\Delta n_r - \Delta n_\varphi = \psi(\sigma_r - \sigma_\varphi)$$

and by definition: $\delta = 2S\psi$

For glass $\delta$ has a value: $\delta_{glass} \approx (19 \times 10^{-6})$ cm/W.

We note that the change in polarisation which this effect produces is of special significance when the intracavity beam is already polarised, as occurs when the polarising element is positioned within the resonator. It can be seen that the magnitude of birefringence has a radial dependence. For a rod of length $L$ placed between parallel polarisers, (which is equivalent to a rod in a cavity containing a polariser) the loss per pass due to depolarisation must be calculated by integration over the area (see Figure 11.2). Thus, we have:

$$\frac{I_{out}}{I_{in}} = \frac{1}{\pi r_0^2} \int_{\varphi=0}^{2\pi} \int_{r=0}^{r_0} 1 - \sin 2\varphi \, \sin^2\left[ \frac{\pi L}{\lambda} \delta Q r^2 \right] r \, dr \, d\varphi$$
$$= \frac{1}{4} \left[ 3 + \mathrm{sinc} \left( \frac{2\delta}{\lambda} P \right) \right] \qquad (11.5)$$

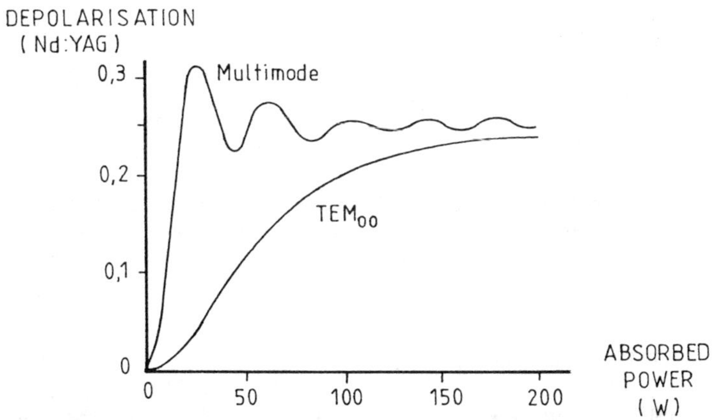


Fig 11.3 Single pass depolarisation loss versus absorbed power.

We note that the birefringent loss depends only on material parameters ($\delta$), wavelength, and absorbed power ($P$). Birefringence effects on multimode laser beams are rather severe because the intensity profile extends out to the surface of the rod where the stress is of greatest magnitude (refer to equation 11.4); fundamental mode ($TEM_{00}$) beams have reduced intensity in the wings and are therefore relatively less distorted. Figure 11.3 shows a plot of the depolarisation as a function of absorbed power for a Nd:YAG rod, for both multimode and single mode operation, calculated using equation 11.5. The level of output power from a Nd:YAG laser is approximately half of the level of absorbed power shown in Figure 11.3; thus, in order to achieve only low losses (for example less than 10% per pass), only low output powers are permitted.

### 11.2.2 Thermal Focussing

Thermal focussing is a problem caused by the variation in the refractive index of the laser rod material with temperature, which may be expressd as $(\partial n/\partial T)$.

For the temperature profile described by equation (11.1), we can write an expression for the radial variation, $\Delta n$ in the refractive index as follows:

$$\Delta n = n(r) - n(0) = \frac{\partial n}{\partial T} [ T(r) - T(0) ]$$

$$= - \frac{Q}{4K} \frac{\partial n}{\partial T} r^2$$

Kogelnik (1965) has shown that if the refractive index variation has a parabolic form, then the effective focal length, $f_1$ (assumed to be long in comparison to the rod length) is given by:

$$f_1 = \frac{2K}{QL \; \partial n/\partial T} \tag{11.6}$$

which gives rise to spherical focussing. Writing $f_1$ in terms of absorbed power in the rod $P_a$, we have:

$$f_1 = \frac{2\pi k}{\partial n/\partial T} \frac{r_0^2}{P_a} \tag{11.7}$$

By way of example, we consider a Nd:YAG laser crystal of 150mm length by 10mm diameter pumped by 10 kW average input power. In this case where $P_a$ represents approximately 7% of the input power, the laser rod exhibits a focal length as short as 50 cm.

## 11.3 THE BASIC RESONATOR

The simplest resonator which may be used with a Nd:YAG laser consists of a rod placed between a pair of plane mirrors is shown in Figure 11.4. The output coupler usually has a reflectivity in the range 30% to 60% for fixed–Q operation and between 4% and 15% when Q–switched output is required.

The achievement of stability for such a "plane–plane" resonator is dependent on the thermal development of a finite focal length in the laser rod.

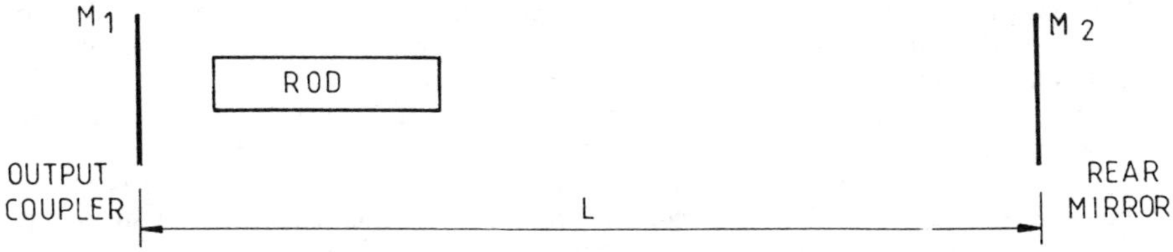


Fig 11.4 A simple resonator.

Considering the case where the rod is close to one of the mirrors, the resonator can be re-drawn as shown in Figure 11.5. Here, the effect of the lensing in the rod has been incorporated in the curvature of the nearby mirror.

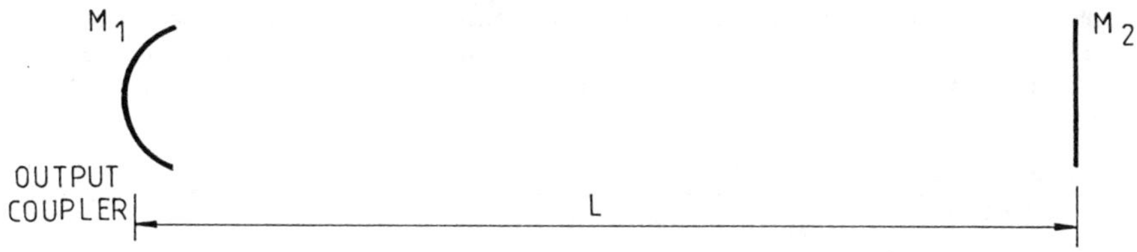


Fig 11.5 The "Equivalent" to that shown in Fig 11.4 with the thermal lensing incorporated in the curvature of $M_1$.

Now, using the stable resonator theory developed in Chapter 1, the above resonator can be analysed and the variation of the $TEM_{00}$ spot size with rod focal length deduced. The variation with rod focal length of the fundamental mode spot size is shown in Figure 11.6. It should be noted that the spot size at mirror $M_2$ decreases continuously as the equivalent focal length of the rod decreases, while the spot size at Mirror $M_1$ has a minimum value when the focal length is equal to twice the resonator length ($L$).

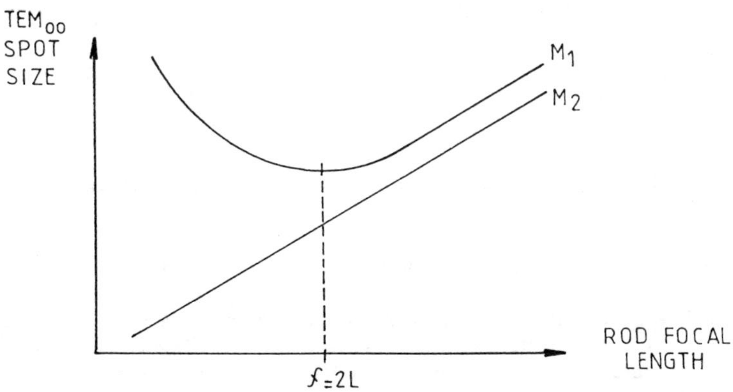


Fig 11.6 $TEM_{00}$ Spot size versus rod focal length.

With a wavelength as short as 1 $\mu$m and a resonator length in the region of one metre, then the $TEM_{00}$ spot size achieved at typical values of rod focal length is approximately 1 mm. Since the rod diameter usually lies in the 4 to 10 mm range, it is clear that such a laser resonator will support a multimode beam; fundamental mode output can only be achieved with the use of internal aperturing, reduced mode volume and consequent reduction in output power.

In many practical cases the use of a multi–transverse mode beam is adequate for the intended application. In this context, *three* important points can be made. *Firstly*, for such systems, it is possible to calculate quite accurately an important quality parameter known as the "*divergence – diameter product*", which is a useful measure of the propagation characteristics of a Gaussian beam. *Secondly*, although the indices $(n,m)$ of the highest order lasing mode are not known, it is reasonable to assume that the diameter of the propagating mode in the rod is equal to the rod diameter. *Finally*, knowing that the ratio of the diameter of a high order mode $(n,m)$ to the diameter of the $TEM_{00}$ mode remains constant with propagation, then it follows that the multimode beam divergence must be proportional to that of the $TEM_{00}$ mode. Combining these three last facts then it can be deduced that:

$$\begin{matrix}\text{Multimode}\\ \text{Divergence}\end{matrix} = \frac{\text{Rod-diameter}}{TEM_{00}\ \text{diameter in rod}} \cdot TEM_{00}\ \text{Divergence} \qquad (11.8)$$

Now, using equation (11.8) and the information given in Figure 11.6, the behaviour of the multimode divergence of the output beam for mirror $M_1$ can be determined. The dependence of the multimode beam divergence on the focussing power of the rod is shown in Figure 11.7.

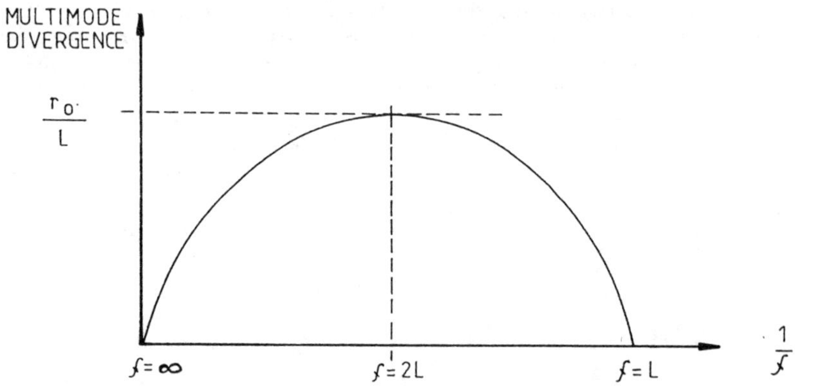


Fig 11.7 Multimode divergence versus rod focal length.

The half angle divergence has a maximum value of approximately $r_0/L$ which occurs for the condition that $f=2L$. It can be shown that beam quality as represented by the divergence–diameter product is dependent on the rod focal length, which is in turn dependent on average input pump power. As the laser output power is also directly related to the input power, then it is clear that, in general, beam quality varies with output power. It should be noted that in the region of the extremum value of beam quality, that the variation with output power is relatively small; many solid–state laser systems are designed to operate at this point.

## 11.4 COMPENSATED RESONATORS

In the previous section it has been shown that there is a limit to resonator stability such that the focal length of the rod cannot be allowed to fall to a value shorter than the resonator length. This constraint can cause serious problems if a design is required which combines high output laser power (leading to a short rod focal length), with a low beam divergence and hence long resonator length. To overcome this problem, convex mirrors may be used to introduce a negative lensing effect which partially counters the positive effect produced by the rod. Alternatively, the end faces of the rod may be ground concave so as to produce a similar result. Referring to Figure 11.7 it can be seen that both of these compensation techniques result in a displacement of the divergence curve along the focal power axis, as shown in Figure 11.8 .

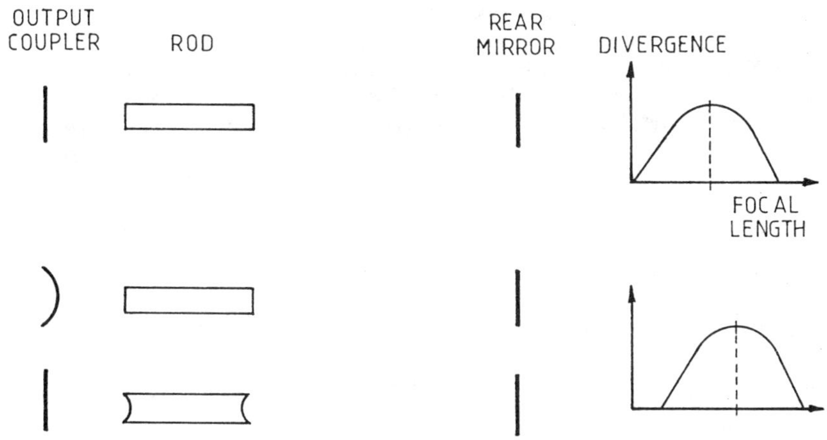


Fig 11.8 Simple compensated resonators.

Unfortunately, the use of these simple techniques involving compensation elements of fixed values is relatively inflexible and cannot be easily adjusted to suit differing situations. There is, however, an alternative design referred to as a "stable telescopic resonator" (Hanna *et al* 1981) in which an intra–resonator telescope is adjusted in focal power to provide the necessary factor to produce a compensated resonator..

Correction of birefringence problems in laser rods is significantly more difficult to achieve than correction for focussing. The polarisation state of a ray is strongly dependent on the position in the aperture from which it emerges. Some degree of success has however been achieved by Scott and de

Wit (1971), who devised a scheme whereby the effect produced by passage through one rod was countered by an opposite effect produced by passage through a second identical element.

## 11.5 SUMMARY

A number of key problems encountered in the design of solid–state laser resonators are a consequence of the variation with input power of the transmission properties of the laser rod. Compensation techniques may be applied, but very careful design is necessary in order to yield acceptable results in terms of beam quality and output power.

## 11.6 REFERENCES

Carslaw H S and Jaeger J C 1948 Conduction of Heat in Solids (Oxford)
Hanna D C, Sawyers C G and Yuratich M A 1981 Opt. and Quantum Electron. 13 493
Koechner W 1973 J. Appl Phys 44 3162
Koechner W 1976 Solid State Laser Engineering (Springer–Verlag)
Koechner W and Rice D K 1970 IEEE J. Quantum Electron. QE–6 557
Kogelnik H 1965 Bell Syst. Tech. J. 44 455
Nye J F 1964 Physical Properties of Crystals (Oxford)
Scott W C and de Wit M 1971 Appl. Phys. Lett. 18 3
Timoshenko S and Goodier J N 1951 Theory of Elasticity (McGraw Hill)

# 12 Resonators for Dye Lasers and Optical Parametric Oscillators

*M H DUNN*

## 12.1 INTRODUCTION

Dye lasers and optical parametric oscillators offer alternative means of obtaining tunable, coherent radiation throughout the near infrared, visible and near ultraviolet regions of the spectrum. Until recently dye lasers fulfilled this role more satisfactorily, particularly in the visible region of the spectrum, and the use of parametric oscillators was more or less confined to regions of the spectrum inaccessible to dye lasers, such as the near infrared. However, recent progress in the development of new nonlinear materials with high optical damage thresholds is beginning to change this situation, and it seems likely that the optical parametric oscillator will find a more widespread usage in the near future. In both types of device, cavities must be engineered to provide narrow linewidth, coherent radiation that is tunable in the presence of an active medium which exhibits a broad gain bandwidth. Cavity design techniques in each case are similar, and so the first part of this chapter considers the design of a dye laser cavity suitable for generating pulsed coherent radiation (pulse durations of the order of 10 ns) with a transform limited linewidth. The second part of the chapter will then review some recent progress in the development of optical parametric oscillators for the visible and near ultraviolet, with emphasis on cavity design, particularly with regard to minimising the effects of the beam walk-off that is associated with the nonlinear optical media used in these devices.

## 12.2 DYE LASER CAVITY WITH TRANSFORM-LIMITED LINEWIDTH

Since this section is tutorial in nature and we wish to emphasise general principles, we will limit discussion to one particular type of pulsed dye laser cavity that has found widespread usage in recent years, and has come to be known as the *'Hansch'* oscillator (Hansch, 1972). We will conclude the section with a brief discussion of more recent designs that are superior in several respects and are hence likely to progressively supersede this earlier design.

A particular version of the Hansch dye laser oscillator is illustrated in Figure 12.1. For the values of parameters given in the figure, linewidths have been determined experimentally. This configuration is a particularly useful one for the comparison of predicted linewidths with those achieved in practice, since this design uses a combination of a diffration grating, preceded by beam expansion, and an etalon so as to achieve linewidths which are close to the transform limit. In order to attain the maximum resolutions of which the various tuning elements are capable, it is first of all necessary that the divergence of the beam within the cavity be determined by diffraction rather

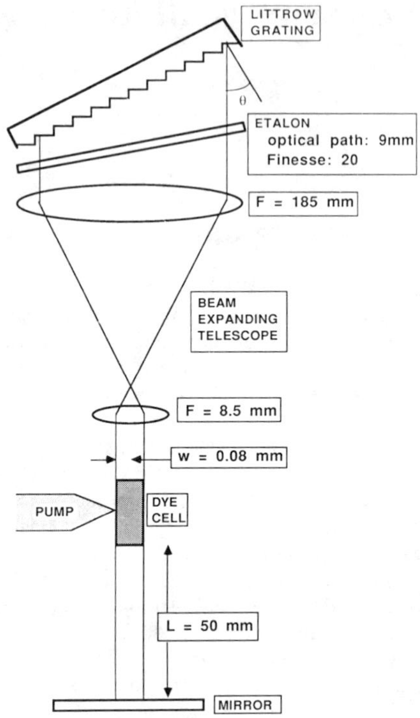


Fig 12.1 Hansch design of dye laser oscillator.

than geometrical optics. If we consider the combination of the dye cell and the plane mirror which terminates one end of the cavity, then in terms of geometrical ray tracing, we can regard the dye cell as being illuminated by its image in the plane mirror. Hence if $w_1$ is the beam radius at the dye cell, (this being determined by the absorption length of the pump light in the dye solution), the approximate angular spread $\theta_g$ of the light propagating towards the grating is given in the limit of geometrical optics by:

$$\theta_g \sim w_1/2L \tag{12.1}$$

where $L$ is the distance between the mirror and the dye cell. On the other hand, the diffraction spreading $\theta_d$ of the beam due to its finite aperture is given approximately by:

$$\theta_d \sim \lambda/\pi w_1 \tag{12.2}$$

For the particular set of parameters given in the figure, namely where $w_1$ is 0.08 mm, this being determined by the absorption of the pump light in a $10^{-4}$ M solution of dye, and $L$ is 50 mm, then the geometrical spreading of the light is of the order of 1mrad, while the diffraction spreading is of the order of 2–3 mrad. So for this particular beam size and cavity geometry, the beam divergence is determined by diffraction and is hence the minimum possible. The cavity is therefore satisfactory in this respect. The importance of the beam divergence is apparent when we come to consider the operation of the tuning elements, for it is beam divergence which determines the spectral resolution that they are able to provide. If we consider first of all the grating as shown operating in the Littrow configuration, then the grating equation is:

$$n\lambda = 2d \sin\theta \qquad (12.3)$$

so that the spectral resolution as determined by the angular spread of the incident and diffracted light is:

$$\delta\lambda = (\lambda/\tan\theta)\delta\theta \qquad (12.4)$$

If the light from the dye cell is allowed to strike the grating directly, then a spectral resolution of 0.7 nm is expected from the above relation. However, this can be significantly improved by first expanding the beam using a telescope in reverse, as shown in Figure 12.1. Such expansion results in a reciprocal reduction in the beam divergence. For the case here, where lenses of focal length 185 and 8.5 mm are used, the beam expands by a factor of 22, so the divergence is reduced to 0.1 mrad, and the resolution of the grating increases to 0.03 nm. In fact, the *observed* linewidth is close to an order of magnitude below this (around 0.003 nm or 3 GHz) because the radiation undergoes multiple passes through the cavity over the duration of the pumping pulse. In the present configuration with a cavity round trip distance of 30 cm, there are close to five round trips during the course of the pumping pulse which lasts for 5 ns.

The transform limited linewidth, $\Delta\nu$, for a pulse of duration $\tau$ is given by:

$$\Delta\nu.\tau = 0.441 \qquad (12.5)$$

for the case of a gaussian pulse shape. In the present case where the pulse duration is 5 ns, the transform limited linewidth is around 90 MHz, and hence there is scope for further improvement. This can be effected by the use of etalons placed inside the laser cavity. The basic equation describing constructive interference of light passing through an etalon is:

$$n\lambda = 2d\cos\theta \qquad (12.6)$$

where $\theta$ is the angle of the plane wave to the normal to the etalon faces (internal angle), and $d$ is the optical thickness of the etalon. The free spectral range can hence be deduced as:

$$\Delta\lambda = \lambda^2/2d \qquad (12.7)$$

and if the finesse of the etalon is $F$, then its resolution is:

$$\delta\lambda = \lambda^2/2Fd \qquad (12.8)$$

We must however, also assess how the resolution of the etalon is diminished by the divergence of the beam passing through it. This we can do by looking at the effect on the selected wavelength (corresponding to constructive interference) of small changes in the angle of incidence about normal incidence. From the above, we see that for an angular spread in the light of $\delta\theta$, the spread in transmitted wavelength is:

$$\delta\lambda/\lambda = (\delta\theta)^2/2 \qquad (12.9)$$

If we now consider the angular spread of the light to be determined by diffraction, then we have:

$$\delta\lambda/\lambda = (\lambda/4\pi w_2)^2 \qquad (12.10)$$

where $w_2$ is the radius of the beam at the etalon. In seeking to achieve maximum resolution, the advantage of placing the etalon in the expanded part of the beam, where the divergence is low, is now apparent.

In choosing an etalon to produce further line narrowing, we must first of all ensure that its free spectral range is greater than the bandwidth initially selected by the grating, otherwise the laser may oscillate on adjacent orders of the etalon. This restricts the etalon thickness. Its finesse must then be adequate to achieve the required line narrowing. In the present example, suppose we consider the use of an etalon with an optical thickness of 9 mm and a finesse of 20. The free spectral range of the etalon is 0.02 nm, and since this is greater than the grating–selected bandwidth of 0.003 nm, the etalon will ensure that only one order is selected. The resolution of this etalon, as determined by the finesse, is 0.001 nm. Further, the resolution is not degraded by the beam divergence, since the limit due to this effect in the present beam geometry is 0.00005 nm. In practice such an etalon is observed to narrow the linewidth of the dye laser to around 0.0004 nm, because its effectiveness is enhanced by multiple passes. This value approaches the transform limit.

The selected wavelength range may be tuned across the bandwidth determined by the grating by either altering the etalon spacing, $d$, or by tilting the etalon and hence altering $\theta$. The former is to be preferred where possible, since tilting the etalon away from normal incidence introduces additional cavity loss. However, alteration of the optical thickness is not normally possible in the case of a solid etalon, which therefore requires tilting in order to achieve the required tuning. We can estimate the loss introduced by considering the reduction in the beam overlap on repeated reflection within the etalon. If we consider the number of reflections in the etalon to be given roughly by its finesse, then the loss introduced by tilting the etalon through a sufficient angle from the normal to scan its pass band through one free spectral range, may be written as:

$$\text{Tilt loss} \propto \lambda F/w \qquad (12.11)$$

Note this is the maximum angle of tilt needed since all wavelengths can then be accessed. The advantage of inserting the etalon in the expanded beam and keeping its finesse low is thus apparent. This latter constraint cannot always be met if the improvement required in the resolution is large. In this case it is better to use two etalons, one of large free spectral range, one of small free spectral range, so allowing the finesse of each to be kept small while still achieving the desired overall improvement in resolution. In the present example it can be seen that this source of loss is small (under 1%), and hence the use of a single etalon to achieve transform limited resolution is quite appropriate.

Although the *Hansch* design has proved to be somewhat of a classic in the context of line narrowing and tuning of dye lasers, there have been several significant improvements and developments in recent years, some of which we now review.

Anamorphic prisms may be used for beam expansion rather than the telescope, since such expansion is required in only one dimension in order to improve grating selectivity (Zhang *et al*, 1981). Another alternative is to use a diffraction grating at grazing incidence, a design which eliminates the need for prior beam expansion altogether. This grating is then followed by either an additional plane mirror, or indeed a second grating in a Littrow configuration (because of the effects of the first grating, the beam reaches this second grating with an expanded cross section) (Shoshan *et al*, 1978, Littman, 1978). By carefully reducing the cavity length of a grazing incidence grating geometry Littman (Littman, 1984) devised a tunable dye laser with a time averaged linewidth <150MHz. The reduction in the cavity length increases the number of round trips and hence the number of passes through the tuning elements within the course of the excitation pulse, so improving their selectivity. A further development is the inclusion of a pair of quarter wave plates within the resonator positioned on either side of the dye cell. This eliminates spatial hole burning in the gain medium, and makes it far easier to retain single (axial) mode oscillation for increased pumping levels of the dye.

The resonator geometries discussed above may be regarded as the conventional geometries for the line narrowing of dye lasers. Mention should however be made of other more novel configurations for line narrowing, such as the modeless, variable bandwidth dye laser developed by Ewart (1985), and improved distributed feedback geometries (McIntyre *et al*, 1988).

## 12.3 VISIBLE OPTICAL PARAMETRIC OSCILLATORS

The recent development of nonlinear optical materials (urea, beta barium borate) with improved nonlinear coefficients and higher optical damage thresholds than classical materials (ADP,KDP) coupled with improvements in beam quality in ultraviolet pump lasers (reduced linewidths, lower divergences in the case of excimer lasers, for example) are together opening up new opportunities for the generation of broadly tunable radiation in the visible and near ultraviolet using *optical parametric oscillators* (Ebrahimzadeh *et al*, 1988(a)&(b), 1989 (a)&(b)). When the nonlinear crystal acts as a gain medium in an optical parametric oscillator, it can be thought of as *'cracking'* each photon of the pump radiation into two photons such that energy and

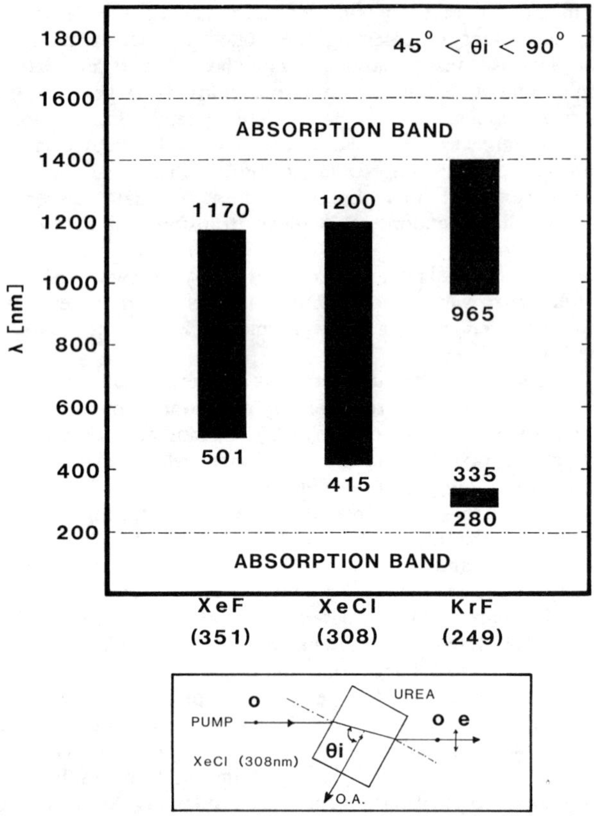


Fig 12.2 Tuning ranges associated with urea optical parametric oscillators (Type II phase matching) when pumped by a range of excimer laser wavelengths and for phase match angles (internal) between 90 and 45°. The identified tuning ranges include both the signal and idler waves.

momentum conservation applies (conservation of the latter quantity is known as phase matching). In other words the frequencies $\nu_1$ and $\nu_2$ of the generated radiation are given by:

$$\nu_1 + \nu_2 = \nu_3 \tag{12.12}$$

where $\nu_3$ is the frequency of the pump photon, together with:

$$n_1\nu_1 + n_2\nu_2 = n_3\nu_3 \tag{12.13}$$

where the $n$'s refer to the respective refractive indices. The selected frequencies can be altered, while still maintaining their sum equal to the pump frequency, by differentially altering the refractive indices. This can be done by altering the direction of propagation of the radiation in the nonlinear crystal and relying on its anisotropic optical properties.

As an example of the tuning ranges possible, Figure 12.2 illustrates the situation in the non-linear material urea when pumped by a variety of excimer wavelengths. The graph shows the tunability achieved as the angle of propagation of the pump wave in the crystal is altered from being orthogonal to the optic axis of the material (non-critical phase matching) to an angle of 45° to the optic axis. For example, in the case of urea pumped with radiation from a XeCl excimer laser, coherent radiation can be generated from 415 nm to 1200 nm, a range that would require some 16 different dyes if it were to be met by a dye laser. Figures 12.3 and 12.4 show the potential of

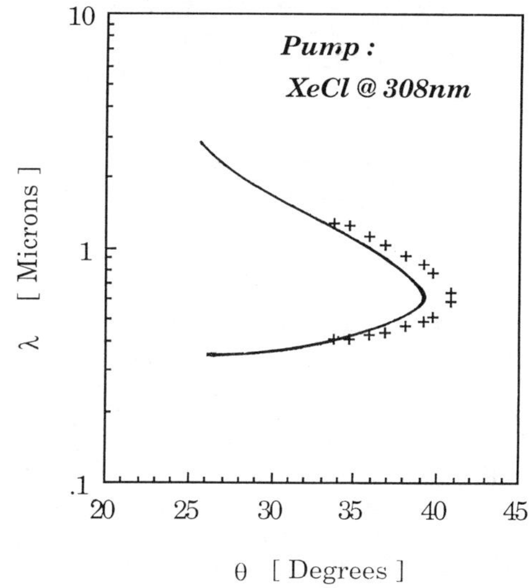


Fig 12.3 Tuning range for the case of a beta barium borate crystal pumped by a XeCl excimer laser (308nm) as the phase match angle is varied. The crosses are experimentaly determined values.

the nonlinear crystal beta barium borate when pumped by the XeCl (308nm) and KrF (249nm) excimer lasers respectively. For example, in the case of pumping at 249 nm, it should be possible to achieve continuous tuning from 280 nm in the near ultraviolet to approaching 3$\mu$m in the near infrared.

Figure 12.5 shows one of the experimental layouts which may be used to realise some of these potential tuning ranges. Optical parametric oscillation is a process which depends critically on the temporal and spatial coherence of the pump laser and hence a line narrowed excimer laser of high spatial quality is needed. This is not the case in dye lasers where the creation of a real population inversion in the gain medium decouples the generation process for the coherent radiation from the properties of the pump radiation. The pump coherence required is accomplished using the technique of injection seeding. Here, a master oscillator is first used to generate radiation in a fundamental transverse mode by use of an apertured stable resonator, and line narrowed through the insertion of intracavity etalons into this resonator (typically the linewidth is less than 0.2 $cm^{-1}$). This radiation is then used to injection seed a power oscillator which has an unstable resonator in order to attain good extraction efficiency and hence high energy from the gain medium.

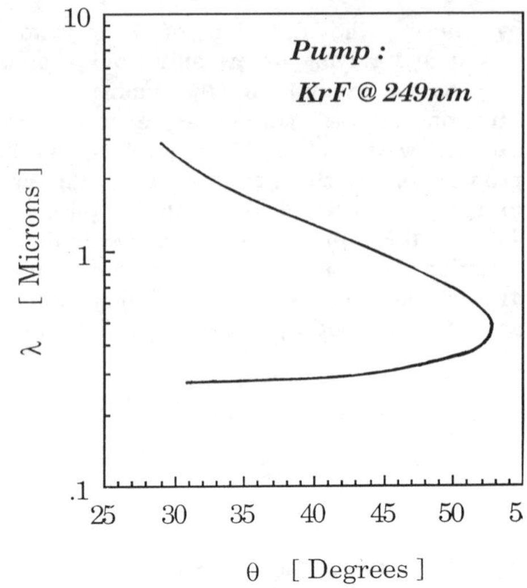


Fig 12.4 Tuning range for the case of a beta barium borate crystal pumped by a KrF excimer laser (249nm) as the phase match angle is varied.

Thus, high energy pulses (>20mJ in this case of duration around 10 ns) can be extracted with a narrow linewidth, and in a low divergence beam (about three time the diffraction limit). Following beam compression using a telescope in reverse, to achieve adequate power densities (>100 MW/cm$^2$), the radiation is then used to pump the optical parametric oscillator. In order to satisfy phase matching requirements, one of the generated waves identified as the *signal* wave, has to be o–polarised, while the other identified as the *idler* wave, has to be e–polarised. The *pump beam* is o–polarised. The cavity of the optical parametric oscillator is such that only one of these waves is resonated (the signal wave), and the oscillator is referred to as being *singly resonant*. Typical operating conditions and performance characteristics of a urea based optical parametric oscillator pumped by a XeCl laser are given in Table 12.1. Evidently, a conversion efficiency of the order of 67% may be achieved under conditions of non–critical phase matching, and that a tuning range of 100nm can be addressed with greater than 15% conversion efficiency.

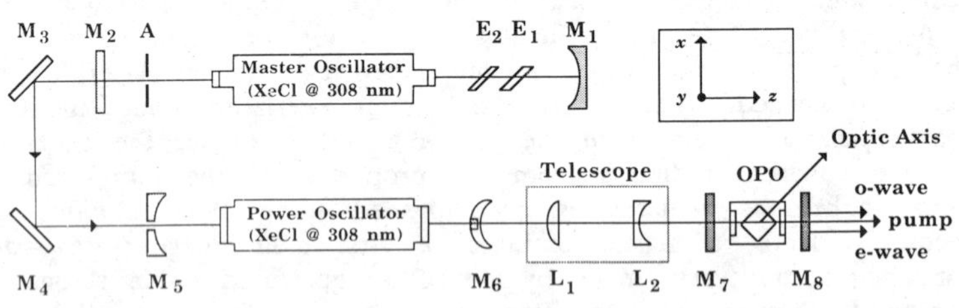


Fig 12.5 Experimental layout for a urea optical parametric oscillator pumped by an injection seeded, line narrowed, XeCl excimer laser. The M's refer to the mirrors, the L's to the lenses, the E's to etalons, and the A to the aperture.

| | | |
|---|---|---|
| | **Conditions** | |
| **Pump Pulse:** | Energy | 16.7mJ |
| | Pulse Duration | 7ns |
| | Spot Size | 1x2mm |
| | (unfavourable aspect ratio) | |
| | Linewidth | <0.2 $cm^{-1}$ |
| | Plane Polarised | |
| **Crystal:** | 25mm Path Length | |
| **Resonator:** | Plane Parallel | |
| | One High Reflector | |
| | 60% Output Coupling | |
| | 108mm Round Trip Length | |
| | Singly Resonant | |
| | **Performance** | |
| **Output:** | Energy | 11.1mJ |
| | $90^o$ Phase Matching | |
| | 67% Overall Optical Conversion Efficiency | |
| **Tunability** | 570–670nm at >15% conversion efficiency | |
| | Total Range to date 537–720nm | |

Table 12.1 Summary of Performance of an Optical Parametric Oscillator based on Urea

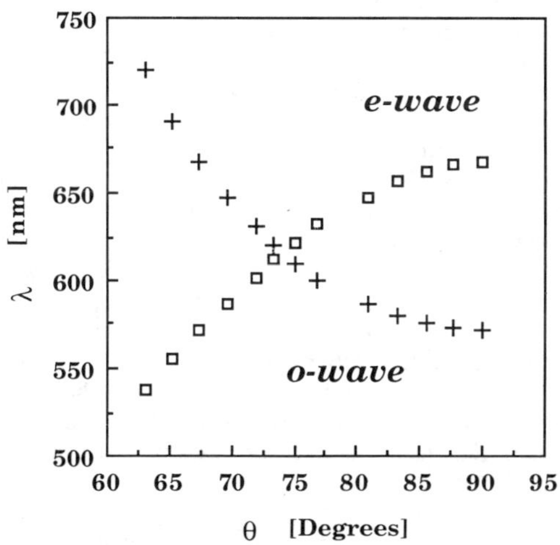


Fig 12.6 Experimentally observed tuning range for the case of an optical parametric oscillator based on urea, pumped by a XeCl excimer laser, as a functon of the phase match angle (internal). The wavelengths of the signal (o–wave) and idler (e–wave) are shown.

Figure 12.6 illustrates the way in which the wavelengths of the idler and signal waves depend on the angle between propagation direction of the pump radiation and the optic axis of the material. Figure 12.7 illustrates how the

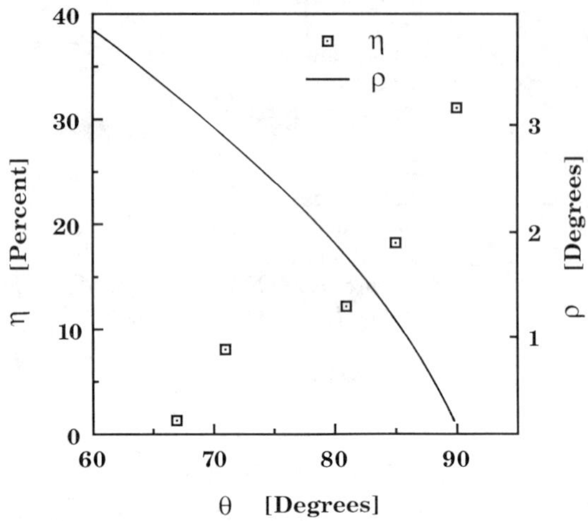


Fig 12.7 Experimentally determined generation efficiency with phase match angle. The solid line shows how the computed walk-off angle varies with phase match angle.

efficiency of generation falls off as this angle decreases from 90° (corresponding to non-critical phase matching), to a value of around 60°. (A crystal of length 15mm was used here, so the maximum conversion efficiency is less than in Table 12.1 above). An explanation for this behaviour lies in the polarisation properties of the generated radiation. Because the idler wave is an e-wave, in general its propagation direction (Poynting Vector direction)is not the same as its wave normal direction. Now, since the wave normal directions of all three interacting waves are colinear in the crystal, it follows that the ray direction of the idler wave walks-off from the ray directions of the other two waves (signal and pump). The angle between the ray and wave normal directions of the idler is called the walk-off angle, and is also plotted in Figure 12.7. As a consequence of this walk-off effect, the generated idler wave no longer fully overlaps the other two waves, and so generation efficiency is expected to reduce with increasing walk-off angle, as indeed is observed. In order to reduce the effects of beam walk-off, and so maintain high generation efficiency out to small values of the phase matching angle, large area beams are to be preferred. However the beam area cannot be arbitrarily increased since this has the detrimental effect of reducing the power density of the pump beam at fixed pulse energy, and as a result the conversion efficiency. So what is required is a means of reducing the effects of beam walk-off while still maintaining the pump power density. This can be done by going to elliptical beam geometries as illustrated in Figure 12.8. The major axis of the ellipse is chosen to lie in the plane in which beam walk-off occurs, so maximising the beam overlap, while the minor axis of the ellipse is orthogonal to this plane and of dimension such as to maintain the

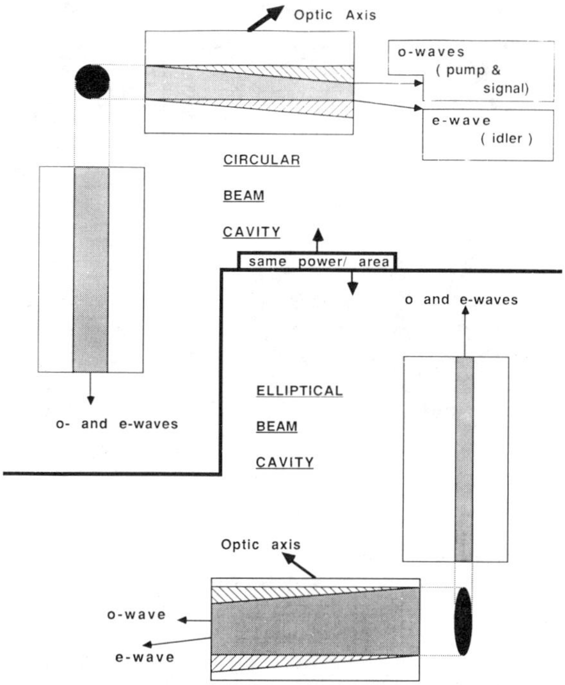


Fig 12.8 An illustration of hov an elliptical beam geometry can reduce the effect of beam walk–off, by maximising the overlap between pump, signal and idler beams, compared to the case of circular beams of the same area.

pump power density at the desired level. A practical method for tailoring the beam profile from the excimer laser is based on the use of an anamorphic prism pair is illustrated in Figure 12.9. The threshold reduction effect expected to result from the use of an elliptical beam geometry as the aspect ratio of the beam is increased from (1:1) circular to (5:1) is illustrated for the particular set of conditions given in Figure 12.10. It can be seen that as the aspect ratio approaches 5:1, the effect of beam walk–off is more or less eliminated for the conditions. Experimental results showing the improvement in conversion efficiency as the aspect ratio of the excimer beam is altered from a ratio of 2:1 (unfavourable from the point of view of minimising walk–off effects) to a favourable ratio of 1:2 are presented in Figure 12.11. In both arrangements the power density at the crystal is the same.

The linewidths obtained in the above arrangements are determined by the phase match requirement associated with the nonlinear process, and are of the order of 0.4 – 0.5 nm. To the extent of providing initial frequency selection, the optical parametric oscillator is superior to the dye laser. Further frequency reduction can be accomplished using techniques similar to those we have already discussed in the context of the dye laser.

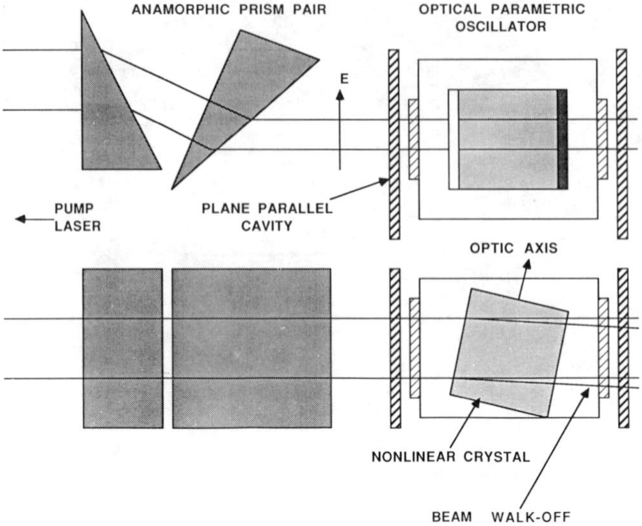


Fig 12.9 A practical method for tailoring the beam profile from an excimer laser by the use of an anamorphic prism pair.

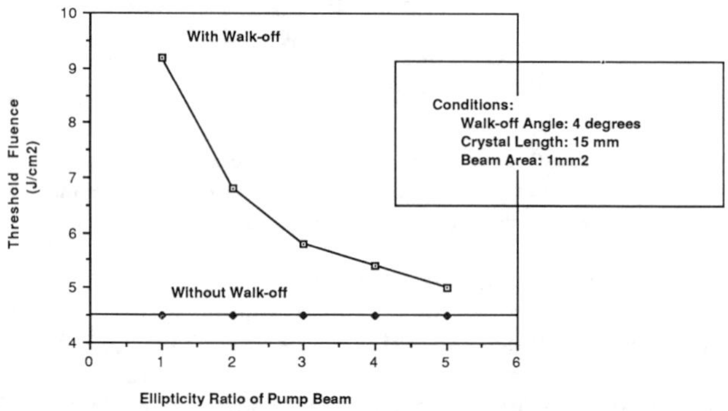


Fig 12.10 Calculated thresholds for an optical parametric oscillator based on urea pumped by the XeCl excimer laser under the conditions stated as a function of the aspect ratio of the beam. The lower trace referes to the situation where the effects of beam walk-off have been disregarded (i.e. the walk-off angle has been put equal to zero). It is seen that an aspect ratio of the order of 5:1 more or less compensates the effect of the walk-off for the given conditions.

The use of beta barium borate in place of urea offers potentially greater tuning ranges, although walk-off is now more pronounced at the small angles of propagation to the optic axis necessary for phase matching. As may be seen from Figure 12.3, a tuning range extending from 405 nm to about 1.3 $\mu$m is obtainable with this material when pumped by a XeCl excimer laser, although the conversion efficiency is somewhat reduced at around 5% ⊗.

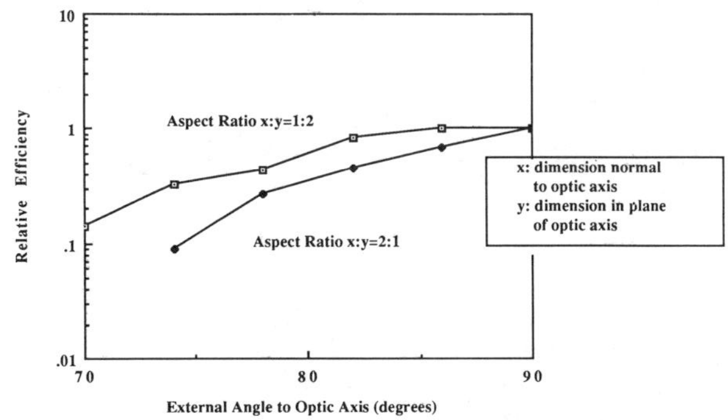


Fig 12.11 Experimentally determined conversion efficiencies for two aspect ratios of the pump beam, one favourable for minimising walk-off (1:2), the other unfavourable (2:1). In the case of the former, the efficiency decreases less rapidly as the phase match angle decreases and the walk-off angle increases.

## 12.4 REFERENCES

⊗ Note addded in proof: This tuning range has recently been increased significantly to extend from 354nm to 2.38$\mu$m, and conversion efficiencies greater than 15% have been observed over much of this range.

Ebrahimzadeh M and Dunn M H 1988a CLEO Technical Digest Anaheim PD30
Ebrahimzadeh M and Dunn M H 1988b Opt. Commun. 69 161
Ebrahimzadeh E and Dunn M H 1989a CLEO Technical Digest Anaheim 11 416
Ebrahimzadeh M and Dunn M H 1989b Opt. Lett. 14 560
Ewart P 1985 Opt. Commun. 55 124
Hansch T W 1972 Appl. Opt. 11 895
Littman M G 1978 Opt. Lett. 3 138
Littman M G 1984 Appl. Opt. 23 4465
McIntyre I A, Lusty M E and Dunn M H 1988 J. Mod. Opt. 35 325
Shoshan I and Oppenheim U1978 Opt. Commun. 25 375
Zhang G W , Grethen H, Kronfeldt H D and Winkler R 1981 Opt. Commun. 40 49

# 13 Resonators for High Power $CO^2$ Lasers

*D J DYSON*

## 13.1 INTRODUCTION

The term "high power $CO_2$ lasers" is taken to mean those carbon dioxide lasers with average powers extending upwards from about 100 watts, currently up to around 10 kilowatts, which are mainly used for industrial processing. Although primarily designed for cw (continuous wave) operation they are often also able to operate in a pulsed mode. (This is essentially chopped cw although some power enhancement at the beginning of the pulse is frequently obtained). Closely related, at the lower end of the power range, are the waveguide lasers which also find use in medical applications. More distantly related are the 'TEA' lasers which operate in a pulsed mode only. Much of what is said in the following sections will apply to these related categories but other aspects may also be involved which are covered in other chapters.

These lasers employ a gaseous amplification medium, usually a mixture of carbon dioxide, nitrogen and helium which is electrically excited (although other methods are possible). For effective laser action to occur, the amplifying medium must be cooled and it is the method of cooling which is usually the most definitive aspect of the design of the laser. Three basic types are recognised i.e. 'conventional', 'fast axial flow' and 'fast transverse flow'. Of these, the first depends on conductive cooling of the gas, while the latter two rely on forced convective cooling. A detailed discussion of the technical aspects of these lasers is inappropriate here and for this, the reader is referred to the relevant literature. Although it is true that particular classes of laser are more likely to be found with a particular type of resonator, the situation is by no means rigid, and it is posible to discuss resonator design to a large extent without reference to type of cooling.

## 13.2 THE REQUIREMENTS OF THE CAVITY DESIGN

The fundamentals of laser resonator design have been discussed in Chapter 1 and it is these features which determine the quality of the laser output beam. Pre-eminent is the requirement for high brightness, or low divergence, allowing the beam to be focused to the smallest possible spot size. The 'best' beam, for this purpose, is normally regarded to be one with a $TEM_{00}$ mode profile from a stable cavity and this type of beam has been widely used, particularly at the lower end of the power range. However, other modes with relatively low divergence have also been widely used. For many applications, a combination of low order TEM modes, and particularly a combination of $TEM_{00}$ and $TEM^*_{01}$ has been found to be satisfactory. Some applications, particularly heat-treating, as opposed to cutting and welding, are less sensitive to beam divergence and higher order TEM modes may be considered. Since higher output power may be available from higher order mode operation, there can be an advantage in being able to operate a laser in more than one mode configuration. For lasers at the top of the power range, where the volume of gain medium is large, unstable resonators can offer considerable advantages. These give a mode with a divergence comparable with that of a $TEM_{00}$ mode, provided the gain is sufficiently high.

Apart from the mode divergence, there are several other important aspects of $CO_2$ laser performance, related to cavity design, which must be considered. The following paragraphs describe some of these.

The coupling between the mode and the amplifying medium must be good, leading to efficient power extraction. This normally means that the shape of the mode within the cavity must be such that all parts of the amplifying medium are exposed to the beam.

The beam mode (or mode combination) should be stable in time. This applies both to the shorter time scale (a few seconds) during which the cavity optical components come into thermal equilibrium and the longer time scale (minutes to hours) during which the system support structure comes into thermal equilibrium, following initiation of the electrical discharge. Inadequate stability in the support structure will result in a failure to maintain correct alignment of the cavity, resulting in the need to 'fine tune' at regular intervals. Although the design of the mechanical structure plays an important part in overcoming this problem, the design of the cavity also has an important effect since this determines the tolerance to misalignment.

The output beam should be reasonably 'clean' in the sense that the proportion of scattered or reflected radiation which is not part of the main oscillating mode should be kept to a minimum. Such scattered light often arises from reflections off the inside surfaces of discharge tube walls, forming a 'halo' around the focused spot, when the beam is focused. Moreover, because of its time-coherence with the main beam, it can cause large perturbations of the beam in the near field, (i.e. away from the focus), due to interference effects. Since even quite small proportions of scattered light can cause large perturbations it is often difficult to distinguish the true operating mode of a laser by observing the beam profile close to the output window. (See also the remarks about bouncing modes in the final paragraph of section 13.8 of this chapter).

The cavity design can have important effects on practical aspects of operating the laser. As well as putting less demand on the stability of the mechanical structure, a reasonable tolerance to misalignment can ease the problem of restoring correct alignment after the optical components have been removed for inspection and cleaning. This is particularly true if it allows the possibility of designing an optical mounting system in which the components can be removed and replaced, without the need for further adjustment.

The design of the cavity can influence the life of the optical components since it determines the power density of the beam at the various optical surfaces. The methods of cooling the optics can also significantly affect their useful life.

In the following sections more detailed consideration will be given to those aspects of the cavity design which bear on the characteristics of the beam outlined above. The analysis of design features presented here, is generally restricted to approximate or 'order of magnitude' calculations. Where situations are too complex for any meaningful simple calculations, relevant experience and empirical data are presented along with rough estimates of how one might scale these known results to cover different situations. Where appropriate, possible methods for a more precise analysis are indicated. Note that it is sometimes the case that precise analysis may not be worthwhile since the situations being considered are inherently too variable or uncontrolled.

## 13.3 THERMAL DISTORTION

The term "thermal distortion" is usually used to refer to the change in shape of a reflecting (or partially reflecting) surface, arising from the heat input from the beam. It may in part be responsible for changes in the output beam from a laser during the first few seconds after the electrical excitation is switched on, since the time constant for thermal equilibrium to become established in a typical component is normally of this order.

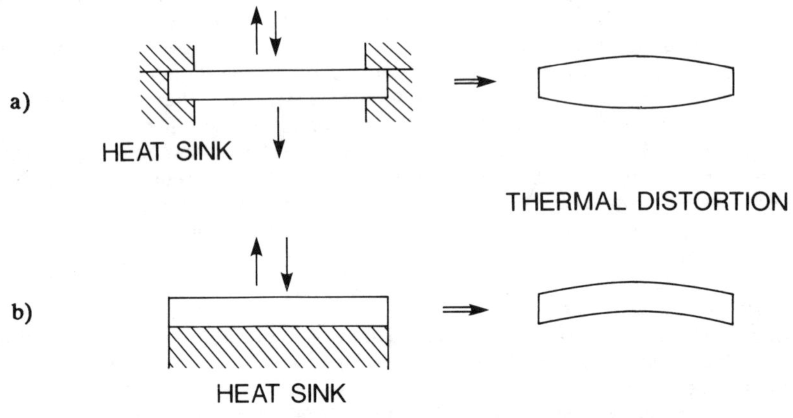


Fig 13.1 Cooling arrangements for high power $CO_2$ laser mirrors, and the resulting distortion due to thermal expansion.

The analysis of thermal distortion depends on identifying the heat flow and temperature rise patterns within the mirror substrate, as heat absorbed from the incident beam is transferred to a heat sink formed by the mirror mount. Figure 13.1 illustrates the two commonly encountered cases. Edge-cooling in case (a) is necessary for transmitting optics such as cavity output couplers and focussing lenses. Heat flow is predominantly radial for thin substrates in the edge-cooling case. Case (b) removes heat through the reverse face of the substrate and is suitable for high reflectance cavity mirrors, cavity folding mirrors and beam–steering mirrors. In this situation, heat conduction is predominantly axial. For the highest power $CO_2$ lasers, heat removal by substrate conduction may be augmented by forced gas cooling across the mirror surface, or air convection.

The simplest case to consider analytically is that when a uniform intensity $I_0$ strikes one face of a thin mirror, and where heat is removed from the reverse face, as in Figure 13.1b, under conditions of *free expansion* where the substrate is not constrained in any way by its mounting. The differential expansion of the heated and cooled faces produces a bowing of the initially, flat substrate whose radius of curvature $R_d$ is:

$$R_d = \frac{K}{eAI_0} \tag{13.1}$$

where $K$ is the substrate conductivity, $e$ is the thermal expansion coefficient and $A$ is the fraction of the incident intensity, $I_0$ absorbed by the mirror.

The edge cooled configuration in Figure 13.1a can also be analysed, but not quite so simply, for distortions resulting from a uniform intensity $I_0$ and purely radial heat flow. As illustrated in Figure 13.2, the uniform heating by absorption produces a parabolic temperature profile across the substrate. Under these conditions, thermal expansion causes a radially varying thermal stress in the substrate, and distorts each surface to produce a radius of curvature $R_d$ given by:

$$R_d = \frac{4K}{e(1+\nu)AI_0} \tag{13.2}$$

where $\nu$ is Poisson's ration. Helpful formulae for deriving these results are given by Roark and Young (1975 a,b).

In practice the beam profile of the incident light will not be a uniform intensity, and the thermally induced curvatures predicted by the above formulae will be modified to some extent when the total incident power occupies less than the whole substrate aperture. Additional uncertainty is also caused by the departure from true free–expansion, caused by mechanical constraints associated with vacuum seals, and clamping forces exerted to attain good thermal contact between the substrate and its mount. Detailed thermo–elastic calculations using numerical techniques may be necessary to assess these effects.

A simple assessment of any given situation may be made by applying the above formulae, using the maximum incident power density for $I_0$, and seeing whether the resultant curvature is sufficient to produce a significant effect. This may be done, for both stable and unstable cavities, by adding the induced curvature, $1/R_d$ to the existing curvatures (taking note of signs since the induced curvature is convex) and re–calculating the cavity mode parameters with the new combined curvatures. One must consider the effect both on the beam intensity profile and on the beam output curvature. Since the output curvature is normally matched to the curvature of the final element in the cavity (e.g. the output window), the latter effect is related simply to the distortion of this component.

If the effects are insignificant one can assume that thermal distortion will have no effect in the real case. If the calculated effort is marginal, it may be necessary to carry out a more detailed calculation of the induced curvature. Ultimately, one could consider solving the general problem of finding a self-consistent solution of a mode supported by a distorted cavity, but one would have to be careful to ensure that this was worthwhile, given the variability of the real situation that was being represented. Numerical methods for cavity analysis are discussed in Chapter 7.

The expression $K/e$ above can be regarded as a figure of merit for the substrate material of a mirror, the larger the value the less the thermal distortions effect. Table 13.1 gives the values for a number of relevant materials.

| | $K$ | $e$ | $K/e$ |
|---|---|---|---|
| Material | $Wm^{-1}K^{-1}$ | $K^{-1}$ | $Wm^{-1}$ |
| Glass | 8.4 | 9 | 0.9 |
| Fused quartz | 10.0 | 0.5 | 20 |
| Copper | 3760 | 18 | 209 |
| Silicon | 1296 | 4.1 | 316 |
| Molybdenum | 1463 | 4.9 | 299 |
| Stainless Steel | 167 | 17 | 9.8 |
| Zinc Selenide | 180 | 8.5 | 21 |
| Germanium | 590 | 5.7 | 103 |
| | $X10^{-1}$ | $X10^{-6}$ | $X10^{5}$ |

Table 13.1 Values of $K/e$ for various materials (figure of merit for thermal distortion).

Some indication of the importance of the thermal distortion effect can be obtained by applying the criteria, widely used in optical manufacture, that a mirror surface must be true to $\lambda/20$ where $\lambda$ is the wavelength. If the beam radius in the situation considered above was $a$; then the condition that the maximum error was $\lambda/20$ would lead to:

$$\frac{\lambda}{20} = \frac{a^2}{2R_d} = \frac{a^2}{2}\,\frac{eAI_0}{K} \tag{13.3}$$

$$I_0 = \frac{\lambda}{10}\,\frac{K}{eAa^2} \tag{13.4}$$

or $$\text{Total power} \approx a^2 I_0 = \frac{\lambda}{10}\,\frac{K}{eA} \tag{13.5}$$

This criterion leads to the conclusion that, for a fixed thermal distortion effect, the total power which can be tolerated remains constant as the beam diameter varies. Putting in typical values, one obtains a critical total power for a copper substrate and $A$ = 0.02 (typical of a gold reflective coating) of ~1kW. Silicon forms a good substrate for low loss multilayer coatings for which values of $A$ of 0.003 are possible, yielding, in this case, a critical total power of ~11kW.

It must be emphasised that this analysis is only strictly applicable where one has a single reflection off a mirror external to a cavity. Where several reflections occur the distortions are additive and the tolerable power is lower.

## 13.4 THERMAL LENSING

The term 'thermal lensing' is usually used to refer to the lens effect induced in a transmissive component as a result of power absorbed from the beam. It arises partly from the change in shape due to thermal expansion effects but, in most materials of significance for $CO_2$ lasers, by far the most important contribution comes from the change in refractive index of the material with temperature. This is an additional effect which changes the output beam parameters of a laser during the first few seconds after switching on the discharge. Although, if the effect is small, it can be regarded as equivalent to a simple lens, the consequences can become complex, causing distortion and degrading the inherent laser beam divergence.

An interesting example of the latter effect arises when one has a spurious beam, perhaps caused by reflection off an internal surface, which causes an interference pattern in the output window. The corresponding variations of refractive index result in a diffraction grating which diffracts some of the main beam, thus enhancing the spurious beam.

(a)

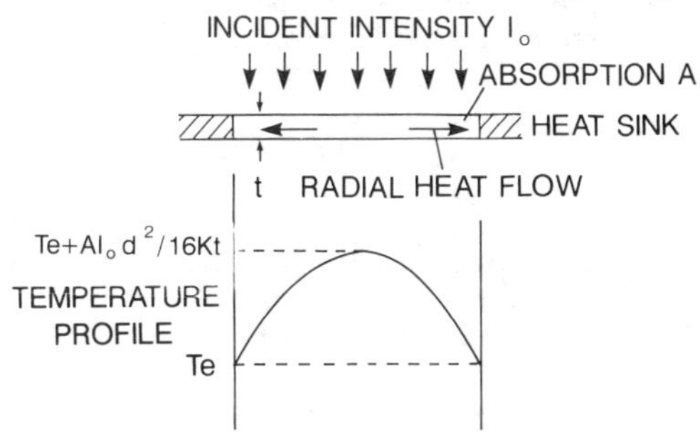


(b)

Fig 13.2 Temperature distribution in an edge cooled mirror of diameter *d* with uniform intensity illumination.

A simple view of thermal lensing can be obtained by considering an edge-coupled, thin window which is heated by absorption of a fraction $A$ of the incident uniform intensity $I_0$, as in Figure 13.2a. The parabolic temperature profile in Figure 13.2b is again obtained, and the beam passing through the window experiences an optical path length variation given by:

$$\Delta x = \Delta n . t = t\frac{dn}{dT} \cdot \Delta T \qquad (13.6)$$

The temperature coefficient of refractive index of the material, $dn/dT$ sets the optical distortion of the transmitted beam. For the idealised geometry of Figure 13.2a, and positive $dn/dT$, the distortion is a converging lens with focal length $f$ given by:

$$f = \frac{2K}{AI_0\,(dn/dT)} \qquad (13.7)$$

The factor $K/(dn/dT)$ can be regarded as a figure of merit of the window material for thermal focussing, and values of $K$, $dn/dT$ and the figure of merit are given in Table 13.2 for materials suitable for $CO_2$ lasers. The thermal expansion distortion of the window also acts to produce a converging lens, but in practice this effect is usually much weaker than that caused by thermal refractive index variations.

| | $K$ | $\frac{dn}{dT}$ | $\frac{K}{dn/dT}$ | $\alpha_0$ |
|---|---|---|---|---|
| Material | $Wm^{-1}K^{-1}$ | $K^{-1}$ | $Wm^{-1}$ | $m^{-1}$ |
| Germanium | 59 | 2.77 | 2.18 | ~ 3 |
| Gallium Arsenide | 48 | 1.49 | 3.22 | 0.8 |
| Zinc Selenide | 18 | 0.64 | 2.8 | 0.05 |
| | | $\times 10^{-4}$ | $\times 10^{5}$ | |

Table 13.2 Material Data for thermal lensing in transmissive optics for $CO_2$ lasers

In the simple model considered so far, any dependence of the optical distortion on window thickness comes from the value of absorption $A$, which has contributions from both the bulk absorption of the substrate and the absorption of surface coatings, including the effects of surface contamination.

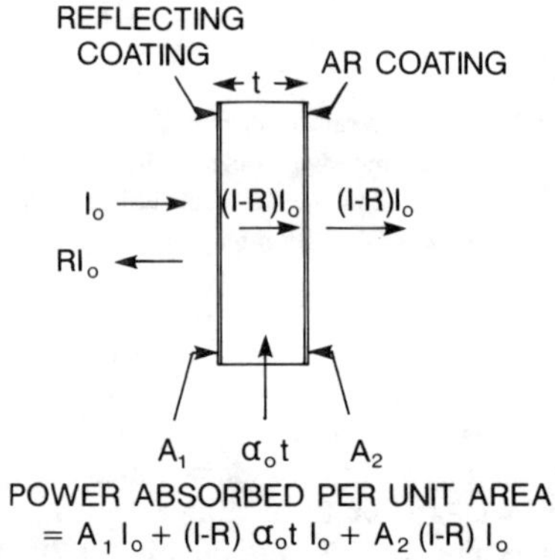


Fig 13.3 Absorption of laser radiation in a partially transmitting cavity mirror, where $I_0$ is the in-cavity intensity and $R$ is the reflectivity.

Figure 13.3 illustrates the situation for the partially reflecting output coupler of a laser cavity. The design of a coupler with a particular material depends on the relative importance of the surface and bulk absorption terms. Surface absorption can be lower than 0.1% for clean dielectric coatings, whilst bulk absorption coefficients at room temperature are indicated in Table 13.2. Unlike the thermal distortion of a total reflector, the figure of merit of the material for thermal lensing in Table 13.2 is not so useful a quantity for assessment of a material for transmitting optics. General conclusions about materials selection cannot be made solely using this figure of merit, since the associated absorption is very dependent on additional effects such as coating quality, material purity and in particular the working temperature.

In assessing the allowable beam size in a $CO_2$ resonator or beam delivery system, one is particularly interested in how the thermal lensing effect varies with beam diameter for a given fixed total power. Suppose one considers the case of a window whose edge is maintained at a fixed temperature and where the effects of convective cooling at the window surfaces are ignored. If the diameter is varied at constant power (and hence constant absorbed power) the central temperature remains constant (since the heat flow out of a cylinder with fixed axial and edge temperatures depends only on the length, not the diameter). The wavefront error therefore remains constant, depending on the total power only.

If, because of the need to withstand a pressure difference, the thickness is also increased, then, if the power absorbed remains fixed due to the absorption being principally at the surfaces of the window, the temperature difference will fall in a way which is inversely proportional to the thickness. The overall wavefront error will still remain the same since it is also proportional to the thickness, i.e. it still depends on the total power only. If, on the other hand, the absorption is dominated by bulk absorption in the material, the temperature difference remains constant and wavefront error increases proportionally with the thickness.

To summarise, two extreme conditions have been identified, depending on whether the absorption is predominantly at the surface or in the bulk of the material. The results are illustrated in Table 13.3 which has been configured to show how the input power would be varied to maintain (i) a constant temperature difference or (ii) a constant wavefront error as the thickness is varied. (Note that although there is no dependence on diameter one is likely to need to change the thickness for practical reasons if the diameter is changed).

| | (i) Const. temp. diff. | (ii) Const. wavelength error |
|---|---|---|
| Dominant surface absorption | $t$ | *Const.* |
| Dominant bulk absorption | *Const.* | $t^{-1}$ |

$t$ = thickness NB – the power is independent of the diameter.

Table 13.3 Scaling of total power to maintain constant temperature difference or constant wavelfront error with dominant conduction cooling.

Real cases will not generally fall into either of these categories and will be further complicated by the additional cooling effect arising from convection at the surface. Germanium has a rather high bulk absorption which increases rapidly with temperature and is probably well represented by the bulk absorption case. It has a high thermal conductivity which makes it useful at medium power levels. Here, one has the interesting situation that the total power that can be tolerated, from a thermal lensing point of view, is inversely proportional to the thickness only. A smaller, thinner, mirror will give a better performance, limited by any pressure differential applied.

Two other commonly used materials are gallium arsenide and zinc selenide. Compared with germanium, these have progressively lower bulk absorptions and lower thermal conductivities. They also have progressively lower coefficients of refractive index with temperature. The net effect is that they perform better than germanium with respect to thermal lensing.

Experience with lasers at the 1kW level operating in the $TEM_{00}$ mode with $1/e^2$ beam diameter of ~7mm shows that relatively little thermal lensing with a clean zinc selenide output window need be expected. However, for long throw applications the lensing is detectable but can be further reduced by the application of a jet of air to enhance the cooling.

The performance of zinc selenide at high power levels is difficult to estimate because of the increasing effect of convective cooling. It is not proposed to tackle this problem in detail here except to say that the convective cooling component does provide a means of tolerating higher power levels at larger diameter. (This is not the case with conduction limited cooling if one is considering the thermal lensing performance only).

It has been assumed in the above arguments that the output window is operating well clear of the regime where the absorption begins to increase with temperature. If this is not the case, then this effect must also be taken into account in estimating the thermal lensing effect. In the first case illustrated in Table 13.2, for example, an increase in thickness could result in a reduced (rather than constant) thermal lensing effect because of the lowering of the temperature. The effects of temperature rise are considered in more detail in the following section.

## 13.5 THERMAL DAMAGE

Thermal distortion and lensing are not the only effects which limit the powers that components can tolerate. As the power is raised, a point is eventually reached where local temperatures reach a critical level and where the absorption begins to further increase, either due to temperature-dependent absorption or permanent physical change (e.g. melting). The increased absorption results in still higher temperatures, leading eventually to catastrophic failure if the incident power is maintained. Local high-absorption areas will normally occur at defects in the component surface or in areas where the surface is contaminated with foreign particles. Failure will therefore occur at power levels below that required to raise the general temperature of the component to the critical level.

One could define a critical power density at which failure would be likely to occur on the basis that the general temperature rise of the substrate was small compared with that required to reach the critical temperature. In cases where the cooling was not good enough to achieve this condition, the maximum permissible power density would be lower.

It should be pointed out here, that the failure mechanisms are time-dependent. Much work has been done to assess the ability of components to withstand very high-energy short-duration pulses of power. The case under consideration is rather one of cw, or quasi cw operation, where the surface temperature is in an equilibrium state determined by the local heat input and conduction into the substrate. Failure mechanisms in the two situations may be different.

It is apparent from the above that the maximum power density which a component can tolerate before failure becomes likely, depends on factors such as: the adequacy of the cooling (although there may be little to be gained by improving the cooling beyond an adequate level); the temperature that the component substrate or coating can withstand before the absorption increases or physical damage occurs; the thermal conductivity of the substrate; and the nature and size of defects in, or foreign particles on, the component surface and their probability of occurrence. It is difficult to generalise about the occurrence of defects and foreign particles in optical components. Defects may vary from production batch to production batch and the nature and occurrence of foreign particles will depend on the operating environment. Consequently, it is not possible to quote precise values of maximum power density to cover all situations.

Considering first the case of mirrors, if one compares the two types of mirror already mentioned, namely gold–coated copper and multi–layer dielectric coated silicon, it is generally accepted that the higher critical temperature and better thermal conductivity of gold coated copper makes it superior for use in the open atmosphere where foreign particles are likely. For cleaner environments, such as the inside of a laser cavity, the dielectric coated silicon can readily withstand the same power densities. Dielectric coatings on copper are available but experience seems to suggest that they are more damage–prone than a well–cooled gold coated copper mirror. One form of mirror which is particularly rugged is plain molybdenum. This has an absorption of ~ 0.03.

Experience over several years indicates the feasability of the use of dielectric coated silicon inside the laser cavity and gold–coated copper outside the laser cavity working at a peak power density ~ $5 kW/cm^{-2}$. In general, it has been found that no mirror will work satisfactorily at this power density in the worst industrial environments and some form of protection (e.g. a clean–air jet) is necessary. Within this limitation, the failure rates observed at this power density have been acceptable.

In the case of transmissive components, cooling is more difficult since direct (e.g. conductive) cooling cannot be applied locally near the centre of the component and it is necessary to rely either on conductive cooling to the edge or convection from the surface. Moreover, unlike the case with fully reflecting mirrors, one is not able to choose materials with necessarily the best thermal conductivities. One is often therefore, in the situation where the effect of the general temperature rise of the component is significant.

With semiconductor materials such as germanium, gallium arsenide and zinc selenide, the critical temperature is determined by the effect of increasing absorption of the substrate material with temperature, leading to the possibility of thermal runaway. In germanium, this effect is particularly critical, the onset of increasing absorption occurring at ~$30^0$C. The critical temperatures for gallium arsenide and zinc selenide are much higher. As already indicated, the effects of coating defects or foreign bodies require for safe operation that there should be a safety margin between the general substrate temperature and the critical runaway temperature. This margin must be larger for the materials with lower thermal conductivity.

The results in Table 13.3 summarise how the temperature rise depends on the geometry of the component in the case where conductive cooling is dominant. As has already been indicated, germanium may be well characterised as showing dominant bulk absorption. In this case, it is seen that the input power for a constant temperature rise is independent of both diameter and thickness. If one assumes a given edge temperature and a fixed maximum safe operating temperature, the transmitted power that can be tolerated is then also fixed, being independent of the component dimensions. However, if one goes to a smaller diameter the power density will increase and a greater safety margin should be allowed for defects and foreign particles. Provided one keeps below the above mentioned power density $\sim$5kW/cm$^2$, experience shows that germanium can be used at powers up to about 200W.

Gallium arsenide and zinc selenide can be used at much higher powers than germanium, although, as already indicated, the calculations are more difficult because of the possible effects of convection. The conditions mentioned above as giving just detectable thermal lensing with zinc selenide (1kW in a 7mm $1/e^2$ diameter $TEM_{00}$ beam) are also usually satisfactory from the thermal damage (i.e. life) point of view for this material.

Experience over many years also indicates that at least in the cases of zinc selenide and gallium arsenide, the onset of increased thermal lensing acts as a warning that the absorption is increasing (usually as a result of surface contamination) before thermal damage or catastrophic failure actually occurs. Normally this will become apparent due to a failure of the process being carried out by the laser. Catastrophic failures normally only occur in cases where this warning is not recognised in time.

It should be pointed out here that, if a catastrophic failure of one of these materials does occur, the result may be the production of the toxic oxides of selenium or arsenic, normally deposited as powders on nearby surfaces. The equipment manufacturers should give warnings of this and instructions as to how to deal with such a failure – normally the oxides should be removed with an excess of water taking care not to inhale or touch the deposits.

## 13.6 ALIGNMENT TOLERANCE

Ideally, the effect of misalignment of one of the mirrors of a cavity should be determined by carrying out a diffraction calculation on the mode of the misaligned cavity, and varying the amount of misalignment so as to establish the level at which the variation of the original properly aligned cavity ceases to be tolerable. This also involves estimating, or postulating arbitrarily, what is defined as tolerable. There are today, many computer codes available which readily enable such calculations to be made and some relevant numerical techniques have been discussed in Chapter 7. However, various methods are also available for making rough estimates, depending on the precise nature of the cavity.

A particularly simple method which can often be applied to cavities involving curved mirrors consists of monitoring the motion of the optical axis when mirror misalignment occurs. The optical axis is defined as the path of a light ray which retraces its path backwards and forwards through the cavity. In the case of cavities consisting of only two mirrors, this is the path which strikes each mirror surface perpendicularly, and is readily located as the line which joins the centres of curvatures of the two mirrors. If one of the mirrors is plane it is the line which passes through the centre of curvature of the curved mirror and is perpendicular to the surface of the plane mirror. Given that one can readily calculate the position of the optic axis, it is possible to calculate how it moves if one misaligns one of the mirrors. The alignment tolerance is then calculated by finding out how much the mirror can be misaligned without causing the optical axis to move by more than a small proportion of the beam diameter.

The situation is slightly more complicated in the case of folded cavities. The procedure for finding the optical axis in this case is to launch rays from either end of the cavity which are perpendicular to the end mirrors, but not necessarily in the centre, and adjust the positions of launch until two paths are found which coincide. The coincident path is the optical axis. Two simple examples of this procedure are discussed in Chapter 5.

If one wishes to find the tolerance to misalignment of one of the fold mirrors, it is necessary only to find two such rays which coincide in position on the mirror in question. The amount by which the mirror has to be tilted to make the rays coincident in direction can then be calculated. This leads to a relationship between a given angle of tilt of this mirror and the movement of the optical axis. Since this relationship is linear, the maximum tolerable tilt of the mirror can be easily calculated.

In general, although the above arguments lead immediately to the result that long radii of curvature produce critical alignment tolerances, one must look carefully at individual cases. For the case of the most commonly used form of unstable resonator, both the centres of curvature lie in the same direction from the cavity and when estimating the effect of moving the centre of curvature of one of the mirrors the movement of the optical axis in the cavity is magnified by a 'lever' effect, as compared with the similar situation in a stable cavity.

## 13.7 REFRACTION EFFECTS

Temperature variations in the gaseous amplifying medium can cause refraction effects on the laser beam inside the cavity. For example, in a transverse flow laser the temperature will rise as the gas travels through the gain region causing an 'optical wedge' effect which would be similar to a misalignment of the laser. In the case of a conventional slow flow laser, on the other hand, the high temperature along the tube axis will result in a negative lens. The effect is probably least in a fast axial flow type of laser where the main temperature gradient is in a direction parallel to the laser beam.

The effects may be estimated by using the figures for the individual refractive indices of the He $N_2$ and $CO_2$ to calculate a refractive index for the gas mixture. The values are, for atmospheric pressure and $0\,^{0}$C:

| | |
|---|---|
| $He$ | 1.000036 |
| $N_2$ | 1.00030 |
| $CO_2$ | 1.00045 |

Assuming a mixture of $70\%He$, $20\%N_2$ and $10\%CO_2$ and a typical working pressure of 50 millibar, the refractive index exceeds unity by $6.5\text{x}10^{-6}$.

In most $CO_2$ lasers, the maximum temperature which occurs is about 200$^0$C since this is the maximum temperature for efficient laser action. If it is assumed that the minimum temperature of the gas is about 30$^0$C, then the difference in refractive index between these two temperatures becomes $2.1 \times 10^{-6}$. The maximum path length change in the gas for this temperature difference amounts to $\lambda/10$ in 0.5 metres.

It can be seen from the above argument that the effect can be significant. The correct procedure to follow, as was the case with thermal distortion, is to calculate the effect on the particular cavity in question. An example of the effect of gas thermal lensing is given in Chapter 5. Note that the effect differs from thermal distortion in that it occurs instantly. If it does not vary, i.e. if the laser is always operated with the same gas temperatures, and provided the effect is small, it can be allowed for in the resonator design. A problem can occur when the temperatures vary as power into the discharge is varied in order to vary the output power.

## 13.8 DESIGNING A LASER CAVITY

Having reviewed most of the factors which have to be considered in the practical design of high power $CO_2$ laser cavities, we now illustrate some of the arguments and trade-offs which arise in typical design projects. It is well established that in conventional 'slow flow' lasers the available output power per unit length is independent of the tube diameter if one maintains pressure-diameter product constant. This result is what is expected from simple scaling considerations in a system where cooling is dominated by conduction to the walls. In effect one has a 'free choice' of tube diameter as far as the generation of laser power is concerned and the actual choice of diameter is governed principally by cavity design considerations. A clear first priority is to ensure that the tube diameter is large enough to keep the power density to an acceptable level from the point of view of thermal damage.

If one considers the simplest case of a symmetrical stable cavity, the requirement that the mode shape couples well to the gain medium normally implies that the cavity mirror radius of curvature is considerably greater than the distance between the mirrors. This leads to a mode of approximately uniform diameter within the cavity. A radius of curvature of at least 3 times the spacing between the mirrors is normally adequate. This condition, itself, determines a minimum size of mode, and hence tube diameter, for a given mirror separation.

It is not necessary for the mirror separation to be equal to the overall length of gain medium if a folded configuration with curved mirror folds is employed. From most points of view, such a design can be treated as being equivalent to a simple unfolded cavity using the same mirror curvatures but with a length equal to a single folded section. This situation is covered in more detail in the Chapter 5.

As well as power density considerations, one also has to consider the problems of thermal distortion and alignment tolerance. These latter problems become more critical as the radius of curvature of the mirrors is increased, in order to obtain a larger mode diameter, with a view to maintaining an acceptable power density as the output power is increased. One therefore has conflicting requirements which become more difficult to satisfy at higher powers. Eventually, a point is reached, depending on the optical materials available, where it is no longer possible to design a $TEM_{00}$ stable cavity which both operates at an acceptable power density and is free from unacceptable thermal distortion effects.

To solve this problem, one may consider using an unstable cavity if, at the diameter required to maintain an acceptable power density, the gain is sufficiently high. (To extract power efficiently, using a cavity with a magnification of 3, a small–signal gain of at least ~100 times per double pass would be desirable). Since the radii of curvature of the mirrors used do not depend on the beam diameter, and are generally much shorter than the very long radii that often occur in large diameter stable cavity designs, thermal distortion effects should be less. The main effect will be on the output beam curvature, rather than the mode profile.

Once the type of cavity and beam diameter have been established, it is necessary to check that the design is adequate from all the points of view that have been considered in this chapter. The alignment tolerance should be estimated and this information used as the basis for the design of the mechanical support structure for the mirrors. The possible effects of refraction effects in the gas, and how these will be dealt with, will have to be considered. The material for the output window must be selected and an estimate made of the thermal lensing effects and temperature rise. If a suitable output window cannot be found it may be necessary to consider an aerodynamic output window.

The possible effects of bouncing modes supported by reflections off the tube walls or "clutter", also caused by reflections off the tube walls, should be considered. Note that bouncing modes can be most effectively supressed by having 'dead' sections where the internal diameter becomes larger. Designs where the length is compressed to the minimum by having active discharge tubes nearly filling the space between the mirrors are particularly prone to these effects. The only remedy in this case is to make the walls non specularly reflecting. This is very difficult because of the long wavelength and oblique angle of incidence; simple roughening is generally not sufficient. Designs incorporating 'fluted' discharge tubes or tubes with metal ring inserts have been employed to help solve this problem.

## 13.9 CONCLUSION

This chapter has been a broad discussion of some of the main effects that have to be considered in the practical design of high power $CO_2$ laser cavities. Because of the wide range of powers and laser types considered, it has had to be very general. Much detail has necessarily been left to the reader who is going to be involved in carrying out such design work.

## 13.10 REFERENCES

Roark R J and Young W C 1975a Formulas for stress and strain 5th Edition (Tokyo: McGraw Hill, Kogakusha) 366

Roark R J and Young W C 1975b *ibid.* 584

# 14 Diode Laser Resonators

***M J ADAMS* and *I D HENNING***

## 14.1 INTRODUCTION

The three basic types of resonator used in diode lasers are the Fabry–Perot (FP), distributed feedback (DFB), and distributed Bragg reflector (DBR). However, extra degrees of sophistication have been incorporated into these three types in efforts to achieve spectral purity, high power, and fast modulation of diode lasers. In many cases the motivation for these efforts has stemmed from the use of lasers as sources for optical fibre communication systems. In this chapter the development of diode laser resonators is traced from the basic double heterostructure to the complicated designs now used for optical integration. In order to set the scene, a brief preliminary discussion is given of diode laser fundamentals; a more detailed account will be found in a recent book (Adams *et al*, 1987).

## 14.2 DIODE LASER FUNDAMENTALS

Efficient electroluminescence in semiconductors is based on radiative recombination of electrons and holes in a structure designed such that an adequate supply of both types of carrier may be injected by the application of a current. The requirement to simultaneously satisfy energy and momentum conservation implies that a direct-gap semiconductor such as GaAs is needed. In order to achieve sufficient electrons in the upper energy levels (usually in the conduction band) and holes in the lower levels (valence band), a p–n junction may be used. However, for most present-day sources for optical communications a more sophisticated structure involving two heterojunctions is employed (Figure 14.1). This 'double heterostructure' provides a well-defined layer – the central 'active' layer – where recombination occurs and carrier wastage by diffusion to other regions is prevented by the potential barriers at the hetero-interfaces. The most common devices of this kind consist of a GaAs active layer between n– and p–type layers of AlGaAs, so that the outer layers are of larger band-gap and therefore transparent to the emitted radiation.

Figure 14.1 shows a double-heterostructure laser (in this case fabricated in the InGaAsP system). The structure makes use of a light-guiding effect which occurs in double heterostructures and which is commonly used in semiconductor lasers. This effect stems from the fact that the narrower-gap active material has a higher refractive index than that of the wider-gap passive layers. For GaAs at a wavelength of 0.85 μm the refractive index is 3.6, whilst for $Al_{0.3}Ga_{0.7}As$ the value is about 3.4. For InGaAsP at 1.55 μm, 1.3 μm and 1.1 μm the figures for refractive indices are about 3.55, 3.4 and 3.3, respectively (see Figure 14.1). For InP at these wavelengths there is very little dependence on wavelength, the index being approximately 3.2. Thus the double heterostructure forms a dielectric waveguide.

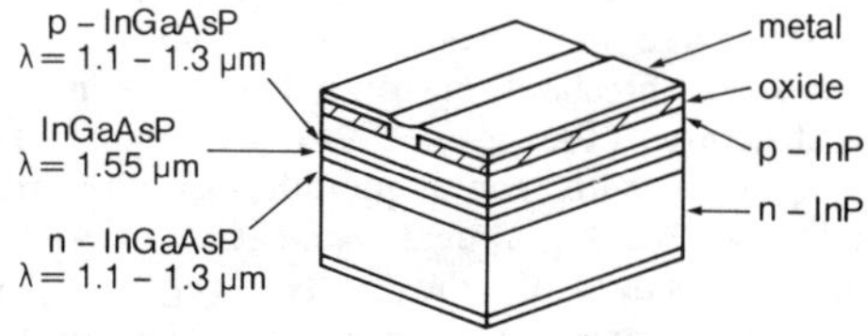


Fig 14.1 **Double-heterostructure laser in the InGaAsP/InP system.**

The emission wavelength of a laser is determined primarily by the band-gap of the active semiconductor. The band-gap of GaAs at room temperature is about 1.43 eV so that the emission wavelength lies at about 0.85 μm. Optical fibre losses and material dispersion are minimised in the range 1 - 1.6 μm with special emphasis on the wavelengths 1.3 and 1.55 μm, and to make sources for this range means using materials other than the AlGaAs/GaAs system. The most promising materials system - InGaAsP/InP - has already been mentioned above.

## 14.3 DIODE LASER STRUCTURES

The semiconductor laser uses the double heterostructure discussed above to provide confinement of carriers and radiation in the direction normal to the junction plane. The thickness of the active layer is dictated largely by the conflicting requirements of minimising the distance which must be filled with high densities of electrons and holes (to ensure that emission exceeds absorption - the so-called *population inversion* condition) and of providing a sufficiently strong waveguide to prevent the optical field from encountering undue loss outside the active layer. The optimum thickness is usually around 0.1 μm. In addition to the active and passive (confining) layers, the structure also includes a substrate, possibly one or more buffer layers, and a final capping layer to ease contacting problems. The entire structure may then look like the example illustrated in Figure 14.1 for the InGaAsP/InP system.

In the structure shown in Figure 14.1 the definition of the active light-emitting region in the direction parallel to the junction plane is rather vague, relying only on an oxide-insulated stripe contact. Current is then free to spread over some distance beyond the stripe width as it flows through the active layer. The radiation that is generated is not confined in the direction parallel to the junctions by any real waveguide structure as in the case for the direction perpendicular to the junctions. Instead a virtual waveguide may exist since there is a region of gain under the stripe, and regions of loss on either side. This *gain-guiding* effect is obviously a function of the current and of the optical intensity in the device. As a result it can give rise to a number of undesirable effects in laser performance, including strong nonlinearity in the light output–current curve, and pulsations in the output of a nominally cw laser. For this reason, laser designs have evolved which provide built-in lateral confinement for the optical field and the carrier distribution.

Two such designs are illustrated in Figure 14.2. The ridge-guide laser (Kaminow *et al*, 1979) of Figure 14.2(a) incorporates a built-in waveguide as a result of the large semiconductor/air refractive index step close to the active layer. Although some degree of current confinement occurs, lateral diffusion in the active layer leads to threshold currents higher than optimum (~ 35 mA). The buried heterostructure (Tsukada, 1974) of Figure 14.2(b) requires two stages of epitaxial growth with some photolithography and etching stages in between. It includes a lateral optical waveguide through the use of InP cladding layers at the sides of the active region. To achieve electrical confinement, p–n blocking structures are incorporated in these cladding layers. The position of these p–n blocking structures must be accurately controlled in order to achieve optimum performance. Typical threshold currents of this class of devices are 10–25 mA. They exhibit well-behaved light–current characteristics with output powers in excess of 20 mW; operating temperatures of above 100°C have been achieved and long life demonstrated.

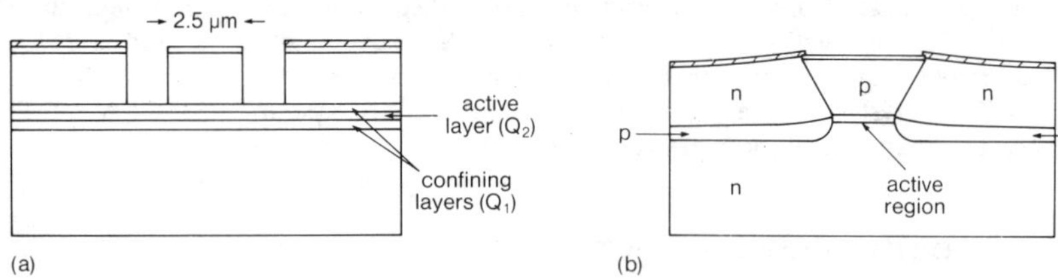


Fig 14.2 Laser structures: (a) ridge-guide laser, (b) buried heterostructure (BH).

## 14.4 BASIC DIODE LASER RESONATORS

In order to obtain laser action, the gain produced by stimulated emission must be sufficient to overcome sources of loss in the device before the photons can escape to the outside world. This loss may include scattering and absorption in both the active and passive layers. Threshold conditions for laser action occur when the gain equals these internal losses plus that due to radiation being emitted from the device (the *useful* loss). When the threshold conditions are satisfied, the output from the device increases strongly in magnitude, the slope efficiency also increases dramatically, and there is considerable spectral narrowing due to the dominance of amplified stimulated emission over spontaneous radiation. The feedback required to give lasing action is provided by some form of resonant cavity, either using mirrors as in the case of the Fabry–Perot (FP) laser, or using a grating as in the case of the distributed Bragg reflector (DBR) and distributed feedback (DFB) lasers. These three types of cavity are illustrated schematically in Figure 14.3; typical cavity lengths are of order a few hundred microns. FP lasers are conventionally fabricated by cleaving the semiconductor along a crystal plane to form the mirrors (the high refractive index of the semiconductor produces a reflectivity of about 30% at the surface. In the DBR laser (Wang, 1974), the mirrors of the FP cavity are replaced by passive gratings. In the DFB case (Kogelnik and Shank, 1972) the grating extends over the full length of the gain region in the cavity. Clearly, for lasers which are to be integrated with other optical devices, it is not possible to cleave facets in the conventional way, so that in such cases, the DFB and DBR structures have obvious advantages.

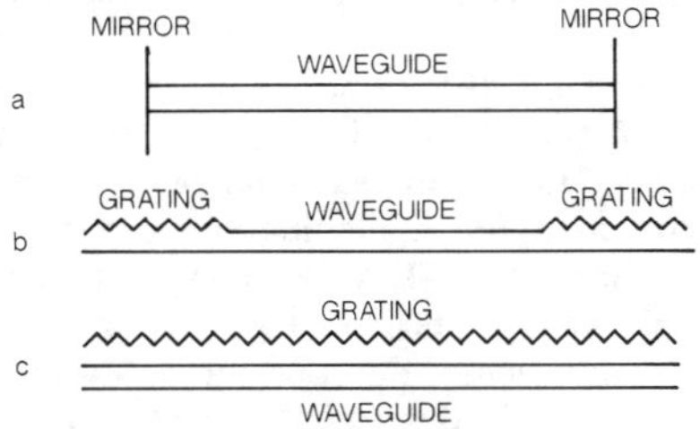


Fig 14.3 Laser cavities: (a) Fabry-Perot, (b) distributed Bragg reflector, (c) distributed feedback.

## 14.5 MORE COMPLICATED RESONATORS

The requirements of long-distance optical communication systems impose severe demands on the spectral purity of diode laser sources. In general, long lengths of silica fibre exhibit considerable dispersion, except at the zero-dispersion wavelength of 1.3 $\mu$m, or when using special dispersion-tailored fibres (Ainslie and Day, 1986). Hence a prime requirement is for single-longitudinal-mode (SLM) operation of the lasers. Since a typical device length is 200 $\mu$m, the longitudinal mode spacing in a 1.5 $\mu$m FP laser is about 1.5 nm. The gain spectrum has a total width of 100 nm or more, but near the peak it varies slowly with wavelength for several nm. It is thus clear that a number of longitudinal modes near the maximum of the gain spectrum can exhibit lasing action. In order to provide some discrimination against unwanted modes, a number of different resonator strategies have been used, and some of these will be discussed here.

Perhaps the simplest approach to SLM operation is to use a short FP cavity. Since the mode spacing varies as the inverse of the cavity length, the hope here is that modes on either side of the preferred mode will not experience sufficient stimulated emission to reach lasing threshold. The shortest length which can be cleaved by conventional means is around 50 $\mu$m (Lee *et al*, 1982), giving a mode separation of about 6 nm. However, this is not sufficient to ensure that a subsidiary mode will not lase, especially during large-signal modulation of the laser. Shorter cavities can be constructed, either by microcleavage (Koren *et al*, 1983) or by the use of surface-emitting designs (Iga *et al*, 1988). The latter approach, which is also of interest in view of the possibility of making two-dimensional arrays of emitters, can give cavity lengths as short as 6 $\mu$m, but the relatively short length over which gain is available means that achieving cw operation at room temperature has been elusive. However, a recent report (Ogura *et al*, 1987) has demonstrated room-temperature cw operation with a threshold current of 2 mA in a surface-emitting DFB structure, so it may be that this option will become more attractive in the future.

An alternative approach to SLM operation is to use compound cavities formed by the introduction of extra reflecting surfaces. The principle here is to superimpose two or more sets of FP resonances in order to favour certain modes (where the resonances coincide) at the expense of others (where the FP modes are at different wavelengths). A number of such configurations have been demonstrated in diode lasers, perhaps the most famous of which is the cleaved-coupled-cavity ($C^3$) device, where two FP lasers of different cavity lengths are coupled together on the same heat sink (Tsang *et al*, 1983). This structure has the virtue of tunability (via the two currents) over a wavelength range of 15 nm, although the complex control circuitry required for stable SLM operation is not attractive for implementation in practical systems.

A simple means of obtaining SLM behaviour, together with tunability, is the use of an external cavity incorporating a grating. In this resonator configuration, the laser facet facing the grating is anti-reflection coated so as to suppress the FP modes of the diode laser, and tuning of the SLM wavelength is achieved by rotating the grating. A tuning range up to 135 nm has been reported (Wyatt *et al*, 1985), and the linewidth of the mode can be below 10 kHz for an external cavity length of 10 – 12 cm. A miniature packaged version of this device has been developed (Mellis *et al*, 1988) for use as a source in coherent transmission systems where tunability and narrow linewidth are key requirements.

Gratings also form an integral part of the basic DBR and DFB resonators discussed in the previous section. For SLM operation of a DBR laser, one grating should have low reflectivity for optimal output coupling and the second grating must have narrow bandwidth and high reflectivity for mode selectivity. The gratings should be located in low-loss waveguides, and two different strategies have been adopted to achieve this, as illustrated in Figure 14.4. In the first, the integrated-twin-guide (ITG) laser (Kobayashi *et al*, 1981), the radiation emitted from the active layer is coupled to the underlying passive guide which contains the gratings. In the butt-jointed version of Figure 14.4 (b), the radiation is coupled directly to the grating (Abe *et al*, 1981).

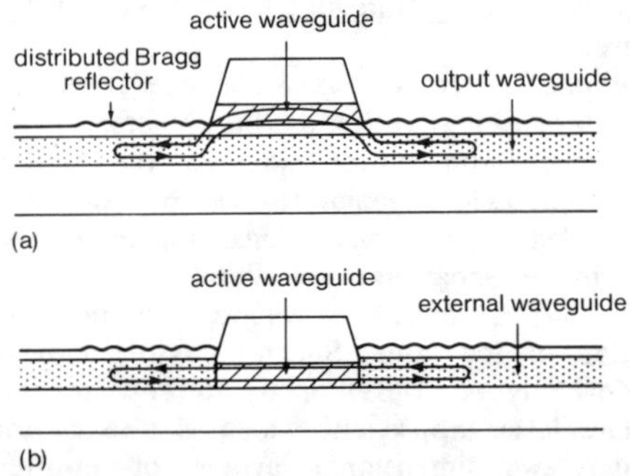


Fig 14.4 **DBR laser structures: (a) ITG, (b) butt-jointed.**

For the DFB laser, oscillation does not occur at the Bragg wavelength, since there are in fact a number of resonant wavelengths located symmetrically on either side of the grating stop-band. This means that in principle the two modes closest to the Bragg wavelength have an equal chance of lasing. In order to overcome this potential disadvantage and achieve true SLM operation, Haus and Shank (1976) proposed the introduction of a $\lambda/4$ phase shift between the two halves of a DFB grating. This has the effect of creating a pass-band in the centre of the grating stop-band, and thus permitting lasing at the Bragg wavelength. A resonator structure of this type has been realised and SLM operation demonstrated in 1.5 $\mu$m DFB lasers (Sekartedjo et al, 1984).

## 14.6 RESONATORS FOR OPTICAL INTEGRATION

The semiconductor laser as a source for optical communications still suffers from a number of limitations. One of these is the difficulty encountered in achieving large-signal modulation at high bit-rates (of order several Gigabits or above). Another is the spectral linewidth of a single laser mode which is rather too high to yet be of use in coherent systems as a source or a local oscillator. These and other limitations can be overcome by the use of other optical components, e.g. a modulator for high-speed operation, or an external cavity for narrow linewidth, and there are considerable advantages in integrating these components with the laser. These advantages include reduced loss due to improved coupling, rugged and reliable packaging, as well as potential cost savings.

The basic unit required to begin this type of integration is a laser-waveguide combination, and several laboratories have now achieved some progress in this direction. A few of the recently-reported devices are illustrated in Figure 14.5.

The DFB laser waveguide shown at the top of Figure 14.5 has been reported by Murata *et al* (1985) to give a minimum linewidth of 1.2 MHz for a passive cavity length of 1.43 mm. Similar results in the range 1 - 2 MHz have been reported by Fujita *et al* (1985) using an etched-facet Fabry-Perot laser, and by Dutta *et al* (1986) with a more complicated structure which included a Fabry-Perot laser and a grating in the passive cavity for frequency selectivity. In another report (Lee *et al*, 1987), a DFB laser was integrated with a passive cavity which was provided with an electrode to change the refractive index via the injection of carriers. The minimum linewidth achieved was 18 MHz, but the FM response of this device was flat from 100 kHz to 500 MHz. This is in contrast to the situation for FM in a conventional laser where the response shows considerable variation over this range due to the different mechanisms responsible (thermal effect at low frequencies and carrier-concentration effect at high frequencies). The achievement of a flat FM response is useful in frequency-shift keying (FSK) systems, since it avoids the need for sophisticated compensation schemes to suppress the modulated pulse distortion.

As additional sections are added to the basic laser-waveguide building-block, further device sophistication becomes feasible, and a few possibilities are shown in Figure 14.5. For example, a recent three-section wavelength tunable DFB laser (Kotaki *et al*, 1987) showed single-mode lasing at any wavelength throughout a range of 4.5 nm. The wavelength of a single longitudinal mode was tuned continuously over 3.15 nm under constant output power condition. In another recent report (Nakano *et al*, 1987), a multi-electrode DFB laser showed a linewidth as narrow as 4.5 MHz and a flat FM response up to 200 MHz.

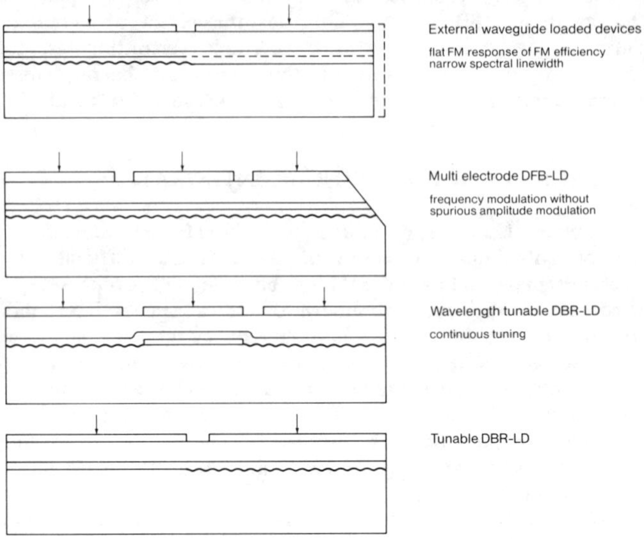


**Fig 14.5 Monolithic integrated laser–waveguide devices.**

Various forms of waveguide modulators have been integrated with DFB lasers. For example, Yamaguchi *et al* (1987) have demonstrated InGaAsP carrier injection modulators integrated with 1.3 $\mu$m and 1.55 $\mu$m lasers, and operated at 450 Megabits per second with NRZ pulse modulation. Kawamura *et al* (1987) have integrated InGaAsP/InP DFB lasers with InGaAs/InAlAs quantum–well modulators. High speed modulation was achieved with a response time of 300 ps, but the modulation efficiency was not as good as that reported for the discrete modulator. We may expect to see improved performance as the growth techniques and design details of these integrated devices are improved.

## 14.7 CONCLUSION

The main resonator types employed in diode lasers have been reviewed with special emphasis on the novel designs used to achieve single longitudinal mode operation and narrow linewidth for use in optical fibre transmission systems. It is envisaged that the fast pace of development of such systems will continue to impose new challenges for the sources which, in turn, will stimulate further innovation in resonator design. Other aspects of diode laser applications which also have a strong impact on resonator technology include (1) optical amplifiers (O'Mahony, 1988), (2) one– and two–dimensional arrays (Scifres *et al*, 1978; Iga *et al* 1988), optoelectronic integration (Forrest, 1985), and bistability (Kawaguchi, 1987). Although these topics have not been discussed here (for reasons of space), it is anticipated that these will be prime areas where novel resonator configurations will emerge in the future.

## 14.8 REFERENCES


Abe Y, Kishino K, Suematsu Y and Arai S 1981 Electron. Letts. 17 945

Adams M J, Steventon A G, Devlin W J and Henning I D 1987 Semiconductor Lasers for Long-Wavelength Optical-Fibre Communications Systems (London: Peter Peregrinus)

Ainslie B J and Day C R 1986 J. Lightwave Technol. LT-4 967

Dutta N K, Cella T, Piccirilli A B and Brown R L 1986 Appl. Phys. Letts. 19 1227

Forrest S R 1985 J. Lightwave Technol. LT-3 1248

Fujita T, Ohya J, Matsuda K, Ishino M, Sato H and Serizawa H 1985 Electron. Letts. 21 374

Haus H A and Shank C V 1976 IEEE J. Quantum Electron. QE-12 532

Iga K, Koyama F and Kinoshita F 1988 IEEE J. Quantum Electron. QE-24 1845

Kaminow I P, Nahory R E, Pollack M A, Stulz L W and DeWinter J C 1979 Electron. Letts. 15 763

Kawaguchi H 1987 Opt. Quantum Electron. 19 S1

Kawamura Y, Wakita K, Yoshikuni Y, Itaya Y and Asahi H 1987 IEEE J. Quantum Electron. QE-23 915

Kobayashi K, Utaka K, Abe Y and Suematsu Y 1981 Electron. Letts. 17 366

Kogelnik H and Shank C V 1972 J. Appl. Phys. 43 2327

Koren U, Rav-Noy Z, Hasson A, Chen T R, Yu K L, Chiu L C, Margalit S and Yariv A 1983 Appl. Phys. Letts. 42 848

Kotaka Y, Matsuda M, Yano M, Ishikawa H and Imai H 1987 Electron. Letts. 23 325

Lee T-P, Burrus C A, Copeland J A, Dentai A G and Marcuse D 1982 IEEE J.Quant. Electron. QE-18 1101

Lee T-P, Menocal S G, Sakano S, Valster V and Tsuji S 1987 Electron. Letts. 23 154

Mellis J, Al-Chalabi S A, Cameron K H, Wyatt R, Regnault J C, Devlin W J and Brain M C 1988 Electron. Letts. 24 988

Murata S, Mito I, Shikada M and Kobayashi K 1985 Proceedings of IOOC - ECOC (Venice) 299-302

Nakano Y, Itaya Y, Fukuda M, Noguchi Y, Yasaka H and Oe K 1987 Electron. Letts. 23 826

O'Mahony M J 1988 J. Lightwave Technol. LT-6 531

Ogura M, Hsin W, Wu M-C, Wang S, Whinnery J R, Wang S C and Yang J 1987 Appl. Phys. Letts. 51 1655

Scifres D R, Burnham R D and Streifer W 1978 Appl. Phys. Letts. 33 1015

Sekartedjo K, Eda N, Furuya K, Suematsu Y, Koyama F and Tanbuk-Ek T 1984 Electron. Letts. 20 80

Tsang W T, Olsson N A and Logan R A 1983 Appl. Phys. Letts. 42 650

Tsukada T 1974 J. Appl. Phys. 45 4899

Wang S 1974 IEEE J. Quantum Electron. QE-10 413

Wyatt R, Cameron K H and Matthews M R 1985 Br. Telecom. Technol. J. 3 5

Yamaguchi M, Emura K, Kitamura M, Mito I and Kobayashi K 1987 Electron. Letts. 23 190

# 15 Diode Pumped Solid State Laser Resonators

*A I FERGUSON*

## 15.1 INTRODUCTION

Diode laser pumping of solid–state lasers was first accomplished in the early 1970's but in recent years has undergone a revival due to rapid developments in the technology of diode lasers. Diode lasers are the most efficient devices to have been devised for converting electrical energy into quasi–coherent optical energy. However the diode lasers themselves are, at present, rather poor laser sources. The beam quality of high power diode arrays is several times diffraction limited, the spectrum is typically several nanometers wide and is temperature dependent. One of the most severe drawbacks of the present generation of diode laser is that they are operated at an optical power level which is very close to the damage threshold of the facet. This means that in any application where pulsed operation is required, the peak power may not greatly exceed the average power available from the diode.

Modern high power laser diodes do however, make excellent pumps for solid–state lasers (Streifer *et al* 1988). They can be tuned to the absorption bands and efficiently coupled into the lowest order mode of the solid–state laser medium. This gives rise to very efficient pumping and eliminates water cooling and its associated plumbing. If we take the case of the diode laser pumped Nd:YAG laser, the conversion efficiency from diode to Nd:YAG laser is typically 40%, while the electrical to optical conversion efficiency in the diode is around 30%, giving an overall conversion efficiency of about 12% (Sipes 1986, Berger *et al* 1987). Many of the widely–used solid–state gain media have long upper state lifetimes with relatively broad emission spectra. This makes them ideal candidates for Q–switching and mode–locking, thereby increasing the available peak power by many orders of magnitude. Since the pump laser can usually be focussed into the lowest order mode volume the lasers almost always operate on the $TEM_{00}$ mode and hence produce diffraction limited output. For this reason diode–pumped solid–state lasers have been called *coherence amplifiers*. The all solid–state (or holosteric) nature of these devices make them highly reliable, compact and efficient.

In many ways, the design of resonators for solid–state lasers pumped by diode lasers is a pure text–book exercise. The low heat load means that thermal effects do not usually have to be taken into account, and lowest order mode operation may usually be assumed. Cavities discussed in this chapter all assume end pumping of the laser rod. This is the preferred technique for moderately low power operation because of its associated high efficiency. Very high power multi–element diode laser arrays are now becoming available, for which the only practical pumping geometry is side pumping. We ignore these in the present discussion. A common feature of the cavities discussed is the requirement for small focal spots in the gain medium, ensuring low threshold operation.

The operation of particular diode-pumped solid-state laser systems will not be discussed in detail as the emphasis will be on resonator design. Details of recent advances in diode lasers and diode-pumped solid-state lasers may be found in recent review articles (Fan and Byer 1988, Byer 1988). Simple monolithic cavities will be discussed followed by a discussion of hybrid cavities which allow the insertion of tuning elements, Q-switches etc. into the cavity. This will be followed by a discussion of cavities for mode-locked lasers including astigmatically compensated cavities. The chapter concludes with a discussion of unidirectional out of plane ring laser cavities for the suppression of spatial hole burning effects.

## 15.2 THE MONOLITHIC CAVITY

The monolithic Nd:YAG laser cavity was the first in the new generation of diode-pumped solid-state lasers and is the system which stimulated much of the recent interest in this field (Zhou *et al* 1985). A diagram of a typical system is shown in Figure 15.1. The laser consists of a solid Nd:YAG rod which is polished flat on one end and curved on the other. The dielectric mirrors are deposited directly onto the ends of the rod. This provides a cavity of exceptional mechanical rigidity. The radius of curvature of the rod end $R$ and the rod length $\ell$ together with the refractive index, n of the material and operating wavelength, $\lambda$ in air completely determine the beam characteristics through the following relationships:

$$w_1 = (\lambda/\pi n)^{\frac{1}{2}}\ \ell^{\frac{1}{4}}\ (R-\ell)^{\frac{1}{4}} \qquad (15.1)$$

$$w_2 = (\lambda/\pi n)^{\frac{1}{2}}\ \ell^{\frac{1}{4}}\ (R^2/(R-\ell))^{\frac{1}{4}} \qquad (15.2)$$

where $w_1$ and $w_2$ are the Gaussian beam spot radii on the plane and curved end respectively.

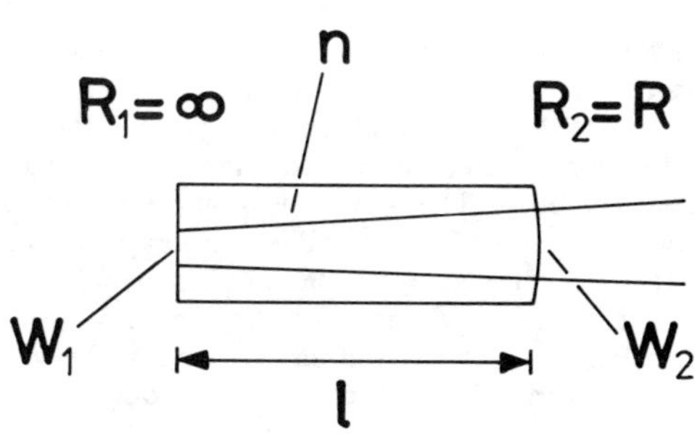


Fig 15.1 The monolithic Nd:YAG laser cavity. This configuration is very robust but the single mode power is very low due to spatial hole burning.

The choice of rod length will be dictated by the need for a large fraction of the pump laser to be absorbed in the rod. For Nd:YAG, pumped at around 810nm, this is in the region of 1mm. The mode spacing will then be determined by this length through the relationship:

$$\Delta\nu = c/2n\ell \qquad (15.3)$$

For Nd:YAG, the rod length is usually chosen to be somewhat longer than the minimum dictated by the absorption coefficient due to handling difficulties. Longer lengths also mean that the constraint on the pump laser wavelength can be relaxed a little. The choice of $R$ will be dictated by the need to ensure that the focussed pump spot is no bigger than the laser spot size. For short absorption lengths the pump beam can be tightly focussed such that it is smaller than the laser spot over the absorption length. For longer absorption lengths the pump and laser modes have to be matched as closely as possible. If the absorption length is longer than the crystal, the pump and laser beam should be mode matched in order to minimise threshold.

Some typical values for a diode–pumped Nd:YAG monolithic resonator are $R$=20 mm, $\ell$=10 mm and $n$ = 1.82, giving rise to a spot radius for the Nd:YAG cavity of $w_1$=23$\mu$m on the plane end and $w_2$=59$\mu$m on the curved end. A design like this would typically have a threshold in the region of 1mW and a slope efficiency of perhaps 20% for an output coupling of about 2% at 1.06$\mu$m.

One of the main applications for the monolithic diode–pumped Nd:YAG laser is as a single frequency narrow linewidth master oscillator to be used for injection seeding high power Nd:YAG lasers. However soon after it had been developed it was discovered that single frequency operation could only be achieved at very modest output powers in the region of less than 0.1mW. This was overcome to some extent by the use of gain–switching techniques (Owyoung *et al* 1985) but has resulted in the virtual abandonment of the monolithic standing wave design in favour of out of plane ring lasers for this particular application (see section 15.5, 4.4 and Chapter 8).

## 15.3 THE HYBRID CAVITY

We refer to the hybrid cavity as one in which the laser rod itself forms one end of the resonator. A typical hybrid cavity is shown in Figure 15.2. It consists of a laser rod of length $t$ and an air gap of length $\ell$. One end of the rod is coated for high reflection at the laser wavelength while the other end is anti–reflection coated. The cavity is completed by an output coupling mirror of radius of curvature $R_2$=$R$. The air gap allows the insertion of various elements into the cavity. These might include etalons for spectral control and single frequency operation or Q–switches (Maker and Ferguson 1988). If the anti–reflection coated end of the rod is curved it is possible, to provide a second waist in the cavity where nonlinear elements such as intracavity frequency doublers may be placed (Baer 1986). A further advantage of the hybrid cavity is that fast frequency servoing becomes possible either by mounting the output coupler onto a piezo–ceramic translator or by inserting a phase modulator into the cavity.

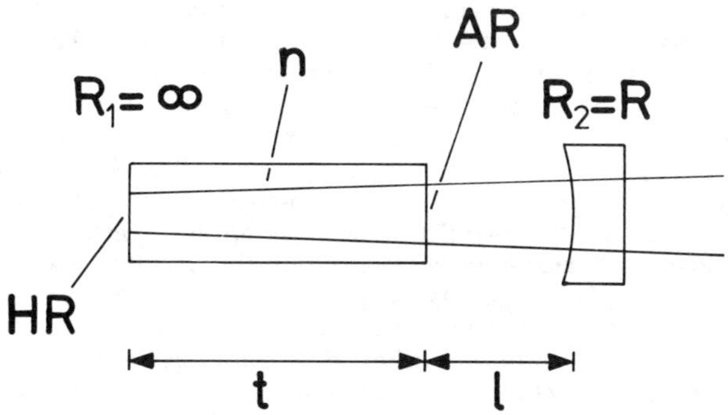


Fig 15.2 The hybrid cavity providing an air gap for the insertion of etalons and Q-switches.

The stability criterion for a cavity of this nature is quite interesting. The criterion can easily be established by replacing the rod by an equivalent thickness in air and applying the standard stability criterion that $\ell'$,the distance in air between the plane mirror and the output coupler of radius of curvature $R$, should be less than $R$. Confusion is often caused by the calculation of the effective distance $\ell'$. The naive approach is to say that the optical path length is $nt$ where $n$ is the refractive index of the laser rod. *This is wrong!* The effective thickness of the rod for the calculation of Gaussian beams parameters is $t/n$. This is the expression which enters the ray matrix for propagation in a medium of index $n$. The reason for this counter-intuitive result is simple when it is realised that when the light enters the medium the wavelength will be reduced by a factor of $1/n$. The diffraction of this light will therefore be less than it would have been in free space and the equivalent distance in air will be reduced by the factor $1/n$. This is quite important in miniature cavities, since solid media with relatively large refractive indices usually form a large fraction of the cavity. The mode spacing is of course given by $(tn+\ell)$.

## 15.4 MODE-LOCKED CAVITIES

One of the major recent advances in the development of diode-pumped solid-state lasers is mode-locking to produce trains of ultrashort pulses (Maker *et al* 1988). This opens up the possibilities of undertaking time-resolved studies and increases the peak power available for nonlinear processes by several orders of magnitude. The design of cavities for mode-locking requires a bit more thought than the simple monolithic or hybrid cavity.

The main requirement for a mode-locked laser cavity is that the mode spacing should be adjustable to the frequency of a modulator placed within the cavity. This modulator couples adjacent modes together and gives rise to short pulses. Typical operating frequencies for mode-lockers are in the region of 100MHz. This means that the effective cavity length must be in the region of 1.5m. A further requirement is that a small spot radius in the region of 20$\mu$m is still needed in order to maintain a low threshold. In principle the hybrid cavity described in Section 15.3 could be extended to give a mode spacing in the region of 100MHz but this is not really practical. In order to see this we can impose the condition on equation (15.1) that $w_1$=20$\mu$m and that $\ell$ be replaced by $\ell'=t/n+\ell$ and set $\ell'$ equal to 1.5m at a wavelength of 1$\mu$m. This sets the condition that $R-\ell'$=2.1 $\mu$m. We may further calculate the spot size on the output mirror and find this to be $w_2$=20.7 mm. The extremely tight tolerance on the matching of cavity length and mirror radius of curvature is totally impractical as it would not allow the subsequent matching of the cavity length to the mode-locker frequency.

A design which overcomes these problems rather elegantly is shown in Figure 15.3a. This consists of a hybrid cavity, as described earlier, but in this case the cavity includes an intracavity lens while the output coupler is now planar. This particular cavity design exhibits the important features that there is a tight focus provided in the rod, and secondly that the beam parameters are not very sensitive to the position of the output coupler. We shall go through the calculation of the cavity parameters in some detail.

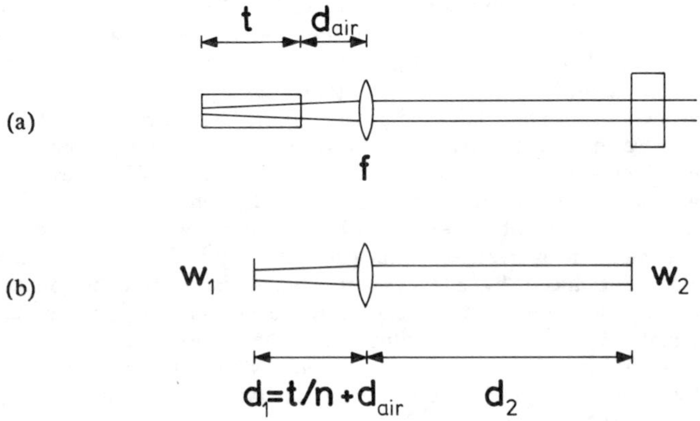


Fig 15.3 (a) An extended cavity suitable for the insertion of a mode-locker. (b) The equivalent cavity in air.

The equivalent cavity is shown in Figure 15.3b. This problem is now almost identical to the problem of focussing by a lens in free space. The plane ends of the rod and output coupler impose the condition that the wavefronts are planar and hence that there are waists at each of these positions. Focussing of a Gaussian beam is treated in almost any text-book on laser beams and resonators (see Chapter 1) and the result of these treatments is the following:

$$d_1-f = (w_1 / w_2)(f^2 - f_0^2)^{\frac{1}{2}} \tag{15.4}$$

$$d_2-f = (w_2 / w_1)(f^2 - f_0^2)^{\frac{1}{2}} \tag{15.5}$$

$$f_0 = \pi w_1 w_2 / \lambda \tag{15.6}$$

where $d_1$ is the distance between the waist $w_1$ and the lens of focal length $f$ and $d_2$ is the distance from the lens to the waist $w_2$.

By taking the product of (15.4) and (15.5) we obtain:

$$(d_1-f)(d_2-f) = (f^2-f_0^2) \tag{15.7}$$

This expression may be compared with the familiar expression for the object and image distance given by Newton's formula which states that:

$$(d_1-f)(d_2-f) = f^2 \tag{15.8}$$

Equation (15.7) reduces to the classical Newton expression if we set $f_0=0$ which corresponds to reducing the spot sizes to zero. A plot of $d_2-f$ against $d_1-f$ obtained using equations (15.7) and (15.8) is shown in Figure 15.4. We can see that for large values of $d_1-f$ the Gaussian beam formula and Newton's formula are essentially the same. When $d_1-f$ approaches the Raleigh range the Gaussian beam and Newton's formula differ substantially.

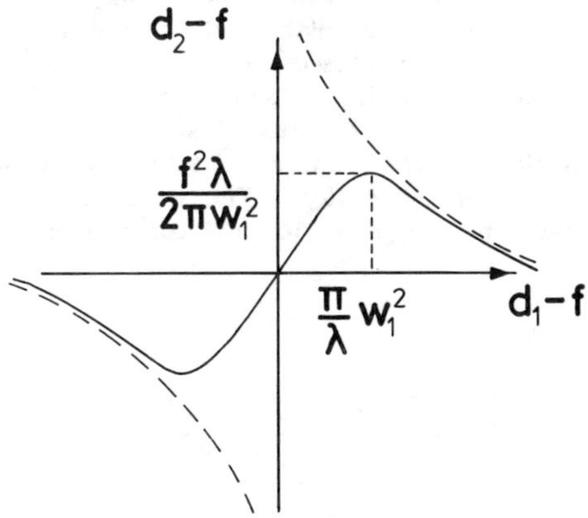


Fig 15.4 A plot of the beam waist positions for Gaussian beam focussing. The classical Newton formula which does not take account of diffraction is shown as a dashed line.

By re-arranging equations (15.4), (15.5) and (15.6) we can solve for the spot radius $w_1$ to give the following expression:

$$w_1^4 = (\lambda^2/\pi^2)[f^2(d_1-f)/(d_2-f)-(d_1-f)^2] \tag{15.9}$$

From this expression it is clear that the spot radius is zero for two values of $d_1-f$, i.e. $d_1-f=0$ and $d_1-f=f^2/(d_2-f)$. Since it is physically impossible to have a zero spot radius this corresponds to the cavity being unstable. This range of values for $d_1-f$ is called the stability region and it corresponds to the region over which a stable cavity can be formed. The spot radius at the centre of the stability region is given by:

$$w_{1\,max}^2 = (\lambda/2\pi)(f^2/(d_2-f)) \tag{15.10}$$

The centre of the stability region corresponds to the region where the spot radius $w_1$ is maximised.

If we take some typical figures we can readily calculate the cavity parameters. Typical values are $d_2$=1.5m, $f$=75mm at a wavelength of 1$\mu$m. This gives rise to a stability region from equation (15.9) which ranges from $d_1-f$=0 to 3.95mm. The centre of the stability region then becomes 1.97mm with the corresponding spot radius of 25$\mu$m. By taking the ratio of equations (15.4) and (15.5), we can calculate the spot size on the output mirror to be 670$\mu$m.

This gives a very satisfactory range of values for the mode-locked cavity. The actual spot size at the pump end of the cavity may be adjusted by careful choice of the spacing between the rod and the lens. Once this has been chosen the cavity parameters are insensitive to adjustment of the distance of the output coupler from the lens. This insensitivity is needed so that the cavity length may be matched to the mode-locker frequency.

Although this cavity still is satisfactory from the point of view of the Gaussian beam parameters, there are some difficulties with this arrangement. The performance of mode-locked lasers is extremely sensitive to the presence of internal reflections which cause etalon effects and upset mode-locking. In this design, the problems are caused by the surfaces of the lens and the internal surface of the laser rod. A simple method of eliminating reflection problems with the rod is to set the internal surface at Brewster's angle. This not only minimises the reflection but also causes any small residual reflection to be directed well out of the cavity. The reflections from the internal lens may also be eliminated by use of an internal concave mirror.

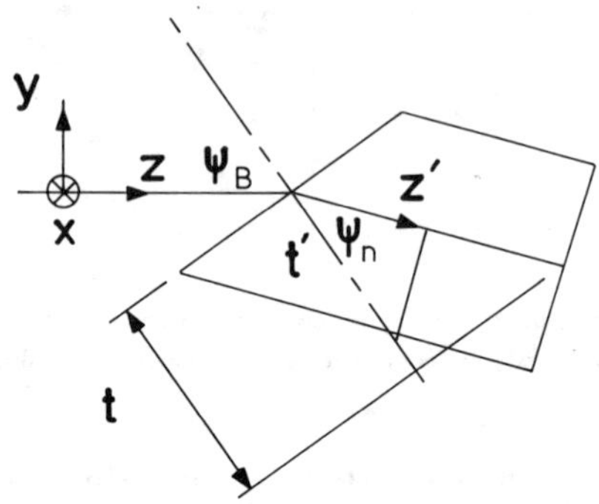


Fig 15.5 The geometry for the calculation of astigmatism on entering a Brewster angled surface.

One problem that is introduced by the incorporation of a Brewster angled surface is that it introduces astigmatism into the cavity (Hanna 1969 and Kogelnik *et al* 1972). This astigmatism is caused by the fact that the effective thickness of the rod is different in the sagittal and tangential planes. To see this consider Figure 15.5. We assume that a Gaussian beam of spot radius $w_0$ is incident on a surface at Brewster's angle. We shall measure the thickness $t$ to be perpendicular to the plane where the beam enters the crystal. If the beam travels a distance $z$ into the crystal the effective distance along the direction in which t is measured is given by:

$$z' = t'/\cos\theta = t'(n^2+1)^{\frac{1}{2}}/n \tag{15.11}$$

On entering the crystal the spot size in the sagittal plane remains unchanged, whereas in the tangential plane it is increased to:

$$w_y = w_0\cos\psi_n/\cos\psi_B = nw_0 \tag{15.12}$$

Inserting these spot sizes into the well-known expression for the change in spot size as a function of distance (see equation 1.18):

$$w(z) = w_0 \left[1 + \left[\frac{\lambda z}{\pi w_0^2}\right]^2\right]^{\frac{1}{2}} \tag{15.13}$$

and allowing for the change in wavelength in enering the medium we find that:

$$w_x(z') = w_0\left[1 + \left[\frac{\lambda t_x}{\pi w_0^2}\right]^2\right]^{\frac{1}{2}} \tag{15.14}$$

$$w_y(z') = w_0\left[1 + \left[\frac{\lambda t_y}{\pi w_0^2}\right]^2\right]^{\frac{1}{2}} \tag{15.15}$$

where we have introduced new effective distances in the medium. The total effective thickness for the sagittal and tangential rays are given by:

$$t_x = t\,(n^2+1)^{\frac{1}{2}}/n^2 \tag{15.16}$$

$$t_y = t\,(n^2+1)^{\frac{1}{2}}/n^4 \tag{15.17}$$

The difference in effective thicknesses given by equation (15.16) and (15.17) represents astigmatism. In the Gaussian beam calculation, the distance $d_1$ now becomes different in the sagittal and tangential planes. This means that the stability region in the sagittal and tangential plane will no longer be coincident. Indeed, it is possible that for a given setting of the air gap between the rod and the focussing mirror or lens, the two stability regions do not overlap at all. In this case the laser would be totally unstable in one plane while being stable in another. Even if the stability regions do overlap, the spot sizes will in general differ in the two planes, leading to elliptical cross–section beams.

The problem of astigmatism in these cavities may easily be overcome by use of an off–axis mirror. It is well–known that the effective radius of curvature of a mirror of radius $R$ at an angle of incidence $\theta$ is given by:

$$R_x = R/\cos\theta \tag{15.18}$$

$$R_y = R\cos\theta \tag{15.19}$$

The condition for astigmatic compensation is that:

$$f_x = R_x/2 = t_x \tag{15.20}$$

$$f_y = R_y/2 = t_y \tag{15.21}$$

Using equations (15.16) through (15.19) we obtain the condition for astigmatic compensation namely:

$$R\sin\theta\tan\theta = 2t[(n^2-1)(n^2+1)^{\frac{1}{2}}]/n^4 \tag{15.22}$$

Inserting some typical values of $t$=10mm, $R$=150mm and $n$=1.82 for Nd:YAG we obtain the result that the angle of incidence $\theta$ should be chosen to be approximately 13.5° for astigmatic compensation. This techniques has been well-proven in the operation of dye lasers and also works well in the case of solid-state lasers.

## 15.5 RING LASERS

The diode laser pumped ring laser has become an important source in the case of the Nd:YAG laser. Despite the fact that Nd:YAG is a homogeneously broadened gain medium, the success at producing single frequency operation in a standing wave configuration has been rather limited. This has been caused by the spatial hole burning effect, which has limited the output power from a monolithic standing wave laser to about 100 $\mu$watts. This may be overcome to some extent by inserting etalons into a hybrid cavity. This is at the expense of output efficiency, caused by the insertion loss of the etalons. Even in this case, the single frequency power is limited to a few mW. Some progress has been made in the hybrid cavity design by placing the rod between two quarter wave-plates (Wallermeroth and Peuser 1988). If the fast and slow axes of the wave-plates are appropriately adjusted it, is possible to establish a single mode within the cavity such that the counter-propagating beams in the laser rod have opposite circular polarization. These beams can therefore not interfere with each other to form a standing wave and spatial hole burning is eliminated. Spatial hole burning may also be completely eliminated by the use a unidirectional ring laser, as discussed in Chapter 4.

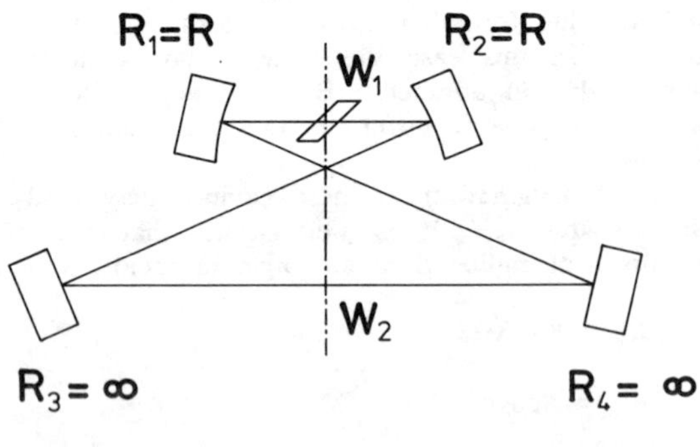


Fig 15.6 An astigmatically compensated ring laser. For the symmetrical geometry shown here the cavity parameters can be calculated in the same way as the standing wave laser described in Section 15.4.

A simple ring laser resonator is shown in Figure 15.6. This resonator is an extension of the kind of resonator described in Section 15.4, for an astigmatically compensated mode–locked resonator. In general, ring laser resonator beam parameters are a little more difficult to calculate than in standing wave resonators because the minimum number of mirrors is larger. However, if a symmetrical arrangement is chosen, the problem simply reduces to the one we investigated in Section 15.4. Referring to Figure 15.6 we see that from symmetry, there must be a waist midway between mirrors $R_1$ and $R_2$ and a second waist midway between $R_3$ and $R_4$. These waists have radii $w_1$ and $w_2$, respectively. The problem is now identical to the extended cavity problem which was solved for the mode–locked laser resonator.

The provision of a ring laser does not in itself ensure single frequency operation, since in general, the laser will oscillate in both directions simultaneously and produce a standing wave field. In order to ensure unidirectional operation, a bias has to be built in to the laser which will make it oscillate preferentially in one direction. The most common way of doing this is to make use of the Faraday effect. It is well–known that the Faraday effect produces a non–reciprocal polarisation rotation whereas other effects such as optical activity, retarders or mirrors produce reciprocal effects. By inserting a Faraday rotator into the cavity together with an optical rotator, it can be arranged that the Faraday rotation and optical rotation add to give a net change in rotation for one direction of propagation while they will cancel for the opposite direction of propagation. If there are polarising elements in the cavity such as Brewster angled surfaces the direction of propagation which experiences no net change in rotation, will experience low loss and will oscillate while the other direction of propagation will be suppressed.

Unidirectional ring operation in a Nd:YAG laser cavity can be conveniently accomplished by applying a magnetic field to the laser rod and using the small Faraday effect in Nd:YAG. The small optical rotation required to compensate for the Faraday rotation can be introduced by operating the ring laser in an out–of–plane geometry (Kane and Byer 1985, Trutna *et al* 1987, Kane and Cheng 1988). It is well known that a geometrical optical rotation can be accomplished simply by reflection (Biraben 1979). For example the polarisation of light can be rotated through $90^{\circ}$ by use of two mirrors at $45^{\circ}$ in two different planes. This is caused by a change in reference planes on reflection. The out of plane ring resonator is an example of the use of this rotation but at very small angles. The ring resonator shown in Figure 15.6 has been made to operate in a unidirectional mode by applying a field of a few Kilo–Gauss to the 4mm rod and coming out of plane by raising the beam on one of the mirrors $R_3$ or $R_4$ by a few mm. Such a laser can be used to provide single frequency output or mode–locked and Q–switched to provide performance free from spatial hole burning effects.

The desirable features of the out of plane ring resonator have been brought together in a monolithic design to provide stable single frequency output. A diagram of such a system is shown in Figure 15.7. The entire crystal length is about 4mm and is placed in a magnetic field. The ring geometry is provided by total internal reflection in the Nd:YAG surfaces. A complication which enters here is the phase shift produced by total internal reflection. This puts constraints on the angles of incidence, in order to ensure that the phase shift in going round the cavity is 180 degrees (Trutna *et al* 1987). The polarisation effect of hitting the output coupler at non-normal incidence together with coming out of plane by only a few degrees provides sufficient suppression to ensure unidirectional operation. This kind of laser system is capable of providing several tens of mW single frequency radiation with very low frequency noise and the system has become widely used for injection seeding high power Q-switched lasers.

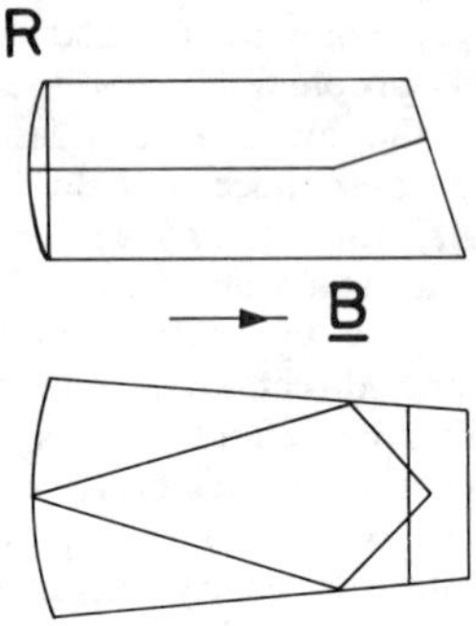


Fig 15.7 A monolithic out of plane resonator. The cavity is formed by three total internal reflectors and a reflection on the curved output mirror. The surfaces are angled such that the first total internal reflection takes the light out of plane. The further two total internal reflections bring the light back into the plane and onto the output coupler thereby completing the cavity.

## 15.6 REFERENCES

Baer T 1986 J. Opt. Soc. Am. 3 1175
Berger J, Welch D F, Scifries W, Streifer W and Cross P S 1987 Appl. Phys. Lett. 51 1212
Biraben F 1979 Opt. Commun. 29 353
Byer R L 1988 Science 239 742
Fan T Y and Byer R L 1988 IEEE J. Quantum Electron. 24 895
Hanna D C 1969 IEEE J. Quantum Electron. QE-5 483
Kane T J and Byer R L 1985 Opt. Lett. 10 65
Kane T J and Cheng E A P 1988 Opt. Lett. 13 970
Kogelnik H W, Ippen E P, Dienes A and Shank C V 1972 IEEE J. Quantum Electron. QE-8 373
Maker G T and Ferguson A I 1988 Opt. Lett. 13 461
Maker G T, Keen S J and Ferguson A I 1988 Appl. Phys. Lett. 53 1675
Owyoung A, Hadley G R, Esherick P, Schmitt R L and Rahn L A 1985 Opt. Lett. 10 484
Sipes D L 1985 Appl. Phys. Lett. 47 74
Streifer W, Scifres D R, Harnagel G L, Welch D F, Berger J and Sakamoto M 1988 IEEE J. Quantum Electron. 24 883
Trutna W R, Donald D K and Nazarathy M 1987 Opt. Lett. 12 248
Wallmeroth K and Peuser P 1988 Electron. Lett. 24 1088
Zhou B, Kane T J, Dixon G J and Byer R L 1985 Opt. Lett. 10 62

# 16 Fibre Laser Resonators

*P URQUHART*

## 16.1 INTRODUCTION

Rare earth doped fibre lasers have recently received growing interest as tunable sources which can be pumped at low power levels to give efficient lasing outputs at useful wavelengths. Optically pumped continuous wave fibre lasers have been demonstrated at important wavelengths such as 1.08$\mu$m, 1.35$\mu$m, 1.55$\mu$m, 2.1$\mu$m and 2.7$\mu$m, showing their potential applicability to current and projected future telecommunications requirements as discussed by Urquhart (1988a). The waveguiding properties of telecommunications-type optical fibre are used to confine lasing waves over a long interaction length. The fibre lasing medium has an outer diameter of typically 100 to 150$\mu$m within which the optical wave is confined to a region of a few microns diameter. The high optical intensity in the fibre, the long interaction length and low losses from scattering and unwanted absorption, combine to ensure that fibre lasers are low threshold power sources.

Optical fibre geometry confers many benefits but it also imposes some special constraints on the design of resonators. The waveguiding action of a single transverse mode fibre restricts the light to axial propagation. Whilst structures such as unstable resonators are not possible when single-mode fibres are used, an important benefit arises from the ease with which complicated designs of resonator can be realised in practice. The long flexible waveguides with diameters of a fraction of a millimetre impose some exacting handling requirements to obtain the necessary quality of performance. Fibre resonators are at an earlier stage in their development than their bulk optical counterparts, although rapid progress has recently been made. This chapter will be devoted to examining some of the achievements.

## 16.2 SINGLE-MODE DOPED FIBRE

Fibre lasers are based on cylindrical glass waveguides, similar in design to telecommunications optical fibre. Optical fibre consists of concentric core and cladding regions and is fabricated so that the core has a slightly higher refractive index than the cladding. A wave which is focused onto the end of the fibre will remain confined mainly to the core as it propagates. If the core of the fibre has a small radius the wave may travel as a single transverse mode, being most intense on the axis of the fibre. Single-mode fibres are most suitable for fibre lasers as they offer the greatest power confinement.

The active lasing species is one of the rare earth ions (such as neodymium, erbium, thulium, holmium or praseodymium) and these determine the wavelengths of lasing emission and pump absorption. The rare earth is added to the fibre core material as a dopant in typical levels of a few hundred parts per million. The fibre itself is a highly transparent glass, which to date has most commonly been a silica/phosphorous pentoxide composite for the cladding and a silica/phosphorous pentoxide/germania composite, along with the rare earth dopant, for the core. Other glass types have been used as fibre laser media and the most important of these is based on the heavy metal fluorides, especially zirconium fluoride as demonstrated by Brierley and France (1988). Fluorozirconate fibres, as they are known, have the advantage of being transparent to wavelengths as long as 3$\mu$m and are very promising lasing hosts but have not at the time of writing been made into components such as fibre directional couplers and grating reflectors.

Fibre lasers offer several advantages over other types of lasers. They use hosts which have very low losses due to scattering and absorption by impurities. If they are single transverse mode at both the pump and lasing wavelengths, they have high intensities and good overlaps between the pump and lasing fields. There is also a long interaction length; the lasing medium can be several tens or even hundreds of metres long if necessary, although lengths of about one metre are typical. An advantage of fibre geometry, which is of particular relevance to the topic of fibre resonators, is the availability of fibre-based components such as directional couplers, which will be considered in sections 16.5 and 16.6.

## 16.3 FIBRE FABRY-PEROT LASER

A fibre Fabry–Perot laser is shown in Figure 16.1 with mirrors mounted directly against the fibre ends. The fibre laser is longitudinally pumped by launching light from a short wavelength external laser through mirror 1. If both the pump and lasing guided waves are single-mode there can be very efficient interaction of the fields. The pumping laser is chosen to be a commercially available source with a wavelength matched to a suitable rare earth ionic absorption band. Ideally, the pump should be a semiconductor laser due to its compact size and high efficiency. Light absorbed by the rare earth ions in the fibre sets up the resonantly enhanced stimulated emission necessary for lasing. The lasing light, together with any unconverted pump light, is coupled out of the fibre through mirror 2. The fibre laser is therefore a means by which power from a short wavelength pump can be converted to provide a source of one of the useful wavelengths mentioned in section 16.1. An example, as discussed by Wyatt *et al* (1988), is to pump an erbium-doped fibre with an 800 nm GaAs semiconductor laser to provide a lasing output at 1.55 $\mu$m.

The fibre Fabry–Perot cavity illustrated in Figure 16.1, with mirrors consisting of flat dielectric coated substrates, is the most basic design of resonator for fibre lasers. Several conditions must be met by the mirrors. Mirror 1 has to be highly transparent at the pump wavelength and highly reflective at the lasing wavelength. The reflectivity of mirror 2 at the lasing wavelength is chosen to satisfy the usual compromise between threshold and slope efficiency: a high reflectivity gives a low threshold, but also a low slope efficiency, while a lower reflectivity improves the slope efficiency at the expense of a high threshold power. Mirror 2 is usually chosen to be low reflectivity at the pump wavelength but in some circumstances it may be advantageous to reflect the pump beam to increase the power absorbed. As with all laser mirrors, both the substrate and the coating should be both low-loss and damage resistant.

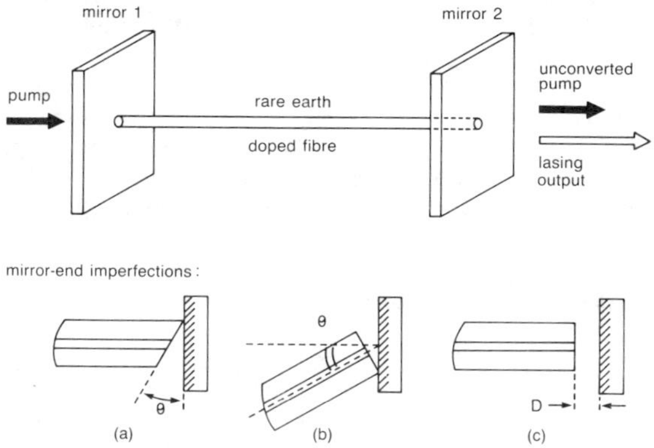


Fig 16.1 Top: Fibre Fabry-Perot laser. Bottom: Three examples of imperfections at the mirror-fibre end interface showing (a) end-angle, (b) tilt and (c) separation.

The fibre ends should be of high quality so that they are both flat and perpendicular to the fibre axis. It is sometimes desirable to polish the ends to remove any irregularities after they have been cleaved. The ends must be mounted perpendicularly against the mirror surface. Small angular misalignments can badly degrade both the reflectivity and the launch efficiency. Three types of mirror-end imperfections are shown in Figure 16.1. The loss caused by imperfectly mounting the fibre end against the mirror may be calculated by approximating the guided modal profile as a Gaussian with a 1/e field width of $w$. (Fibre modes are more accurately described by the weakly guiding analysis, which uses Bessel and Hankel functions). As an example, consider tilt, as illustrated by Figure 16.1 (b). If the fibre axis is at an angle of $\theta$ radians to a front surface mirror of reflectivity R the fraction, $\rho$ of the incident power that is successfully coupled back into the fibre can be calculated to a first approximation using a simple adaptation of the work of Marcuse (1977) and Marcuse and Stone (1986):

$$\rho = R \exp[-k_2^2 w^2 \theta^2] \tag{16.1}$$

$k_2$ is the propagation constant of a wave in the cladding and $w$ is expressed in terms of the fibre core radius $a$ by:

$$w/a = \frac{1}{\sqrt{2}} \left[0.65+1.619V^{-3/2}+2.879V^{-6}\right] \tag{16.2}$$

The so called $V$-value is a normalised frequency, given in terms of $n_1$ and $n_2$, the core and cladding refractive indices respectively, by:

$$V = ka(n_1^2 - n_2^2)^{\frac{1}{2}} \tag{16.3}$$

If $k_2w$ has the typical value of 25 then for a 1° tilt only, 0.83$R$ of the power is returned to the fibre, which could badly degrade the lasing performance. The cavity loss variation of a fibre laser with mirror tilt has been measured by Hanna *et al* (1988). Care has to be taken to ensure that the fibre ends are positioned accurately. By using high quality micropositioners, however, this can be achieved in practice, and erbium doped fibre lasers emitting at 1.55$\mu$m with threshold launched powers of a few milliwatts and slope efficiencies (with reference to the launched power) of 10–15% have been demonstrated by Wyatt *et al* (1988). Neodymium systems lasing at about 1.08$\mu$m have shown even higher efficiencies, as demonstrated for example by Shimitzu *et al* (1987).

## 16.4 CONNECTORS AND SPLICES

It is often necessary to join fibre ends together to achieve a longitudinal power transfer. The two methods of doing so, connectorisation and splicing, are necessary to make some types of resonator. They are also needed in circumstances where a doped fibre end is to be joined directly to an undoped fibre end. The methods are quite straightforward to follow. In a connector, as shown in Figure 16.2, two high quality fibre ends are located in a precision machined ferrule so that their axes are aligned.

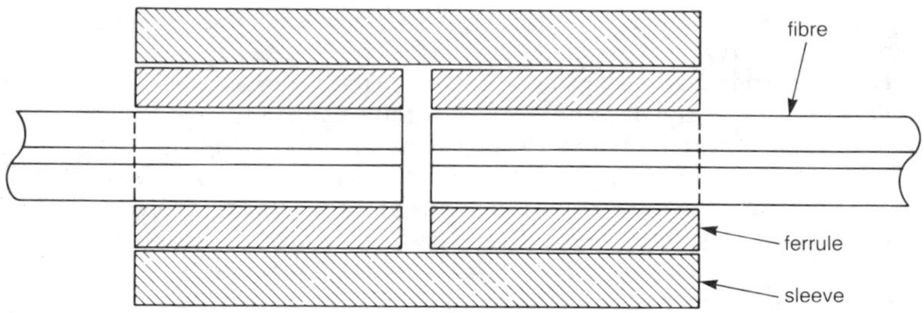


Fig 16.2 Schematic cross section of a connector. Not to scale.

There are various methods of splicing, the most important of which is fusion splicing, illustrated in Figure 16.3. The splice is formed by bringing together two ends between two electrodes. A high voltage is applied across the electrodes and the resulting discharge between them causes local fusing of the glass in the vicinity of the fibre ends. Obviously, a connector is demountable while a splice is not. In both cases the lowest losses are achieved when the two fibres are identical, well aligned and have high quality ends. A small proportion of the incident light is sometimes reflected from connectorised ends, with R ~ 4%. Their use within a resonator can cause a low finesse sub-resonance and if possible should be avoided. The theory of loss due to connectorisation using dissimilar single-mode fibres, which may have misaligned fibre ends, has been calculated in terms of the Gaussian mode approximation and is presented by Marcuse (1977). The problem of misaligned fusion splices has been covered by Wright (1986).

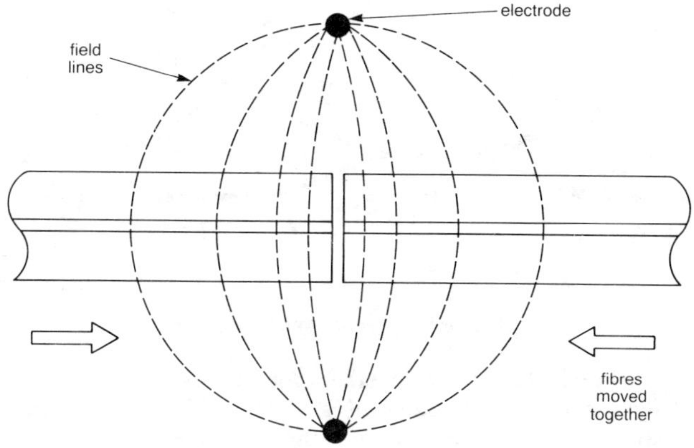


Fig 16.3 Fusion splicing using an arc between two electrodes (schematic).

## 16.5 DIRECTIONAL COUPLERS

Beam–splitting in fibre–based resonators is most efficiently carried out using fibre directional couplers. The cores of two fibres in a parallel configuration are brought into close proximity so that their evanescent fields overlap and there is a transverse exchange of power between them. The input ports are designated 1 and 2 and the output is from ports 3 and 4. The coupler has a coupling ratio $K$ and an excess loss $\gamma$, both of which may have values ranging between 0 and 1. When $K \approx 0$ there is little power transfer across the coupler and when $K \approx 1$ there is nearly complete transfer. The loss comes about as a result of scattering and absorption. Usually $\gamma$ is small and is about 0.01. If coherent waves are launched into the coupling region from ports 1 and 2 and the complex field components propagating towards the coupler are $E_1^t$ and $E_2^t$, then the output waves travelling away from the coupler at ports 3 and 4 have field components given by:

$$E_3^a = jK^{\frac{1}{2}}(1-\gamma)^{\frac{1}{2}}E_1^t + (1-K)^{\frac{1}{2}}(1-\gamma)^{\frac{1}{2}}E_2^t \tag{16.4}$$

$$E_4^a = jK^{\frac{1}{2}}(1-\gamma)^{\frac{1}{2}}E_2^t + (1-K)^{\frac{1}{2}}(1-\gamma)^{\frac{1}{2}}E_1^t \tag{16.5}$$

The imaginary coefficients of $E_1^t$ and $E_2^t$ indicate that in crossing over the coupler from port 1 to 3 or port 2 to 4 there is a phase lag of $\pi/2$, whereas there was no phase change associated with the output that does not cross over. The output intensity terms are calculated by multiplying the field terms by their complex conjugates.

There are two main types of directional couplers, known as the fused and polished couplers, which have been discussed in Lamont *et al* (1985) and Digonnet and Shaw (1982), respectively. Both types offer low excess loss beamsplitting whilst maintaining fibre wave–guidance. Both are four port devices to which equations (16.4) and (16.5) apply. The fused coupler is shown in Figure 16.4 from the side and in cross-section. It is made by

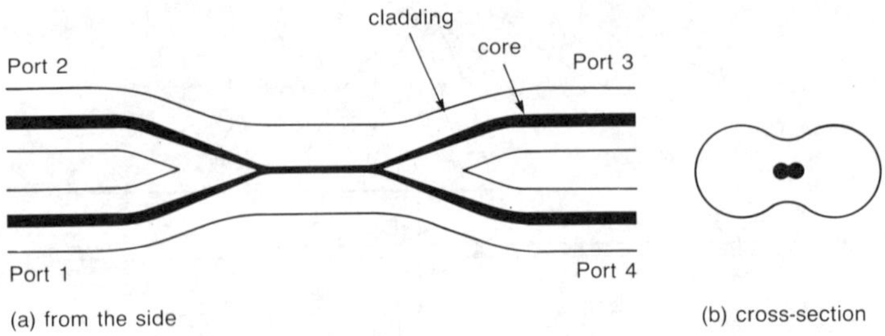


**Fig 16.4 Fused tapered coupler.**

bringing two single-mode fibres together for a region of a few millimetres along their length and heating them in a flame which is hot enough to cause the glass of the two fibres to fuse together. Wave-guidance in the coupling region is performed by the air-glass interface at the outer surface of the cladding. The coupling ratio $K$ is governed by the interaction length, the distance between the two cores and factors such as the core radii and refractive indices.

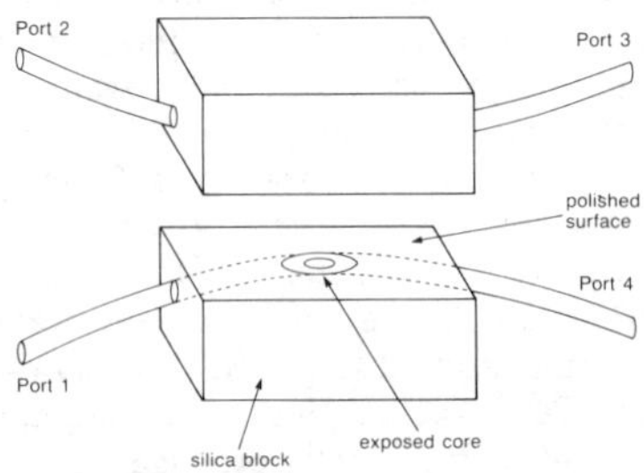


**Fig 16.5 Polished coupler.**

A polished coupler is illustrated in Figure 16.5. As can be seen, it is constructed in two halves. A fibre is laid and cemented into a curved groove in a silica block so that a small portion protrudes above the surface. The surface is then polished flat so that, as shown in the lower block of Figure 16.5, the core of the fibre is exposed. Two polished blocks are placed back-to-back to form a coupler. It is usually necessary to introduce a few drops of an refractive index-matching fluid between the blocks to ensure good optical contact. The coupling ratio is determined by factors such as the length of the interaction region, the longitudinal and lateral offset, the relative orientation of the exposed cores and the refractive index of the intermediate fluid layer.

The coupling ratio of both types of couplers can be varied. Adjustment of the coupling ratio of the polished coupler is brought about by moving the two blocks by a few microns. Variation of the coupling ratio of the fused coupler has been achieved by surrounding the coupling region with a silicone potting compound to give good thermal contact and changing the temperature with the use of a Peltier element, as described by Millar *et al* (1988).

## 16.6 OTHER INTRA-CAVITY COMPONENTS

Fibre lasers have been made to operate Q-switched and mode locked. Such performance is obtained with the use of some form of intra-cavity element, such as an acousto-optic modulator. To date these have all been bulk optical components and so it has been necessary to couple light out of the fibre and collimate it using a lens (usually a microscope objective) with a numerical aperture closely matched to that of the fibre. A discussion of the requirements is given by Alcock *et al* (1986). The problem with this method is that additional loss is introduced into the cavity. Ideally, one would like a modulator in fibre form. An all-fibre Q-switching modulator has not yet been demonstrated.

The fibre Fabry-Perot resonator illustrated in Figure 16.1 gives relatively broad band output with no potential for tuning. Wavelength selection and tuning may be achieved by replacing the output mirror with a diffraction grating to reflect only in a narrow bandwidth. In this design, light must be coupled out of the fibre and collimated, so that the problem of intra-cavity loss is again apparent. An all-fibre grating reflector, based on a single polished half-coupler block with a grating overlay has been shown to overcome this difficulty by Bennion *et al* (1986). A phase grating is formed by holographic means on the short exposed section of fibre on top of the block with the grating perturbations running perpendicular to the fibre core. The fibre optical version of a distributed Bragg reflector was thus demonstrated. Using this technique, Jauncey *et al* (1988) have demonstrated feedback sufficiently narrow to obtain single longitudinal mode lasing at 1082 nm from a very short length of neodymium-doped fibre. The method has not, at the time of writing, been applied to give single-mode output at the potentially more useful erbium transition at 1550nm.

## 16.7 ALTERNATIVE RESONATOR DESIGNS

There are several difficulties with Fabry-Perot resonators, as depicted in Figure 16.1. Longitudinal pumping requires launching light through a mirror on a fibre end. The three problems with doing so are *firstly*, as discussed in section 16.3, the system is susceptible to loss due to angular misalignment, *secondly* the mirror coatings are found to be prone to damage with pump powers of a few Watts cw focused down onto a small spot and *thirdly*, the launch efficiency is lower than for an unmirrored end. An associated limitation is that if a fibre laser is to be used as a telecommunications source, it is undesirable to locate a mirror between the lasing fibre and the transmission fibre. Fibre laser cavities that permit the pump light to be coupled directly into unmirrored ends are therefore required. Various designs of resonator, based on directional couplers, have been demonstrated and should overcome these problems. These will be discussed in this section.

Figure 16.6 illustrates in tabular form a number of ways in which directional couplers can be incorporated into fibre lasers. It is a compilation of most of the designs of fibre interferometers that have been demonstrated to date, showing the fibre-based devices and their equivalents in bulk optics. Only three components are used in these devices. These are the mirrors, the directional couplers and the fibre itself. All of the devices shown in Figure 16.6 can be fabricated entirely from rare earth doped fibre and all except one of them can be made without any splicing or connectorisation being necessary. Alternatively, they can be made so that a portion of the device incorporates doped fibre and is spliced to the remainder, which is undoped.

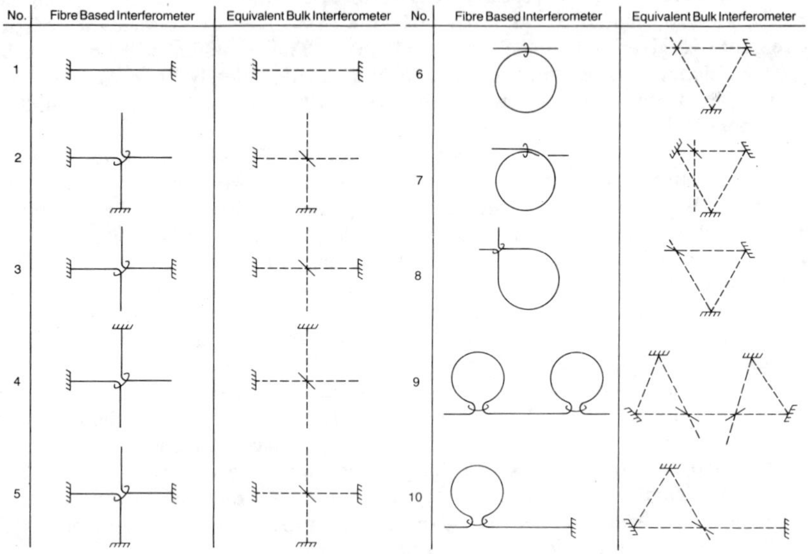


Fig 16.6 Tabulation of the interferometer architectures discussed in section 16.7, together with the equivalent devices using bulk optical components.

Diagram 1 in Figure 16.6 shows the fibre Fabry–Perot resonator. This is followed by four designs of four-port directional coupler interferometer. Diagrams 2 and 3 are known as the types I and II transversely coupled fibre Fabry–Perot (TCFFP) resonator, respectively and have been discussed theoretically by Urquhart (1987) and experimentally by Brierley and Urquhart (1987). They have been used as laser cavities by Barnsley (1987). High finesses are possible in the type I TCFFP resonator when the coupling ratio is low and the type II TCFFP resonator when the coupling ratio is high. Pump light can be launched into the cavities through the unmirrored ends provided that the coupling ratio is chosen appropriately. Diagram 4 of Figure 16.6 is a fibre Michelson interferometer, as demonstrated by Kashyap and Nayar (1983) and is the third design of four port directional coupler interferometer with two mirrors. It is a non-resonant device, demonstrating a broadband frequency response. It has not to date found direct application in fibre lasers but is important as a fibre sensor. The fibre Fox–Smith resonator, shown in diagram 5 of Figure 16.6, has been discussed theoretically by Urquhart (1988b) and experimentally by Barnsley *et al* (1988). It is a directional coupler resonator with three mirrors (consisting of two coupled cavities) and has been used as a method for line narrowing fibre lasers to obtain single longitudinal mode output. Each of the cavities operating individually will produce a set of resonances. The transmission of the complete resonator is high when both of the constituent cavities are on resonance and low when only one cavity is on resonance. If the cavity lengths are chosen to be in the ratio of two integers which are close, but not equal in value, the frequency spacing of the high transmission peaks will be made to be significantly further apart than the spacing of either of the transmission peaks of the cavities acting individually. This is an example of the Vernier effect. When a gain medium, such as erbium dopant, is incorporated in the fibre, lasing will take place preferentially at the frequencies which simultaneously satisfy the resonance conditions of both cavities. The lasing modes are then sufficiently far apart that one of them may be selected with the use of a diffraction grating. The fibre Fox–Smith resonator is interesting in that it is the only example in Figure 16.6 of a compound resonator, in which two cavities are coupled together in the same structure. Other compound fibre-based resonators have been proposed by Urquhart (1988b) and these all demonstrate the possibility of mode suppression by the Vernier principle. One of the benefits of fibre-based resonators is the ease with which such structures can be assembled.

Two types of fibre ring resonator are shown in diagrams 6 and 7 of Figure 16.6 and these have been demonstrated by Mears *et al* (1985) and Stokes *et al* (1982), respectively. The first of these is made by joining together two ends of a directional coupler and the second can be constructed by folding over the fibre in an appropriate manner before the coupler is made. The first type of ring, known as a directly coupled ring resonator, is the only design of resonator shown in Figure 16.6 which cannot be made without using a splice or a connector. The indirectly coupled ring resonator of diagram 16.7 can be constructed from a continuous length of fibre. The directly coupled ring can be constructed so as to demonstrate high finesse performance with the use of a low coupling ratio, while the indirectly coupled ring gives a high finesse when the coupling ratio is high.

Diagrams 8–10 of Figure 16.6 show the fibre loop reflector and two ways that it can be incorporated into a resonator. The loop itself, shown in diagram 8 is a non-resonant interferometer which acts as an all-fibre mirror. Light entering the loop from one of the two open ports is split at the coupler, circulates both clockwise and counter clockwise and coherently recombines at the coupler. A reflected portion of the power exits by the port into which it was originally launched and the remainder, which is the transmitted power, exits by the other port. Both pathways round the loop are of identical length. Two such loops can be concatenated in series, as shown in diagram 9 and described by Miller *et al* (1987) to form an all-fibre resonator and laser. Such a structure can be made without a splice. Alternatively, as shown in diagram 10, only one of the reflectors need be in the form of a loop and such a configuration was used by Millar *et al* (1988).

It is interesting to consider what constitutes the cavity length of a resonator composed of two loops. To do so, one has to trace the path of a photon as it undergoes a round trip: the cavity length is half the round trip length. Let the photon enter the system at the left hand coupler. It will undergo a circuit of the left hand loop, followed by a transit of the connecting section of fibre, followed by a circuit of the right hand loop and finally a return transit of the connecting fibre section. The cavity length must therefore be half this length, which is the sum of half the lengths of the loops plus the length of the connecting section of fibre. The effective length determines the free spectral range of the Airy function and is also required to predict the lasing threshold power.

A range of fibre resonator types has been briefly discussed in this section. Figure 16.6 illustrates the scope for using basic components to enable the design of a whole class of resonators for fibre lasers. It can be seen how fibre, mirrors and directional couplers can be concatenated in a large number of ways to form useful devices.

## 16.8 CONCLUSION

Fibre lasers use the waveguiding properties of optical fibres, together with the gain produced by an inverted population of rare earth ions to produce the amplification necessary for lasing. Fibre lasers are most efficiently operated with longitudinal pumping in circumstances when the optical fibre is single transverse mode at both the pump and lasing wavelengths. In this chapter, the constraints that the fibre geometry places on the design of laser resonators have been discussed. Fibre Fabry–Perot resonators must be carefully constructed to avoid losses at the interfaces between the fibre ends and the mirrors.

Beamsplitting in fibre optics is carried out with low excess losses using fibre directional couplers. The opportunity to carry out low–loss beamsplitting means that a large range of fibre resonator designs are available. Various types of optical fibre resonators have been reviewed. These include four–port directional coupler resonators, ring resonators and structures incorporating fibre loop reflectors. There are a large number of ways that one could envisage incorporating mirrors and directional couplers to design novel resonators. It is therefore to be expected that the range of designs discussed here will be extended to satisfy the growing number of operating requirements of fibre lasers.

## 16.9 REFERENCES

Alcock I P, Tropper A C, Ferguson A I and Hanna D C 1987 IEE Proc. Part J (Optoelectron.) 134 183–186

Barnsley P 1987 MSc Thesis University of St Andrews

Barnsley P, Urquhart P, Millar C and Brierley M 1988 J. Opt. Soc. Am. A 5 1339–1346

Bennion I, Reid D C J, Rowe C J and Stewart W J 1986 Electron. Lett. 22 341–343

Born M and Wolf E 1980 6th Ed (Oxford: Pergamon)

Brierley M C and France P W 1988 Electron. Lett. 24 935–937

Brierley M and Urquhart P 1987 Appl. Opt. 26 4841–4845
Digonnet M J F and Shaw H J 1982 IEEE J. Quantum Electron. QE–18 746–754
Hanna D C, Smart R G, Suni P J, Ferguson A I and Phillips M W 1988 Opt. Commun. 68 128–132
Jauncey I M, Reekie L, Townsend J E, Payne D N and Rowe C J 1988 Electron. Lett. 24 24–26
Kashyap R and Nayar B K 1983 IEEE/OSA J. Lightwave Technol. LT–1 619–624
Marcuse D, 1977 Bell Syst. Tech. J 56 703–718
Marcuse D and Stone J 1986 IEEE/OSA J. Lightwave Technol. LT–4 377–381
Mears R J, Reekie L, Poole S B and Payne D N 1985 Electron. Lett. 21 738–739
Millar C A, Miller I D, Mortimore D B, Ainslie B J and Urquhart P 1988 IEE Proc. Part J (Optoelectron.) 135 303–309
Miller I D, Mortimore D B, Urquhart P, Ainslie B J, Craig S P, Millar C A and Payne D B 1987 Appl. Opt. 26 2197–2201
Shimitzu M, Suda H and Horiguchi M 1987 Electron. Lett. 23 768–769
Stokes L F, Chodorow M and Shaw H J 1982 Opt. Lett. 7 288–290
Urquhart P 1987 Appl. Opt. 26 456–463
Urquhart P 1988a IEE Proc. Part J (Optoelectron.) 135 385–407
Urquhart P 1988b J. Opt. Soc. Am. A 5 803–812
Wright J V 1986 SPIE 630 110–119
Wyatt R, Ainslie B J and Craig S P 1988 Electron. Lett. 22 1362–1363

# 17 $CO^2$ Waveguide Laser Resonators

*I E ROSS*

## 17.1 INTRODUCTION

Over the past 20 years or so the waveguiding principle and waveguide resonators have been applied to a number of laser types, including dye lasers, diode lasers, helium–neon, helium–xenon and carbon monoxide lasers, excimer lasers and far infrared (FIR) lasers. However, it is the *carbon dioxide* waveguide laser, the subject of this chapter, which has attracted by far the most research and development activity during this period.

The reason for this is largely due to the wide range of possible system application for these devices. They may offer a combination of high specific power, high efficiency, comparatively wide frequency tuning range, compact size, rugged construction and low cost. Applications fall into two main categories, those where the laser is the transmitter (and/or local oscillator) in a remote sensing or laser radar system, and a second class of applications in which the laser represents a highly controllable heat source for certain industrial and medical applications involving, for example, the cutting, drilling or perforation of non-metallic materials.

The concept that a section of optical waveguide could be incorporated within a gas laser was first proposed by Marcatili and Schmeltzer in 1964. They calculated the modes of hollow dielectric guides and determined the propagation characteristics for the situation where the guide diameter/width is large compared to the wavelength of the light. Based on this theory and on the known gain for the helium–neon laser, they speculated that *waveguide* He–Ne lasers would be feasible. Several years later, and without knowledge of the work of Marcatili and Schmeltzer at Bell Labs, Steffen and Kneubühl (1968) utilised waveguiding concepts to explain the apparently anomalous observation by Schwaller *et al* (1967) on transverse mode frequencies in a hydrogen cyanide laser operating in the far infrared portion of the spectrum.

The first observation of laser action in a laser actually *designed* as a waveguide laser was reported by Smith (1971) who, confirming the predictions of Marcatili and Schmeltzer, successfully operated a waveguide He–Ne laser consisting of a small bore (430 $\mu$m), 20 cm long glass capillary tube and external mirrors. Best performance with this laser was obtained when the 30 cm radius of curvature mirrors were positioned at a distance from the guide entrance very slightly smaller than the mirror radii. From the standpoint of conventional Gaussian beam analysis described in Chapter 1, this geometry is clearly unstable, with very high Fresnel losses, so that a new resonator theory, which takes account of an intra-cavity waveguiding section (Chapter 3), must be invoked to explain the properties of this laser. Applications of the waveguiding principle to cw $CO_2$ lasers using DC discharge excitation of the gain medium, followed fairly rapidly (Bridges *et al* 1972, Abrams and Bridges 1973), and since that time, the waveguide laser literature has been almost completely dominated by work on $CO_2$ devices.

During the remainder of the 1970s, the main driving force for research and development on waveguide $CO_2$ lasers was the military requirement for lasers to be used in infrared ranging/surveillance/measurement systems. Germane to this application is the fact that these lasers may be designed as physically very compact packages, constructed from rugged metal/ceramic materials, allowing vacuum sealing technology consistent with long operational lifetimes prior to gas degradation. At the same time, they may exhibit high specific power and efficiency coupled, in some designs, with good beam quality and frequency stability.

However, the $CO_2$ waveguide laser did not break into the commercial market to any significant extent until the beginning of the 1980s, when an additional important new feature became available for incorporation in $CO_2$ laser design. This new development involved the use of transverse radiofrequency discharges (Laakmann 1980) as a preferred alternative to longitudinal DC discharges for laser excitation. The additional laser design flexibility led to the commercial availability of lower cost lasers to meet the requirements of the scientific, industrial and medical markets.

This chapter reviews some of the important practical developments in waveguide $CO_2$ lasers, emphasizing those issues which relate to the *resonator design*. The practical characteristics of waveguiding in the 10$\mu$m spectral region are outlined in Section 17.2, and this section also includes a discussion of the main options for resonator design, giving practical examples where appropriate. A working knowledge of the contents of Chapter 3 on waveguide resonator theory is assumed. Section 17.3 covers some of the materials–related and technological implications of $CO_2$ laser design and construction. Section 17.4 discusses the optical design implications of waveguide lasers which incorporate multiple discrete gain lengths, with the beam optically *folded* from guide to guide. The chapter concludes with a brief review of some recent work on resonators for slab–shaped gain media, which combine features of both waveguiding and unstable resonators.

## 17.2 RESONATORS FOR WAVEGUIDE $CO_2$ LASERS

### 17.2.1 Design of the Waveguide

In the overall design process, an early decision is required on the selection of material(s) for the guided wave section of the laser. Although the choice of resonator parameters is a crucial aspect of the design, there are, as will soon be apparent, other important factors, which may on occasion dominate the choice of materials and geometry in such a way as to limit the choice of resonator configurations. Nevertheless, one vitally important factor concerns the attenuation characteristics of the guide itself. The waveguide power attenuation coefficient, $\alpha_{nm}$ for circularly symmetric guides, derived by Marcatili and Schmeltzer (1964) is given (see Chapter 3) by:

$$\alpha_{nm} = \frac{1}{4\pi^2}\, \rho_{nm}^2 \,\frac{\lambda^2}{a^3}\, \mathrm{Re}\{\tfrac{1}{2}(\epsilon+1)(\epsilon-1)\}^{-\frac{1}{2}} \qquad (17.1)$$

where $\rho_{nm}$ is the $m^{th}$ root of the $n^{th}$ order Bessel function $J_n(x)$ (i.e. a number which is different for the various modes, has the lowest value for $\alpha_{11}$ and increases with $n,m$), $a$ is the guide radius, $\lambda$ is the wavelength of the propagating radiation and $\epsilon$ is the complex dielectric constant of the wall material. Examination of equation (17.1) indicates that in considering the question of power loss, one is faced with specifying both a material–dependent factor ($\epsilon$) and the guide diameter. Three materials have been in common use for 10$\mu$m waveguides, namely pyrex, alumina ($Al_2O_3$) and beryllia (BeO). As regards waveguiding losses, equation (17.1) predicts that beryllia should be between one and two orders of magnitude less lossy than alumina, which in turn should have slightly higher transmission than pyrex. Experiments on guide attenuation confirm this order of preference with respect to the materials, if not the exact scale of the differences. Of greater significance in overall laser design, is the fact that the ceramic materials have considerably superior thermal properties to those of pyrex. In addition pyrex, although inexpensive, is fragile, less easy to machine and more difficult to seal effectively to metal components. On the other hand, handling BeO is more complicated because of the health risk, and on balance, unless there is a particular reason to use very small bore waveguides (when guide losses become more important because of the $a^{-3}$ factor), alumina is almost always the preferred material. One case where use of BeO may be justified is in waveguide lasers designed for wideband tuning, where high pressure operation and a narrow bore gain section are necessary.

Returning to equation (17.1), we note that the value of the Bessel function root increases with $n,m$, which suggests that differential guiding losses would be an effective basis for discrimination against unwanted higher order waveguide modes. However, as noted in Chapter 3, this approach must be viewed with some caution. Given that the gain spectrum of the $CO_2$ laser medium extends from about 9.2 $\mu$m to 10.8 $\mu$m, the $\lambda^2$ factor suggested by equation (17.1) has significance in the context of line–tunable $CO_2$ lasers. This mode and wavelength dependence often shows itself in real devices with the observation of mode hopping during line tuning. However, the situation is complex and this topic is an area of current active research (Jenkins *et al* 1985).

Finally, on the question of the dependence of losses on guide diameter, experimental results show approximate verification of the $a^{-3}$ factor (Hall *et al* 1977). For the $CO_2$ laser wavelength, and for typical gain/guide lengths of 10–50 cm, waveguide diameters are in the approximate range 1.0 – 3.0 mm, while representative practical values of $\alpha_{11}$ are in the range 0.1–0.01% cm$^{-1}$.

### 17.2.2 Waveguide Resonator Design

Of the three low coupling loss optical configurations reviewed in Chapter 3, Cases I and III are the most interesting from the practical point of view. They are shown schematically in Figure 17.1 (a) and (b).

Comparing the two arrangements, Case I has the virtues of simplicity, lower loss and the fact that it produces the most compact resonator. However the degree of transverse mode discrimination is poor. Case III resonators on

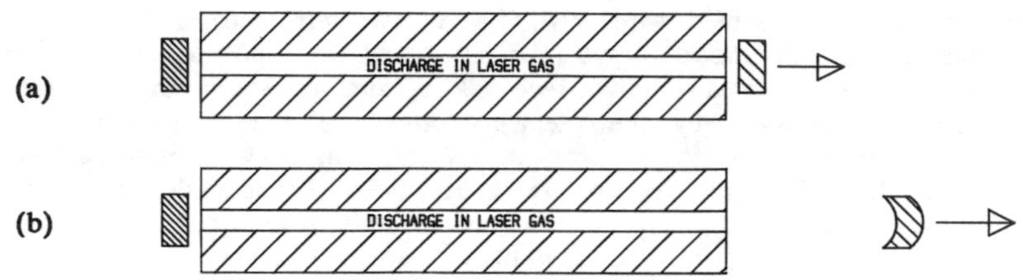


**Fig 17.1 Waveguide laser resonator designs which include (a) Dual Case I (b) Case III reflectors.**

the other hand, require an increase in cavity length, but, as described in Chapter 3, offer a much higher level of discrimination against unwanted higher order modes.

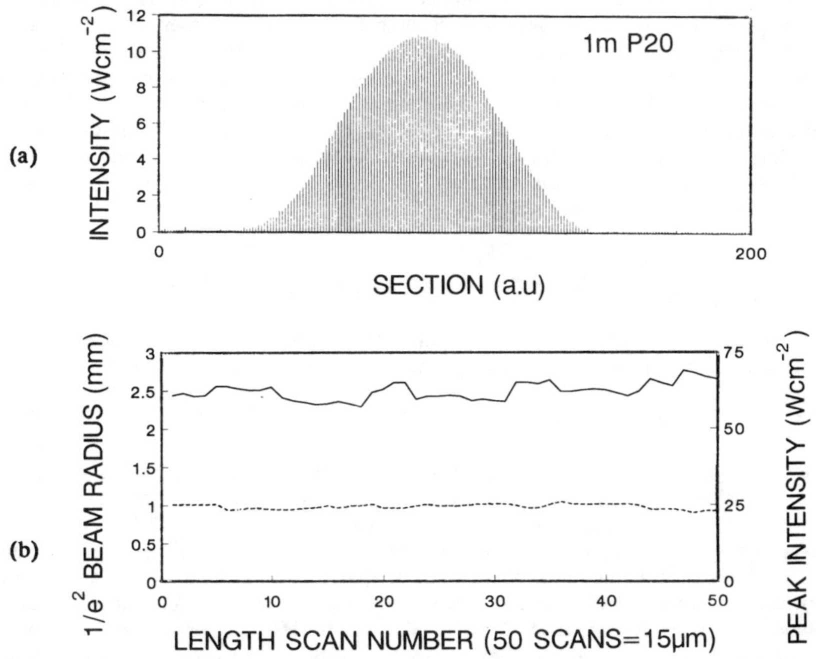


**Fig 17.2 (a) Beam intensity profile of a 4W dual Case I resonator (b) Dependence of beam intensity (upper trace) and radius (lower trace) laser trace on cavity length for a 20W Case III resonator.**

Figure 17.2 shows some experimental measurements of the transverse beam profile, indicating the level of mode purity of waveguide lasers of the two types mentioned above, when optimized for lowest order mode throughout the signature. These curves are generated by repeatedly scanning the laser output beam across a fast detector. Figure 17.2a shows the transverse intensity profile of a laser designed with dual Case I reflectors, measured at a distance of 1 metre from the laser output mirror. Generally, this performance quality can be achieved with an active feedback system to select the highest gain transition, for example by piezoelectric control of the length of the laser cavity. The more difficult problem of assuring this mode performance throughout the full cavity free spectral range, with the spatially–economic Case I reflectors, has been the subject of considerable research effort over the years. Among the approaches investigated are the technique of seeking to ensure the "co–incidence" of transverse mode frequencies described in Chapter 3 (Section 3.4.2), and the exploitation of mode–dependent differential waveguiding losses, where geometric design and the control of manufacturing tolerances are important.

Figure 17.2b shows some results obtained with a 20 Watt laser designed with a Case III waveguide coupler at the output mirror end of the laser. The results shown are from many measurements similar to that shown in Figure 17.2a. Here the same beam profile assessment technique is employed, in which the beam is tuned over a fast detector while the laser length is slowly scanned over the cavity free spectral range. The upper trace shows the peak intensity of the beam, where the variation with cavity length reflects the characteristic laser "signature" as different transitions are selected. The lower trace indicates the stability of the radius of the beam profile (measured at the $1/e^2$ intensity points), as the cavity length is tuned. The excellent consistency of these results clearly indicates the benefits of utilising the Case III reflector in the waveguide resonator design.

## 17.3 THE TECHNOLOGY OF $CO_2$ WAVEGUIDE LASERS

No account of waveguide $CO_2$ lasers would be complete without a brief discussion of some of the technology which characterises these lasers.

### 17.3.1 Excitation Techniques

The laser gas, usually a mixture of carbon dioxide, nitrogen and helium, with other possible additives such as xenon, must be excited with an electrical discharge. Two possible schemes are shown in Figure 17.3. The first waveguide $CO_2$ lasers were excited in conventional fashion using a high voltage DC discharge with the current flowing longitudinally along the waveguide tube between an anode and cathode positioned in side arms. The inter–electrode voltage is typically about 5–7 kV, with most lasers having several sets of electrodes to achieve longer gain lengths without excessively high voltages. The cathode is a chemically active region and a source of problems for long life device operation, requiring attention to the choice of material and geometry.

Since about 1980, there has been a growing tendency, both in research and for commercial laser sources to use radiofrequency (rf) discharges to excite the gas. The frequency of the rf field is usually in the range 30 MHz to several hundred MHz. Interestingly, the very first gas lasers (He–Ne and $CO_2$) were rf excited, but, due to the problems of coupling rf power to the discharge and the fact that the alternative technique was technologically more

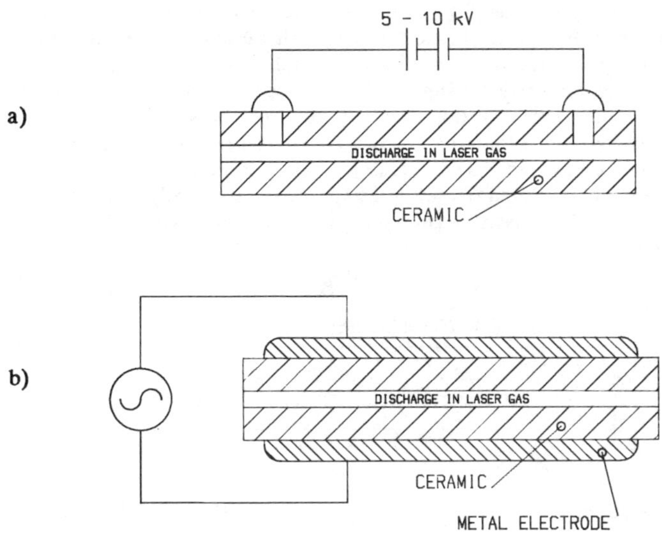


Fig 17.3 Waveguide $CO_2$ laser excitation techniques.
(a) DC excitation (b) Rf excitation.

familiar to most researchers, DC discharges became accepted for gas laser excitation. After a decade of research on rf discharges for laser excitation and significant advances in rf semiconductor technology, the situation has completely reversed for waveguide lasers. Many of the advantages of rf excitation derive from the positive impedance character of the discharge in the region of operation relevant to laser excitation. By contrast, DC and low frequency discharges exhibit a negative differential impedance and are unstable under certain operating conditions. Moreover, because the rf discharge is transverse to the optic axis, it can be sustained with voltages as low as a few hundred volts. The high frequency oscillatory nature of the electric field largely prevents the adverse chemical effects which are associated with the cathode in uni-directional DC discharges. In addition, the rf power can be capacitively coupled into the discharge (as in Figure 17.3b) so that it is possible to avoid the use of metal electrodes, thereby reducing chemical interaction with the laser gas.

### 17.3.2 Waveguide Laser Construction

Modern waveguide lasers are generally based on a guide constructed from a ceramic-metal combination, which is inherently rugged, compact and readily sealed using high vacuum techniques, without the use of organic sealants.

Two construction techniques are shown in Figure 17.4. The first is shown in 17.4a, and consists of a waveguide assembly mounted inside an outer metal, gas-sealed enclosure; the so called 'ship-in-a-bottle-method'. The second technique, shown in 17.4b, is to seal the ends of the waveguide ceramic to enclose the laser gas. The waveguide, which guides the light between the cavity mirrors, can be either an all–ceramic assembly or a combination of metal and ceramic. Both square and circular waveguide cross sections can be manufactured using well–developed ceramic processing techniques. Thus, circular cross sections with bore sizes between about 1 and 3 mm can be extruded or preferably precision drilled, while square cross section guides can be fabricated using ceramic channels and covers. These can either be enclosed in a sealed metal housing, or alternatively, the ceramic surfaces may be sealed by glaze/thermo–diffusion bonding, or soldering of metallised ceramic surfaces.

(a)

(b)

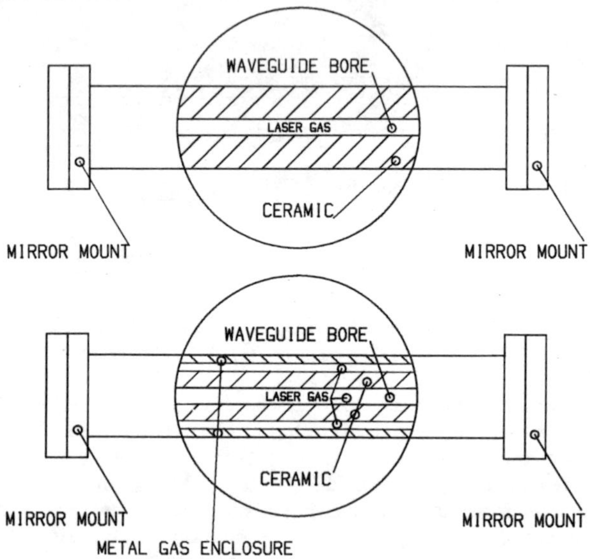


**Fig 17.4 Waveguide laser construction techniques. (a) Metal enclosure on the gas envelope (b) Ceramic gas envelope.**

For rf–excited devices, heterogeneous structures with a combination of ceramic and metal walls can be fabricated from slab technologies with the entire structure housed in a sealed gas–tight enclosure.

This type of construction with its widespread use of metal and ceramic components owes much to the sealing/vacuum technology developed for the manufacture of high power valves (tubes). The use of these technologies makes possible the manufacture of devices which are ideal for long life and permit sealed operation. Whatever envelope design is employed, there is one additional interface between the output laser mirror or window (usually coated zinc selenide or germanium) and its metallic mount, where a high quality seal is required and which has in the past been a problem, with some manufacturers having to resort to unreliable organic seals. However, in recent times, the use of indium, metal or glass frit seals has become standard in commercial devices (Conder *et al* 1985).

### 17.3.3 Influence of the Gain Medium on Waveguide Mode Control

As with most types of laser, the problems of resonator design do not end with an analysis based on the "empty" cavity, since the gain medium may have significant effects on the transverse mode characteristics of the laser. For waveguide lasers, this can be a particularly thorny problem, because of the close proximity of the waveguide walls and the gain medium. No general solution exists. The gas discharge gain medium behaves as an "optical element" which may have spatial variations in both gain and refractive index. The magnitudes of these variations depend on electrical power input, gas pressure and composition and display different spatial symmetry for DC and rf excitation. For DC discharges, there is cylindrical symmetry so that, on the axis of the tube, the gas temperature peaks while the density and gain experience a minimum. Thus, higher-order transverse modes, which have axial minima, are favoured. In addition, the temperature-induced density variations produce a refractive index profile, the effect of which is a negative lens. Generally, it is necessary to select a slightly smaller bore diameter than might be otherwise calculated for such DC excited lasers.

Although rf excited devices have become the commercial market leader for waveguide $CO_2$ lasers, the effects of the transverse nature of the rf are actually somewhat more complex than for DC discharges. Although to first order, the use of high frequency excitation prevents the build up of a cathode fall region, He and Hall (1984) showed that space charge regions do exist, leading to symmetry about a plane equidistant between the two electrodes, resulting in a non-uniformity in the discharge and associated gain profile. In addition, the spatial extent of the space charge regions varies with the rf frequency in such a fashion that the adverse conditions decrease as the rf frequency increases. This effect is illustrated in Figure 17.5, which shows the transverse beam profile for several values of the generator frequency (Miller and Williams 1985). As can readily be seen, the best mode is achieved at lower rf frequency. The laser power tends to peak at higher rf frequencies so that a trade-off is required between mode performance and power.

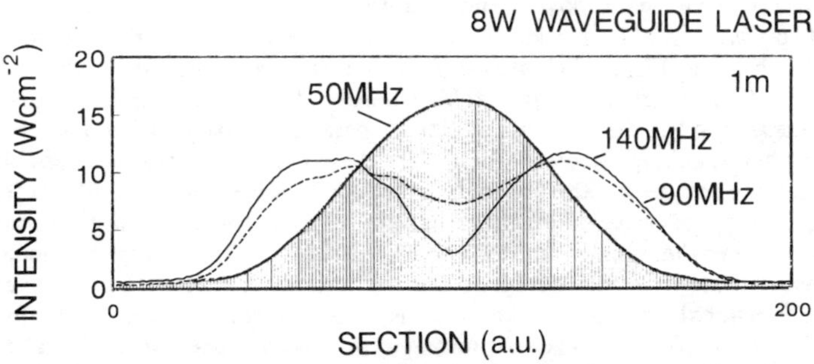


Fig 17.5 Transverse mode variation with generator frequency - due to changes in spatial inhomogeneities within the gain medium with frequency.

## 17.4 FOLDED $CO_2$ WAVEGUIDE LASERS

There are many light industrial, medical and military applications which require intermediate power levels up to the 100W level. Since the highest specific power from a $CO_2$ waveguide laser under static gas conditions is approximately 1W per cm of discharge, this implies an active resonator length of at least 1m. There are often operational or system reasons why devices of this length are inconvenient or impractical. Furthermore, there are length limitations in the fabrication of waveguides with cross-sectional dimensions of 0.5–2.5mm. Presently, waveguide construction methods such as ultrasonic drilling and glass/glaze or thermo-diffusion bonding of polished ceramic slabs are capable of producing guides within the stringent requirements of bore finish and straightness to a maximum length of about 50cm (Conder *et al*, 1985). Longer guide lengths may be achieved by butting together individual sections but the problems of pre-alignment and alignment drift make this option unattractive. However, the machining of parallel waveguide channels in a rectangular ceramic block and the provision of a thin ceramic plate as the "lid" is a relatively easy task. These individual waveguide channels may then be optically coupled in series to produce a long equivalent guide length with rigid alignment. A suitable gain medium can then be provided for this *folded* structure using either multi-electrode longitudinal DC discharges, or transverse rf discharge excitation between common electrodes.

The folded waveguide laser may thus concentrate perhaps a metre of discharge gain medium into a compact laser head not much larger than that required for a single element waveguide laser. However, folding for waveguide lasers is a more difficult problem than for conventional Gaussian resonators (Chapter 5). In both types of resonator there are optical losses associated with the additional intracavity reflectors, but for folded waveguide resonators, there is the additional feature of coupling losses at the ends of each guide element, as well as the complexity associated with the effects of tilted mirrors on the resonator mode. As a result of these additional losses, the increase in overall gain length in folded waveguide devices does not produce a *pro rata* increase in laser power. To minimise this effect, and maximise the benefits of the additional gain length, considerable efforts have been made to design low loss folding configurations, and to study the effects of folding mirror misalignment on laser beam power and quality.

The two basic fold geometries are the U-fold and the V-fold, as illustrated in Figure 17.6. Comparing the two, we note that the U-fold requires two reflecting surfaces per fold whereas V-folding requires only one. On the other hand, the U-fold allows parallel waveguide channels and therefore, higher packing density than for the V-fold. Also, it should be noted that as the angle in the V-fold is reduced, with the aim of increasing the packing density of waveguide channels, there is a corresponding increase in the distance from the guide to the fold mirror. This leads to increased coupling losses, since the self-coupling loss of the fundamental $EH_{11}$ waveguide mode is proportional to $d^{3/2}$, where $d$ is the distance between guide and reflector. These sources of loss assume greater importance since, in a folded resonator, each fold is involved twice per round trip, so that the cumulative effect of even an apparently modest single-pass folding loss may result in a significant reduction in the laser output power.

In principle, the U-fold optics may be placed at an arbitrary distance from the ends of the waveguides, but for low loss folding they are usually as close as possible to the guide end. This is most conveniently achieved using the prism reflector exploiting total internal reflection, as shown in Figure 17.6b. The U-fold two-mirror alignment problem is then reduced to one involving a single element. Single-pass folding losses for the fundamental $EH_{11}$ mode as low as 3.5 ± 0.5%, have been obtained using AR coated zinc selenide prisms (Miller and Williams, 1987).

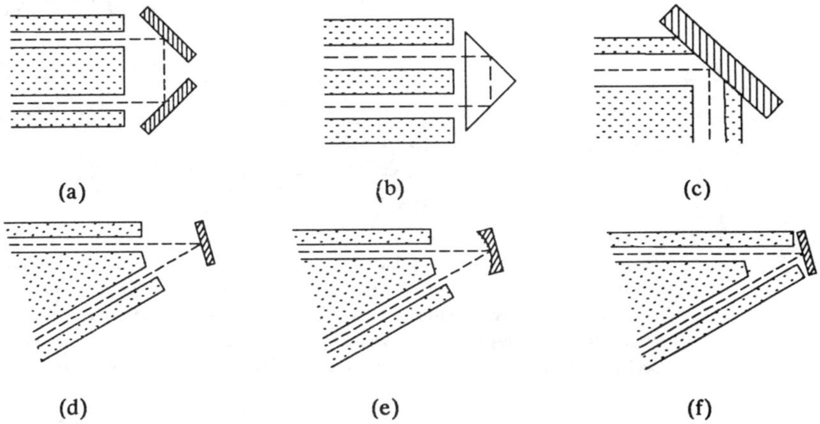


Fig 17.6 Waveguide folding geometries a) Conventional U-fold b) Prism U-fold with total internal reflection c) Right-angle fold d) Conventional V-fold e) V-fold with curved (astigmatic) reflector f) Partial waveguide V-fold.

Smaller V-fold angles (and therefore larger values of *d*) may be employed with reduced penalty in terms of folding/coupling losses by the use of properly selected curved reflectors as shown in Figure 17.6e. Here, the divergence of the incident light is compensated by the curved reflecting surfaces. It should be pointed out that the low-loss and highly mode-discriminating Case III reflector, described in section 17.2.2, may also be used to advantage in this situation. Also shown in Figure 17.6f is the partial waveguide V-fold reflector, in which the light experiences three-walled or partial-waveguiding almost to the mirror itself. Using laser power measurements and Rigrod analysis for a Z-fold laser, Jackson *et al* (1989) have estimated single pass folding losses as low as 1.2 ± 0.1%.

These results and others (Miller and Williams 1987, Newman *et al* 1987) have demonstrated folded $CO_2$ waveguide resonators with output powers in excess of 50W with specific power extraction of typically ½W per cm of discharge.

Some more complex multiple fold waveguide resonator configurations which have been investigated are shown in Figure 17.7. There are however certain practical limitations to the output power achievable by length scaling with multiple folds. For example, as the laser power approaches the 100W level the flux incident on the folding mirrors, which is twice that incident on the end reflectors, becomes large enough for optical damage to be a major concern. However, length scaling by folding the resonator is not the only way to produce higher output power from waveguide $CO_2$ lasers. An alternative approach is described in the next section.

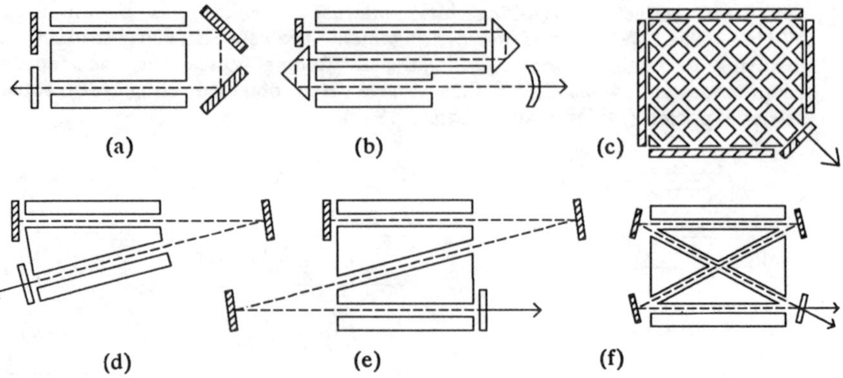


Fig 17.7 Various configurations for folded waveguide resonators. a) Conventional U-fold resonator b) Double U-fold resonator with prism U-folds and Case III output coupling (Miller and Williams 1987) c) Multiply folded intersecting waveguide lattice (Cantoni 1989) d) Conventional V-fold resonator e) Z-fold resonator (Jackson *et al* 1989) f) X-fold resonator (Newman and Hart 1987).

## 17.5 RESONATORS FOR SLAB WAVEGUIDE LASERS

The previous section discussed the virtues of parallel or near-parallel waveguide channels coupled in series as a means of increasing the active gain length to provide higher laser output powers. An alternative approach to the exploitation of an array of parallel waveguide gain media involves the coupling of radiation from guide to guide in parallel, so as to *phase-lock* the individual waveguide laser oscillators (Newman *et al*, 1987). The outputs from the multiple elements in the array are mutually *coherent*. This represents a powerful technique in the quest for compact high power IR sources with good beam quality, and is an area of current research interest.

A further development of these ideas involves the complete removal of the walls which separate the individual parallel waveguides. This results in a rectangular, slab-like cross section. Assuming that the electrodes are cooled then the gas discharge itself may be effectively diffusion-cooled, except that heat flow is now 1-dimensional whereas in the case of square/circular guides it is 2-dimensional. One of the attractive features of rf discharges is that transverse operation over large areas may be achieved in a stable fashion. This opens the way to high power outputs from waveguide lasers, where the power scales with the *area* of the discharge. Measurements by Abramski *et al* (1989), using an rf excited discharge slab with aspect ratios between 1 and 8, demonstrated area scaling of laser power output with multimode values up to 20 $kWm^{-2}$. The resonator, which is necessarily hybrid because of the large aspect ratio rectangular geometry, incorporated reflectors which appear as dual Case I in one dimension, and free space, multimode plane-plane in the orthogonal dimension.

This simple dual Case I resonator arrangement has the benefits of high power extraction at the expense of mode quality. As with most lasers, the

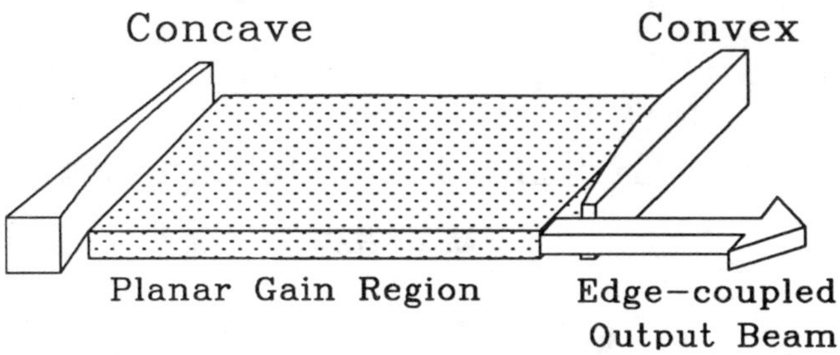


Fig 17.8 Unstable resonator for slab waveguides.

challenge is to maintain high power extraction while providing excellent beam quality for propagation and focussing purposes, i.e low values of $M^2$ (Chapter 9). One approach to this problem is to use a hybrid combination of waveguide and unstable resonator geometries. Using this type of resonator, with a waveguide slab of aspect ratio 20:1, Jackson *et al* (1989) obtained an output power of 240W from an compact laser head. The hybrid resonator, part waveguide and part confocal unstable is shown in Figure 17.8. Beam profile measurements indicated a beam with near diffraction-limited divergence.

## 17.6 REFERENCES

Abrams R L, Redding J R and Jenkins R M 1973 IEEE J. Quantum Electron. QE-9 940

Abramski K M, Colley A D, Baker H J and Hall D R 1989 Appl. Phys. Lett. 54 1833

Bridges, Burkhart and Smith 1972 Appl. Phys. Lett. 20 10 403

Cantoni A 1989 US Patent 4815094

Conder P C, Redding J R and Jenkins R M 1985 Infrared Phys. 25 61

Hall D R, Gorton E K and Jenkins R M 1977 J. Appl. Phys. 48 1212

He D and Hall D R 1984 IEEE J. Quantum Electron. QE-20 5 509-514

Jackson P E, Hall D R and Hill C A 1989 Appl. Opt. 28 5 935-41

Jackson P E, Baker H J and Hall D R 1989 Appl. Phys. Lett. 54 1950-52

Jenkins R M and Devereux R 1985 IEEE J. Quantum Electron. QE-21 1722

Laakmann K D 1980 SPIE 247 74-78

Marcatili E A J and Schmeltzer R A 1964 Bell Syst. Tech. Journ. 43 1783

Miller G M and Williams C A 1985 Proceedings 7th. Nat. Quantum Electron. Conference

Miller G M and Williams C A 1987 SPIE 806 14

Newman L A and Hart R A 1987 Laser Focus/Electro-Optics 80-96

Schwaller P, Steffen H, Moser J F and Kneubühl 1967 Appl. Opt. 6 827

Smith P W 1971 Appl. Phys. Lett. 19 132

Steffen H and Kneubühl F K 1968 Phys. Lett. 27A 612

# 18 Excimer Laser Resonators

*M J SHAW*

## 18.1 INTRODUCTION

Excimer lasers, which are finding uses in applications ranging from materials processing to inertial confinement fusion, represent an increasingly important area of laser technology. Although the scope of excimer lasers is generally considered to include both the rare gas dimers, such as $Xe_2$ which lase in the vacuum ultra violet (VUV) region of the spectrum, and also the mercury halides, which lase in the visible, by far the most important and frequently encountered class of excimer lasers are the rare gas halides (RGH), which operate in the near UV. In this chapter, discussion will be restricted to these RGH lasers, which include ArF (193nm), KrF (248nm), XeCl (308nm) and XeF (351nm).

RGH lasers are characterised by high gain, relatively large gain bandwidth (but not as large as dye lasers) and significant non-saturable losses. RGH lasers which are excited by electric discharges (by far the most commonly-used technique) have excitation durations which are limited by discharge stability to times corresponding to only a few round trips of the laser cavity. This fact, together with the very short excited state lifetime (typically a few ns), are the dominant influences in the design of excimer laser resonators.

In section 18.2 we will discuss simple plane resonators and in section 18.3 techniques which are used to reduce the linewidth. Section 18.4 discusses unstable resonators and injection seeding while in section 18.5 the topic of short pulse amplification is briefly explored. We conclude, in section 18.6 with some information on optical materials which are used for RGH laser resonators and beam transfer optics.

## 18.2 PLANE RESONATORS

The simple plane-plane resonator is probably the most commonly used design for excimer lasers. The resultant highly multimode, divergent, broadband output is perfectly acceptable for many applications. Moreover, the analysis of the plane-plane resonator is instructive in showing how non-saturable losses affect energy extraction.

### 18.2.1 Extraction Efficiency

Figure 18.1 shows a typical plane-plane resonator having an output coupler of

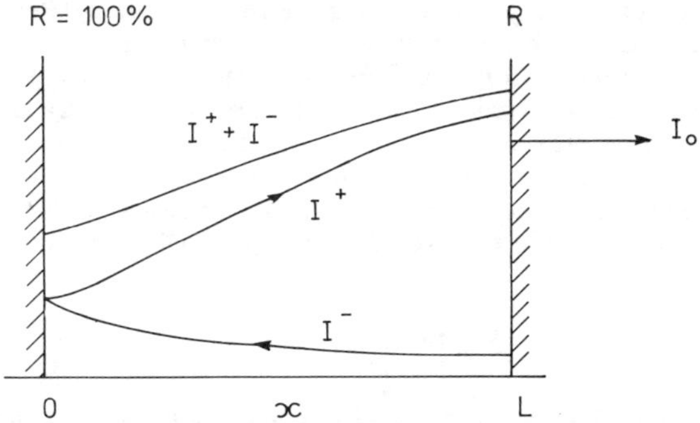


Fig 18.1 Circulating intensities inside a plane-plane resonator having an output coupler reflectivity *R*.

reflectivity *R* and a totally reflecting rear mirror. The circulating intensities within the cavity are $I^+$, directed towards the output coupler, and $I^-$ directed away. The saturation of gain involves the *sum* of $I^+$ and $I^-$, so that the increment in intensity of $I^+$ or $I^-$ at any point within the cavity is governed by the equation:

$$\pm \frac{\partial I}{\partial x} = I^{\pm} \left[ \frac{g_0}{1+ \frac{I^+ + I^-}{I_S}} - \alpha \right] \qquad (18.1)$$

where $g_0$ is the small signal gain coefficient, $\alpha$ the non-saturable loss coefficient and $I_S$ the saturation intensity. Separating (18.1) into individual equations for $I^+$ and $I^-$ we have:

$$I^- \frac{\partial I^+}{\partial x} + I^+ \frac{\partial I^-}{\partial x} = 0 \qquad (18.2)$$

Thus, the product of the left and right going intensities, (illustrated in Figure 18.1) is a constant, $\delta$ given by:

$$I^+ I^- = \delta \qquad (18.3)$$

Boundary conditions are imposed by the cavity mirror reflectivities. In terms of the output intensity $I_0$, we have at the coordinate $x = L$ the values:

$$I^+ = \frac{I_0}{1-R} \quad , \quad I^- = \frac{RI_0}{1-R} \qquad (18.4)$$

Which gives: $$\delta = \frac{R}{(1-R)^2} I_0^2 \qquad (18.5)$$

and similarly at $x = 0$ we have: $$I^+ = I^- = \sqrt{\delta} = \frac{\sqrt{R}}{1-R} I_0 \qquad (18.6)$$

With these boundary conditions, Equation (18.1) may be integrated. The case where $\alpha$ can be neglected leads to an analytical result (Rigrod 1965). This is definitely not the case for excimer lasers where $\alpha$ is typically about 10% of $g_0$. Formal solutions for this case have been given by Rice et al (1980) but they are complex and require numerical evaluation of transcendental equations. A very useful approximation for the extraction efficiency which gives results which are accurate to within a few percent, except at very large output coupling, is obtained by using the average intensity within the cavity. From (18.4) and (18.6), we find that the average total intensity within the cavity is given by:

$$I_{ave} = \tfrac{1}{2}\left[\frac{1+\sqrt{R}}{1-\sqrt{R}}\right] I_0 \tag{18.7}$$

Applying this to the threshold condition for oscillation, we have:

$$\frac{g_0 L}{1+I_{ave}/I_S} = \alpha L - \tfrac{1}{2}\ln R \tag{18.8}$$

We can now determine the cavity extraction efficiency, $\eta_{ext}$:

$$\eta_{ext} = \frac{I_0}{g_0 L I_S} = \frac{2(1-\sqrt{R})}{(1+\sqrt{R})}\left\{\frac{1}{\alpha L - \tfrac{1}{2}\ln R} - \frac{1}{g_0 L}\right\} \tag{18.9}$$

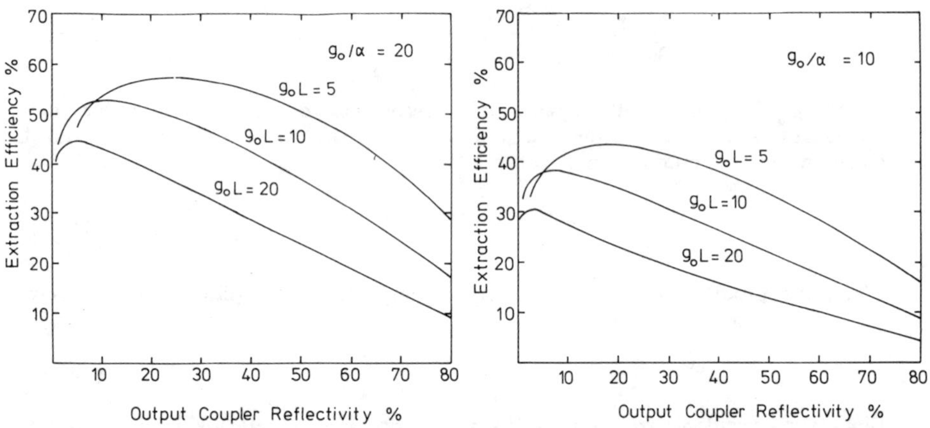


Fig 18.2 **Extraction efficiency from the average intensity approximation as a function of output coupler reflectivity for various values of gain-length product. Gain-to-non-saturable loss ratio $g_0/\alpha$ = 20 and $g_0/\alpha$ = 10.**

This expression is plotted in Figure 18.2 as a function of output coupler reflectivity for the cases of $g_0/\alpha$ = 10 and 20. This range of gain-to-loss ratios, together with the range of gain-length product, $g_0 L$ probably cover most of the parameter space of interest for RGH excimer lasers. The maximum achievable extraction efficiency is given by the expression (Shaw, 1983):

$$\eta_{ext} = 1 - 2(\alpha/g_0)^{\frac{1}{2}} + \alpha/g_0 \tag{18.10}$$

which results in values of 47% for $g_0/\alpha$ = 10 and 60% for $g_0/\alpha$ = 20. Thus, the extraction efficiency at optimum output coupling for $g_0L$ = 5 is very close to the theoretical maximum. In practice, cavity build-up time effects (see next section) dictate the need for higher values of $g_0L$ and typical output couplers for excimer lasers have reflectivities of approximately 10%.

We should note that the overall laser efficiency, $\eta$ is given by the product:

$$\eta = \eta_{ext} \cdot \eta_f \cdot \eta_{el} \tag{18.11}$$

where $\eta_f$ is the fluorescence efficiency, which is a measure of the fraction of the total energy deposited in the gas by electrical pumping which ends in the upper laser level, and $\eta_{el}$ is the electrical efficiency, i.e. the efficiency with which electrical power is deposited in the gas. Typical values are $\eta_f$ = 0.2 and $\eta_{el}$ = 0.3, which gives typical overall efficiencies in the region of 2–3%. We should also note that the output intensity, $I_0$ of the laser is given by:

$$I_0 = \eta_{ext} g_0 L I_s \tag{18.12}$$

which, since $g_0L$ is not arbitrarily extendable without reducing $\eta_{ext}$ significantly, is mainly determined by the saturation intensity $I_s$. The latter is independently variable since $I_s = h\nu/\sigma\tau$, where $\tau$ is the quenched lifetime of the upper laser level. Thus, high output intensities are achieved by operating excimer lasers at high pressure, which by increasing the quenching rate, reduces $\tau$ and hence increases $I_s$.

### 18.2.2 Cavity Build-Up Time

Most discharge excited (and some electron-beam excited) excimer lasers have a pumping pulse duration which allows only a few round trips in the laser cavity. It is important for efficient operation that the intracavity intensity builds up in a time short compared to the excitation time. Thus, build-up time is an important issue in excimer laser resonator design.

We will assume that the intensity, $I$ within the resonator builds-up in an exponential fashion described by the equation:

$$I = I_0 e^{\gamma t} \tag{18.13}$$

Thus, in a time equal to a cavity round trip, $t = 2L/c$ we have

$$I = I_0 \exp(2\gamma L/c) = I_0 R_1 R_2 \exp(2g_0 L) \tag{18.14}$$

Hence:

$$\gamma = \frac{c}{2L}\left\{\ln R_1 R_2 + 2g_0 L\right\} \tag{18.15}$$

We define the cavity build-up time, $t_b$ as the time taken for the intensity within the cavity to increase from the spontaneous emission intensity $I_e$ to the saturation intensity $I_s$. Thus:

$$t_b = \frac{2L/c \; \ln(I_s/I_e)}{2g_0 L + \ln R_1 R_2} \tag{18.16}$$

To obtain the intensity ratio we start from the spontaneous power $dP$ emitted by a volume element $dV$ into a solid angle $d\Omega$ and bandwidth $\Delta\nu$. This is given by:

$$dP = h\nu\frac{N}{\tau_{rad}}\left\{\frac{\Delta\nu_g}{\Delta\nu_s}\right\}\frac{d\Omega}{4\pi}\ dV \tag{18.17}$$

where $\Delta\nu_g$, is the homogeneous gain bandwidth and $\Delta\nu_s$ is the spontaneous emission bandwidth and $\tau_{rad}$ is the radiative lifetime of the upper laser level with population density $N$.

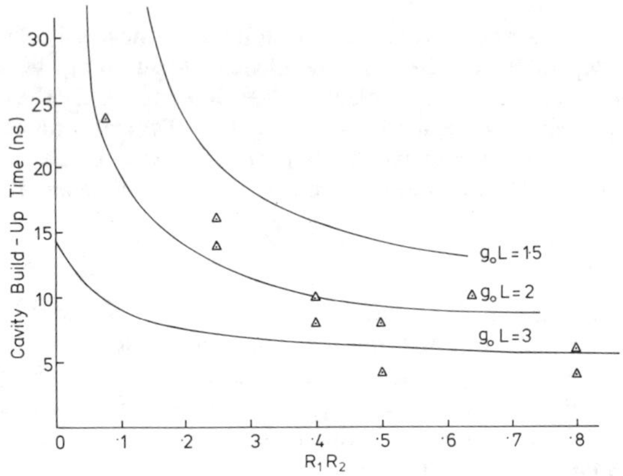


Fig 18.3 Calculated cavity build-up time from Equation 18.16 as a function of mirror reflectivity. Triangles are experimental points from an e-beam pumped KrF oscillator.

Integrating equation (18.17) over the gain length $L$ and dividing by the beam area we have

$$I_e = h\nu\frac{NL}{\tau_{rad}}\left\{\frac{\Delta\nu_g}{\Delta\nu_s}\right\}\frac{d\Omega}{4\pi} \tag{18.18}$$

But $g_0 = N\sigma$ where $\sigma$ is the gain cross section and $I_s = h\nu/\sigma\tau$ where $\tau$ is the quenched lifetime of the upper laser level.

We thus have:
$$\frac{I_e}{I_s} = g_0L\left\{\frac{\tau}{\tau_{rad}}\right\}\left\{\frac{\Delta\nu_g}{\Delta\nu_s}\right\}\frac{d\Omega}{4\pi} \tag{18.19}$$

and so with a knowledge of the laser beam divergence solid angle $d\Omega$ and estimates of the lifetime and bandwidth ratios, we can obtain values for the intensity ratio of sufficient accuracy to enable the build-up time to be evaluated. Under typical excimer laser conditions $\ln(I_s/I_e)$ is in the region of 12 to 15.

Figure 18.3 shows measurements of the build-up time for a short cavity electron-beam pumped KrF laser (Edwards et al 1980) compared with curves from equation (18.16). The agreement between theory and experiment is not too good in this case and may be due to the effects of parasitic oscillations which tend to build up faster than the desired cavity mode intensity, thus robbing the cavity of gain and increasing the build–up time. The expected value of $g_0L$ for this experiment was in the region of 3–4.

## 18.3 NARROW–LINE OPERATION

A typical free running excimer laser will produce a broad spectral output (up to 100cm$^{-1}$ in KrF for example). For many applications it is required to narrow this linewidth considerably. This turns out to be quite difficult to achieve efficiently due to the large mode volume and small number of round trips available in most excimer lasers. One approach to solving this problem for high energy lasers pre–supposes the existence of a line–narrowed, low–power master oscillator, the output of which is either amplified in a power amplifier or is used to injection seed a slave oscillator.

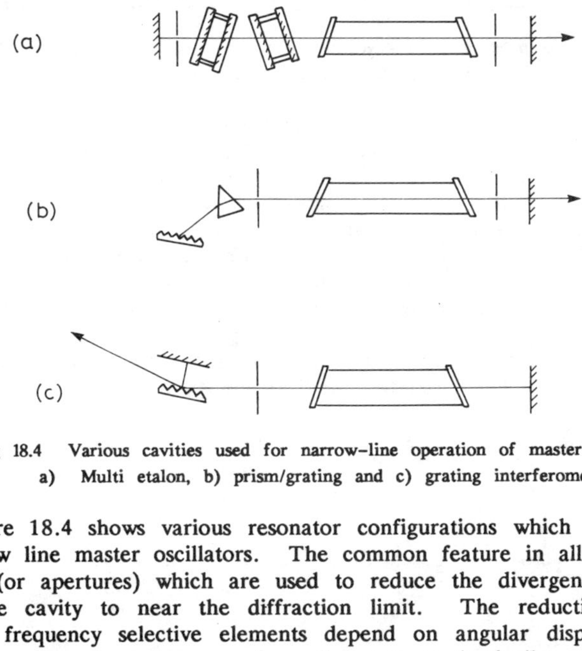


Fig 18.4 Various cavities used for narrow–line operation of master oscillators. a) Multi etalon, b) prism/grating and c) grating interferometer.

Figure 18.4 shows various resonator configurations which have been used for narrow line master oscillators. The common feature in all of them is the aperture (or apertures) which are used to reduce the divergence of the beam within the cavity to near the diffraction limit. The reduction is necessary since all frequency selective elements depend on angular dispersion for their selectivity. However, the use of small apertures (typically ~ 1mm diameter) severely limits the mode volume and hence the output power of the oscillator.

Tilted etalons are perhaps the easiest way to reduce the linewidth of an oscillator. Each etalon needs to have a free spectral range greater than the linewidth produced by the insertion of the previous (coarser) etalon. Finally, the etalon produced by the end mirror of the cavity must also be taken into account. However, with proper choice of free spectral range and finesse, single longitudinal mode operation is possible. Partanen and Shaw (1987) used a cavity with two tilted etalons to obtain single mode operation in a long pulse (120 ns) electron beam–excited KrF laser. A low gain active medium was used, so the cavity build–up time amounted to 9 round trips of the 50 cm long cavity. A coarse, air–spaced etalon with a free spectral range of 70

$cm^{-1}$ was used together with a 3 cm thick solid fused silica etalon with a free spectral range of 0.11 $cm^{-1}$. The finesse of both etalons was 15. Single-mode operation was verified by using a diagnostic etalon having a resolution of 0.005 $\mu m^{-1}$, sufficient to resolve the longitudinal mode structure. An additional useful diagnostic for single-mode operation is to observe the laser output pulse with a fast photodiode. Figure 18.5a shows the output where several modes oscillate which results in a characteristic mode-beating trace. In contrast, Figure 18.5b shows the smooth temporal trace associated with the case of single-mode oscillation.

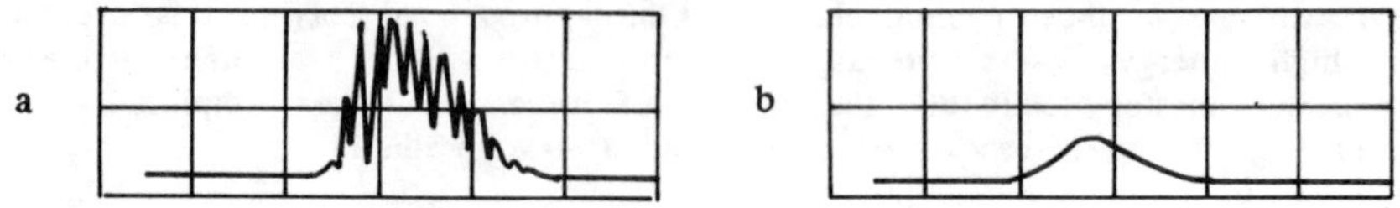


Fig 18.5 a) Output pulse shape observed where a few modes were oscillating (20ns/div) b) Output pulse for a single mode shot (20ns/div).

Combinations of gratings and prisms are frequently used as line narrowing elements in excimer lasers. Prisms are useful in providing beam expansion between the frequency selective element and the gain medium. The reason beam expansion is important is that it reduces the divergence angle (by the magnification factor) and so the same angular dispersion can produce a narrower linewidth. Using a high degree of beam expansion, prisms alone can produce significant line narrowing. Figure 18.6a shows the master oscillator of a commercial excimer laser system (Lambda Physik EMG 150) which uses four prisms for beam expansion and three dispersing prisms to give a linewidth of 0.5$cm^{-1}$. Figure 18.6b shows a modification to the same laser involving the addition of an etalon which reduces the linewidth to 0.15$cm^{-1}$. A simple,

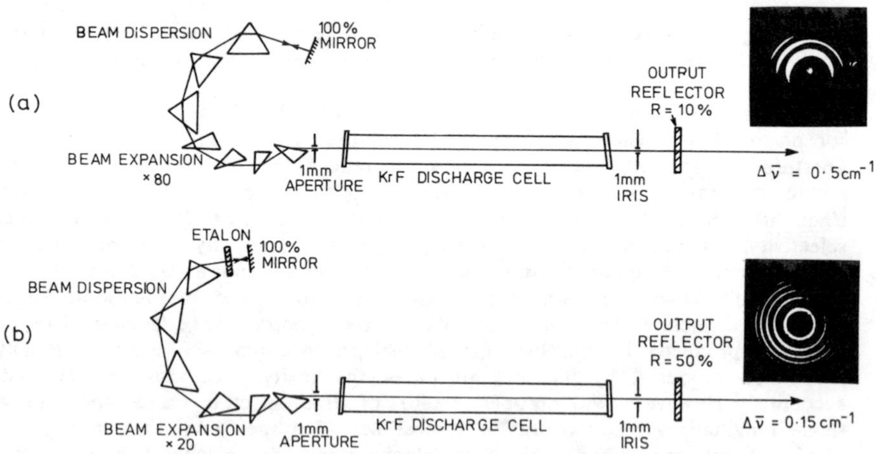


Fig 18.6 a) Line narrowed KrF master oscillator employing prism dispersion only. Output linewidth = 0.5 $cm^{-1}$. b) The same oscillator modified by the insertion of a Fabry Perot etalon. Output linewidth = 0.15 $cm^{-1}$.

but extremely effective way of obtaining very narrow line operation is the grating interferometer (Figure 18.4c). The grating is used near grazing incidence, and the first diffraction order is normal to the grating surface. The resolution of the grating is enhanced when it is used near grazing incidence because the beam size on the grating surface becomes larger. The mirror is aligned exactly parallel to the grating. The mirror and grating form an interferometric device which has similar properties to a Fabry–Perot etalon. However, it does not suffer the drawback of walk–off losses and finesse reduction caused by tilting of a thick intracavity etalon. Because of the etalon effect, losses caused by this device are reasonably small although the grating is used near grazing incidence. If the output is taken as a zero–order reflection from the grating as shown in Figure 18.4c the overall cavity losses are small.

Armandillo *et al* (1984) have used such a cavity with an XeF laser to get single–mode operation. The grating used, however, did not have the resolution of the order of the free spectral range of the grating/mirror etalon. This was not necessary because the line structure in an XeF laser consists of many narrow lines, and the grating is required only to select one of them.

In a KrF discharge laser, Partanen (1986) has used a grating interferometer with an additional intra cavity etalon to achieve a linewidth of $0.03\text{cm}^{-1}$.

## 18.4 TRANSVERSE MODE SELECTION

### 18.4.1 Unstable Resonators

We next turn our attention to the topic of the control of beam divergence in excimer lasers. Because of the inherent limits on intensity in excimer media, large apertures are employed to increase the laser output power. In discharge–excited excimer lasers, considerable effort is made to maximise the cross–sectional area of the discharge. In electron beam–pumped systems, excitation of very large apertures (up to 1 $\text{m}^2$) is possible. The consequence of these large apertures is that Fresnel numbers are very large (up to $10^6$) and conventional stable resonator designs only produce highly divergent output beams.

Unstable resonators are, as pointed out in Chapter 2, an ideal way of achieving diffraction limited beam divergence from large excited volumes and are extensively employed in excimer lasers. A typical discharge laser having an aperture of a few centimetres may have a beam divergence of several milli–radians with a stable cavity, whereas this may be reduced to a few hundred microradians if an unstable resonator is employed.

Such a beam divergence is still about a factor of 10 larger than the diffraction limit and the reason for this is the old problem of too few round trips during the excitation pulse. The number of round trips required to establish a diffraction limited output can be estimated as follows. Consider a diffracting core on the resonator axis with a Fresnel number equal to unity, it will have a radius of $(L_c\lambda)^{\frac{1}{2}}$ where $L_c$ is the cavity length. After $n$ round trips in a cavity of magnification $M$ it will have grown to equal the output beam radius $a$.

Hence: $$a = (L_c\lambda)^{\frac{1}{2}} M^n$$

and so $$n = \frac{\ln(a^2/L_c\lambda)}{2\ \ln M}$$

An approximately equal number of round trips would be required for converging light scattered into the cavity to contract to the diffracting core in the first place, and so the time taken to achieve diffraction limited output is given by equation 18.20.

$$t = \frac{2L_c}{c}\ \frac{\ln(a^2/L_c\lambda)}{\ln M} \tag{18.20}$$

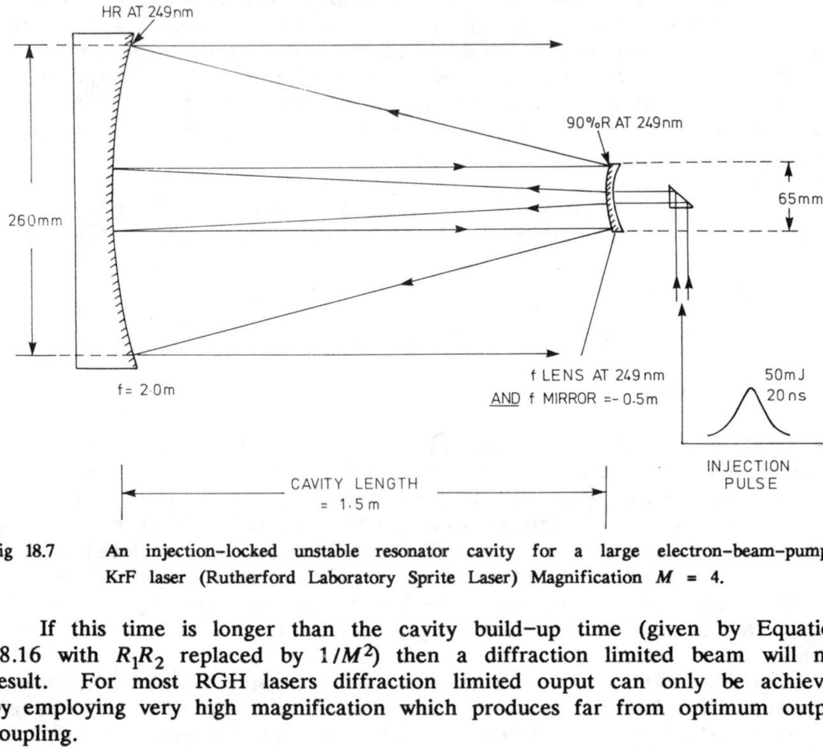


Fig 18.7 An injection-locked unstable resonator cavity for a large electron-beam-pumped KrF laser (Rutherford Laboratory Sprite Laser) Magnification $M = 4$.

If this time is longer than the cavity build-up time (given by Equation 18.16 with $R_1R_2$ replaced by $1/M^2$) then a diffraction limited beam will not result. For most RGH lasers diffraction limited ouput can only be achieved by employing very high magnification which produces far from optimum output coupling.

### 18.4.2 Injection Seeding

The conflicting requirements of a high magnification for greater mode selectivity and the desire for optimum output coupling can be overcome by the use of the technique of injection seeding (sometimes called injection locking). The basis of this technique is the injection of power from a low power master oscillator into a large-aperture, high-power slave oscillator at a time just prior to the onset of the excitation pulse. Provided the spectral brightness of the master oscillator power coupled into the slave cavity far exceeds the spontaneous power (from equation 18.19), then the injected modes will grow at the expense of non-injected modes. This is true of longitudinal as well as transverse modes and so injection seeding can be used to line narrow the slave oscillator without the need for any spectral filtration in the slave oscillator cavity.

Techniques for injecting the master oscillator beam into the slave cavity vary. For small aperture lasers diffraction coupling via a hole in the collimating (rear) reflector is effective (Bigio and Statkine, 1983). For large resonators such a scheme is not practical and coupling via partial reflectors is generally used as shown for example in Figure 18.7. In this case the feedback mirror has a reflectivity of 90% coated onto a diverging lens with a focal length equal to that of the mirror. Four passes of the $M = 4$ resonator are required for the injected beam to fill this cavity.

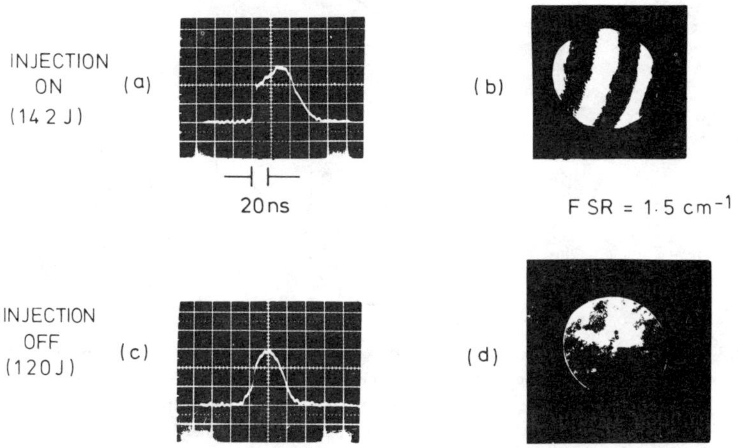


Fig 18.8 Temporal and spectral variation of the output of the resonator shown in Fig 18.7 (a) and (b) with injection, (c) and (d) without injection.

The effects of injection seeding are shown in Figures 18.8 and 18.9. In Figure 18.8 the temporal and spectral behaviour of the slave oscillator output are shown. In this case, the injected KrF pulse was of 20 ns duration and 50 mJ energy with line narrowed operation of linewidth 0.5 $cm^{-1}$. The diagnostic etalon shows the expected narrow line oscillation of the slave oscillator. A considerable side benefit of injection seeding is shown in the photodiode traces. In the case with injection, the output pulse turns on more quickly and the total output energy is considerably increased. This occurs because the build-up time (Equation 18.16) is reduced by the cavity intensity which grows from an injected intensity which can considerably exceed the spontaneous emission. It should however be noted that the effect on build-up time of injected intensity is via a logarithmic term in equation 18.16 and hence, although a relatively small increase in injected intensity over spontaneous emission is necessary to achieve injection control of the slave cavity, the increase in intensity required to substantially reduce the build-up time is very much greater.

Figure 18.9 shows the effect of injection seeding on the near field burn patterns of the laser output as viewed from a distance of some 10m from the laser cavity, so that highly divergent modes show up clearly. In going from patterns E to A in the figure, the injected power in the cavity is decreased by injecting the seed pulse approximately 10 ns earlier in each case.

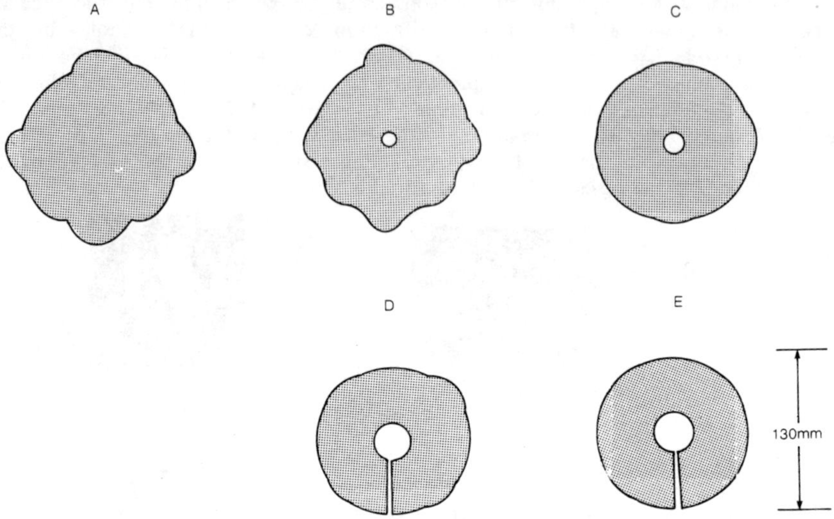


Fig 18.9 Near-field beam patterns produced by the resonator shown in Fig 18.7 at a distance of about 10m. In this case the cavity was adjusted to produce a converging beam. E has the injection pulse optimally timed. Burns D to A are produced by injecting the seed pulse approximately 10 ns earlier.

One problem which is evident with injection-seeded slave oscillators is that of de-locking. This is especially so where the pulse duration of the slave oscillator exceeds that of the injected master. De-locking manifests itself as a tendency for the slave oscillator to oscillate in its lowest loss mode. This will result in the delocking of the oscillator if this lowest loss mode does not happen to be the injected mode. Figure 18.10 illustrates this effect, where a narrow line, master oscillator beam of 20 ns duration is injected into a slave oscillator with a run time of 60 ns. Only in the case where the injected signal is at line centre, i.e. the highest gain condition, does a spectrally pure output result.

## 18.5 SHORT PULSES

There have been some attempts to mode-lock excimer lasers to produce short pulses. None have been particularly successful. The small number of round trips during the excitation pulse is, in general, insufficient to produce significant pulse narrowing. On the other hand, the broad-band nature of excimer gain media, particularly KrF, are such that pulses of less than 100 fs can be amplified successfully. This has given rise to a number of hybrid excimer-dye laser systems for the production and amplification of

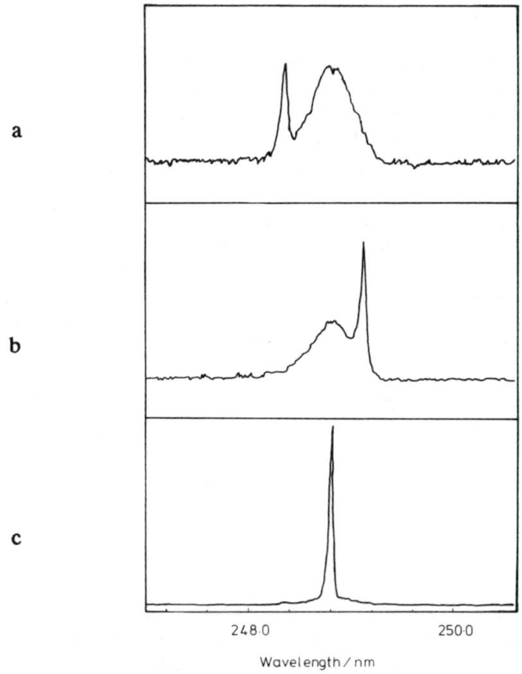


Fig 18.10 Optical Multi-channel Analyser spectra taken of the output of resonator in Fig 18.7 as the injection pulse was tuned across the KrF gain band. The fraction of energy in the narrow line ($\Delta\nu \approx 0.1$ cm$^{-1}$): (a) = 11 (b) = 21 (c) = 78%.

subpicosecond pulses to very high power. In a typical system (eg Szatmari et al 1987), a pair of synchronized excimer lasers is used. One, employing a stable resonator and a XeCl gas fill, is used to pump a short pulse dye laser at 365 nm, which in turn pumps a distributed feedback dye laser at 497nm producing an 80 fs pulse. This pulse is amplified and doubled to 248nm and further amplified to peak powers close to 1 TW in the second (KrF) excimer laser. Many such systems, which combine the short pulse capability of dye lasers with the high power capability and useful UV wavelengths of the excimers, are in existence in laboratories throughout the world. They provide exciting new tools for research into high power, ultra fast phenomena in many branches of science.

## 18.6 PRACTICAL CONSIDERATIONS

Excimer lasers operate with hostile gain media containing highly reactive gases undergoing discharge excitation. The UV output itself is also a source of problems for transmissive or reflective optics. No description of excimer laser resonators would be complete without some consideration of the practical resonator design problems which have to be overcome.

### 18.6.1 Windows

The window in the excimer cavity is in contact with the gain medium and will suffer attack from fluorine in the case of ArF, KrF and XeF lasers or HCl in the case of the XeCl laser. The cheapest and highest optical quality material which is transparent at all four RGH wavelengths is synthetic fused silica. Unfortunately, the lifetime of fused silica windows in most excimer lasers is too short to be of use. The problem of fluorine etching in fused silica can however be considerably improved by the chemical removal of HF from the laser mix. If this is done, bare fused silica windows can survive typical $F_2$ concentrations of 0.5% at 1 atmosphere pressure for several months before etching becomes a problem. Anti-reflection coatings on fused silica substrates employing combinations of alumina and cryolite ($Al_20_3/Na_3AlF_6$) appear to have the same or possibly slightly better survivability.

For small lasers, it is better to use the crystalline materials $CaF_2$ or $MgF_2$ for windows. However, these need to be of single crystal quality and grown from synthetic starting materials rather than from natural materials, in order to avoid colour centre formation. Both these materials resist HF attack well and have the added advantage of being soft and easily repolished by hand.

### 18.6.2 Mirrors

The simplest plane–plane resonator for an excimer laser can be formed by using an uncoated $CaF_2$ etalon as an output coupler giving an average reflectivity of about 8% and an aluminised rear mirror on a $CaF_2$ substrate with the coating away from the laser gas. With the aluminium layer used in this way, the laser damage level is increased compared with its use as a front surface reflector from about 0.2 $Jcm^{-2}$ to about 0.4 $Jcm^{-2}$ (at 248 nm). It is of course also protected from halogen gases. This simple arrangement is not suitable for more complex cavities and is difficult to align, since the mirrors also comprise the gas seal. It is more usual to employ tilted $CaF_2$ windows and to use external optics. For high average power or high energy operation, these optics need to be dielectric multilayers rather than metals, which in general have damage thresholds in the UV which are too low.

The choice of optical materials for multilayers in the UV becomes more restricted as the wavelength decreases. The situation is satisfactory for high damage threshold multilayer coatings for XeF and XeCl lasers, while KrF coatings have much improved in recent years, but there are still problems with ArF reflectors. Most work has been done at 248 nm. Typical coating combinations and approximate damage levels are (eg. Wadt, 1986):–

| | |
|---|---|
| $ThF_4/Na_3AlF_6$ | $\sim 3Jcm^{-2}$ |
| $Sc_20_3/Si0_2$ | $\sim 4Jcm^{-2}$ |
| $Al_20_3/AlF_3$ | $\sim 5Jcm^{-2}$ |
| $Al_20_3/Si0_2$ | $\sim 6Jcm^{-2}$ |

The alumina/silica coating, although hard and of high damage threshold, is not necessarily an automatic choice since due to the similarity in refractive index between $Al_2O_3$ and $SiO_2$, a very large number of layers is required to obtain high reflectivity. The stability of the resulting narrow-band stack is a possible drawback of this design. New combinations of materials are still under development however, with particular applications in mind. For example, in large KrF lasers for fusion, the development of both anti-reflection (AR) and high reflectivity (HR) coatings, which survive in an $F_2$ environment, is of great importance.

## 18.7 CONCLUSIONS

For many applications excimer lasers with simple stable resonators provide useful high power output in the near UV. If the requirement is for narrow line, near diffraction limited beams, then this can be met by using injection seeding master/slave oscillators and most commercial manufacturers offer such systems. The ultimate requirements of single frequency, single transverse mode operation are best met by other lasers more suitable than the excimers. Frequency up-conversion followed by amplification in an excimer amplifier is then the most suitable route to high quality laser beams in the UV.

## 18.8 REFERENCES


Armandillo E, Lopatriello P V M and Giuliani G 1984 Opt. Lett. 9 327

Bigio I J and Slatkine M 1983 IEEE J. Quant. Electron QE-19 1426

Edwards C B, O'Neill F and Shaw M J 1980 Appl. Phys. Lett. 36 617

Partanen J P 1986 Appl. Opt. 25 3810

Partanen J P and Shaw M J 1987 J. Appl. Phys. B43 231

Rice, J K , Tisone G C and Patterson E L 1980 IEEE J. Quantum Electron. QE-16 1315

Rigrod W W 1965 J. Appl. Phys. 36 2487

Shaw M J 1983 J. Appl. Phys. B30 5

Szatmari S, Schafer F P, Muller-Horsche E and Muckenheim W 1987 Opt. Commun. 63 305

Wadt W W 1986 J. Fusion Energy 5 57

# INDEX